CIRCUITS, SYSTEMS, AND SIGNALS FOR BIOENGINEERS: A MATLAB-BASED INTRODUCTION

This is a volume in the
ACADEMIC PRESS SERIES IN BIOMEDICAL ENGINEERING

JOSEPH BRONZINO, SERIES EDITOR
TRINITY COLLEGE—HARTFORD, CONNECTICUT

CIRCUITS, SYSTEMS, AND SIGNALS FOR BIOENGINEERS: A MATLAB-BASED INTRODUCTION

John L. Semmlow

ELSEVIER
ACADEMIC
PRESS

AMSTERDAM • BOSTON • HEIDELBERG • LONDON
NEW YORK • OXFORD • PARIS • SAN DIEGO
SAN FRANCISCO • SINGAPORE • SYDNEY • TOKYO

Elsevier Academic Press
30 Corporate Drive, Suite 400, Burlington, MA 01803, USA
525 B Street, Suite 1900, San Diego, California 92101-4495, USA
84 Theobald's Road, London WC1X 8RR, UK

This book is printed on acid-free paper.

Library of Congress Cataloging-in-Publication Data

Application submitted.

British Library Cataloguing in Publication Data
A catalogue record for this book is available from the British Library

ISBN-13: 978-0-12-088493-3
ISBN-10: 0-12-088493-3

For all information on all Elsevier Academic Press publications
visit our Web site at www.books.elsevier.com

Printed in the United States of America
    06   07   08   09   10   9   8   7   6   5   4   3   2

DEDICATED TO

Susanne Oldham
Who has shown me, and shared with me, so much of life.
Let the adventure continue . . .

PREFACE

This textbook is written to support a bioengineering course covering material traditionally taught in an electrical engineering service course. The course is intended to be taken by second-semester sophomores or first-semester juniors and introduces basic engineering concepts related to signal processing and linear systems analysis. Major topics include the Fourier transform, the Transfer function, the Laplace transform, time and frequency domain representations, and sinusoidal (phasor) analysis.

The primary motivation for this text is to give bioengineering students signal processing and linear systems tools relatively early in their studies. Not only will this allow them to apply these skills earlier in their application courses, but should foster a sense of ownership of these tools. Approaches such as the Transfer function and the Fourier and the Laplace transforms should no longer be considered tools "borrowed" from electrical engineering, but techniques that are used by, and belong to, bioengineering. As long as these tools are taught outside the context of bioengineering (often without enthusiasm) they may not be identified as ours.

With this objective in mind, the textbook contains a number of special features. Most significantly, the text relies heavily on the use of basic MATLAB® in both examples and problems. I have found this software package to be an invaluable educational tool for teaching concepts in linear systems analysis, and believe that not using this resource would be a pedagogical mistake. MATLAB is particularly useful in demonstrating graphically the dynamics of linear systems, the spectra arising from the Fourier transform, and Transfer function frequency plots. It is also very useful in sinusoidal analysis of mechanical and electrical systems. The latter might be more appropriately analyzed using software dedicated to electrical and electronic analysis such as "pSpice," but the educational objectives of this text are served adequately by MATLAB.

The text is written in a casual, almost breezy style in an effort to appear more accessible to younger students. I have tried to develop some of the deeper concepts, such as the Fourier series analysis and the Transfer function, using a highly intuitive approach. For example, the Fourier transform is presented as an extension of basic correlation, but with sinusoids as probing functions. Many of the problems may be considered somewhat dry, but I believe that drill is essential for learning this material. A few more open-ended problems are included in each problem set, usually in the section involving MATLAB. The overriding objective of this text is to give students a solid foundation in the concepts of linear systems analysis. Examples were chosen to be instructive and are not always immediately relevant to bioengineering. Examples of biomedical applications are presented throughout the text, but only if they provide educational support to the topics being covered.

This text is intended to support a one-semester course, although not all of the material can be covered in a three-hour-per-week format. The additional material is included to give the instructor some options and allow the course to be tailored to specific bioengineering programs. One option is to omit the last chapter on electronic devices, and teach only the first nine chapters which fit comfortably into one semester. Alternatively, some of the sections on mechanical systems could be omitted and the chapter on basic electronics included. Finally, a few sections are marked as optional (such as nodal analysis), and these could also be omitted. Ultimately, the instructor may wish to pick and choose within and among the chapters.

The text comes with a number of educational support materials. An enclosed CD contains data and MATLAB functions needed to solve the problems, a few helpful MATLAB routines, and all of the MATLAB examples given in the book. Since many of the problems are extensions or modifications of examples, these CD files can be helpful in reducing the amount of typing required to carry out an assignment. For instructors, a set of PowerPoint files is available that includes all of the figures and most of the equations along with supporting text for each of the chapters. In addition, there are a set of files that contain the solutions to the problems at the end of each chapter. These files are available for download from the publisher at *www.books.elsevier.com*. At Rutgers University, this course is offered in conjunction with a laboratory called "Signals and Measurements"—a manual covering this laboratory is in preparation and should be available shortly.

I am deeply indebted to the assistance provided by two Rutgers undergraduate students who helped with the editing of this text. I especially thank Ms. Jenner Yeh for the outstanding job she did in correcting my many errors, and rewriting particularly obscure passages. In addition, Mr. Mohammad Zia was very helpful in checking the MATLAB problems and other examples. Their contributions were exceptional and any errors or shortcomings that remain are entirely my responsibility. I am also very much indebted to four anonymous reviewers who carefully scrutinized the first three chapters as part of a Whitaker review process. Although funding was not forthcoming, their reviews were invaluable, and I hope that have met at least some of their challenges. Regarding the effort and concern they gave to their reviews, I wish that I could thank them personally, but given the situation this is the best I can do. I also wish to thank Susanne Oldham (to whom this book is dedicated) for her patient editing and unwavering support. Thanks also to Lynn Hutchings and Peggy Christ who demonstrated great understanding during the frenetic preparation of this manuscript.

John L. Semmlow, Ph.D.
New Brunswick, NJ 2004

CONTENTS

3

FREQUENCY TRANSFORMATIONS   69

4

CIRCUIT AND ANALOG ANALYSIS IN SINUSOIDAL STEADY STATE   121

5 ANALYSIS OF ANALOG MODELS AND PROCESSES   161

6 FREQUENCY CHARACTERISTICS OF CIRCUITS AND ANALOG PROCESSES: THE TRANSFER FUNCTION   193

7 RELATIONSHIPS BETWEEN ANALOG ELEMENTS   239

8   THE ANALYSIS OF TRANSIENTS: THE LAPLACE TRANSFORM   289

9   SYSTEM MODELS AND BEHAVIOR   335

1 BIOENGINEERING SIGNALS AND SYSTEMS

1.1 BIOLOGICAL SYSTEMS

A system is a collection of processes or components that interact for some common purpose, although that purpose may only be the invention of human intellect. Many systems of the human body are based on function. The cardiovascular system's function is to deliver oxygen-carrying blood to the peripheral tissues. The pulmonary system is responsible for the exchange of gases [primarily oxygen (O_2) and carbon dioxide (CO_2)] between the blood and air, whereas the renal system regulates water and ion balance and adjusts the concentration of other types of ions and molecules. Some systems are organized around mechanism rather than function. The endocrine system mediates a range of communication functions using complex molecules distributed through the blood stream. The nervous system performs an enormous number of tasks using neurons and axons to process and transmit information coded as electrical impulses.

The study of classical physiology and of many medical specialties is structured around human physiological systems. (The term *classical physiology* is used here to mean the study of whole organs or organ systems as opposed to newer molecular-based approaches.) For example, cardiologists specialize in the cardiovascular system, neurologists in the nervous system, ophthalmologists in the visual system, nephrologists in the kidneys, pulmonologists in the respiratory system, gastroenterologists in the digestive system, and endocrinologists in the endocrine system. There are medical specialties or subspecialties to cover most physiological systems. (Another set of medical specialties is based on common tools or approaches, including surgery, radiology, and anesthesiology, whereas one specialty, pediatrics, is based on the type of patient.)

Given this systems-based approach to physiology and medicine, it is not surprising that early bioengineers applied their engineering tools, especially those designed for the analysis of systems, to some of these physiological systems. Early applications in bioengineering research include the analysis of breathing patterns and the oscillatory movements of the iris muscle. Applications of basic science to medical research date from the eighteenth century. In the late nineteenth century,

Einthoven (Raju, 1998) recorded the electrical activity of the heart, and throughout that century, electrical stimulation was used therapeutically (largely to no avail). Although early researchers may not have considered themselves engineers, they did draw on the engineering tools of their day.

The nervous system, with its apparent similarity to early computers, was another favorite target of bioengineers, as was the cardiovascular system with its obvious links to hydraulics and fluid dynamics. Some of these early efforts are discussed in the sections on system and analog models (Sections 1.3.2 and 1.3.3). As bioengineering has expanded into areas of molecular biology, systems on the cellular, or even subcellular levels, have become of interest.

Regardless of the type of biological system, its scale, or its function, we must have some way of interacting with that system. Interaction or communication with a biological system is done through biosignals. The communication may only be one-way, such as when we attempt to infer the state of the system by measuring various biological or physiological variables to make a medical diagnosis. From a systems analytic point of view, changes in physiological variables constitute biosignals. Common signals measured in diagnostic medicine include electrical activity of the heart, muscles and brain; blood pressure; heart rate; blood gas concentrations and concentrations of other blood components; and sounds generated by the heart and its valves.

Often it is desirable to send signals into a biological system for purposes of experimentation or therapy. In a general sense, all drugs introduced into the body can be considered biosignals. We often use the term *stimulus* for signals directed into some physiological process, and if an output signal is evoked by these inputs we term it a *response*. (Terms shown in *italics* are an important part of a bioengineer's vocabulary.) In this scenario, the biological system is acting like an input–output system, a classic construct or model used in systems analysis (Figure 1.1).

Figure 1.1   A classic systems view of a physiological system that receives an external input, or stimulus, that evokes an output, or response.

Classical examples include the knee-jerk reflex, where the input is a mechanical force and the output is mechanical motion, and the pupillary light reflex, where the input is light and the output is a mechanical change in the iris muscles. Drug treatments can be included in this input–output description, where the input is the molecular configuration of the drug and the output is the therapeutic benefit (if any). Such representations are further explored in the sections on systems and analog modeling (Sections 1.3.2 and 1.3.3).

Systems that produce an output without the need for an input stimulus, for example the electrical activity of the heart, can be considered biosignal *sources*.

(Although the electrical activity of the heart can be moderated by several different stimuli, exercise for example, the basic signal does not require a specific stimulus.) Input-only systems are not usually studied, because the purpose of any input signal is to produce some sort of response: even a placebo, which is designed to produce no physiological response, often produces substantive results.

Because all of our interactions with physiological systems are through biosignals, the characteristics of these signals are of great importance. Indeed, much of modern medical technology is devoted to extracting new physiological signals from the body or gaining more information from existing biosignals. The next section discusses some of the basic aspects of these signals.

1.2 BIOSIGNALS

Much of the activity in biomedical engineering, be it clinical or research, involves the measurement, processing, analysis, display, and/or generation of signals. Signals are variations in energy that carry information. The variable that carries the information (the specific energy fluctuation) depends upon the type of energy involved. Table 1.1 summarizes the different energy types that can be used to carry information, and the associated variables that encode this information. Table 1.1 also shows the physiological measurements that involve these energy forms as discussed later in the chapter.

Biological signals are usually encoded into variations of electrical, chemical, or mechanical energy, although occasionally variations in thermal energy are of interest. For communication within the body, signals are primarily encoded as variations in electrical or chemical energy. When chemical energy is used, the encoding is

TABLE 1.1 Energy Forms and Associated Information-Carrying Variables

Energy	Variables (Specific Fluctuation)	Common Measurements
Chemical	Chemical activity and/or concentration	Blood ion, oxygen, carbon dioxide, pH, hormonal concentrations, and other chemistry
Mechanical	Position Force, torque, or pressure	Muscle movement, cardiovascular pressures, muscle contractility Valve and other cardiac sounds
Electrical	Voltage (potential energy of charge carriers) Current (charge carrier flow)	EEG, ECG, EMG, EOG, ERG, EGG, GSR
Thermal	Temperature	Body temperature, thermography

ECG, electrocardiogram; EEG, electroencephalogram; EGG, electrogastrogram; EMG, electromyogram; EOG, electrooculogram; ERG, electroretinogram; GSR, galvinic skin response.

usually done by varying the concentration of the chemical within a *physiological compartment*, for example, the concentration of a hormone in the blood. Bioelectric signals use the flow or concentration of ions, the primary charge carriers within the body, to transmit information. Speech, the primary form of communication between humans, encodes information as variations in air pressure.

Outside the body, information is commonly transmitted and processed as variations in electrical energy, although mechanical energy was used in the seventeenth and early eighteenth centuries to send messages. The semaphore telegraph used the position of one or more large arms placed on a tower or high point to encode letters of the alphabet. These arm positions could be observed at some distance (on a clear day), and relayed onward if necessary. Information processing can also be accomplished mechanically, as in the early numerical processors constructed by Babbage (Dodge, 2000). More recently, mechanically based digital components have been attempted using variations in fluid flow. Modern electronics provides numerous techniques for modifying electrical signals at very high speeds. The body also uses electrical energy to carry information when speed is important. Since the body does not have many free electrons, it relies on ions, notably Na^+, K^+, and Cl^-, as the primary charge carriers. Outside the body, electrically based signals are so useful that signals carried by other energy forms are usually converted to electrical energy when significant transmission or processing tasks are required. The conversion of physiological energy to an electric signal is an important step, often the first step, in gathering information for clinical or research use. The energy conversion task is done by a device termed a *transducer*, specifically a *biotransducer*.

A transducer is a device that converts energy from one form to another. By this definition, a light bulb or a motor is a transducer. In signal processing applications, the purpose of energy conversion is to transfer information, not to transform energy as with a light bulb or a motor. In physiological measurement systems, all transducers are so-called *input transducers*: they convert nonelectrical energy into an electronic signal. An exception to this is the electrode, a transducer that converts electrical energy from ionic to electronic form. Usually, the output of a biotransducer is a voltage (or current) whose amplitude is proportional to the measured energy. Figure 1.2 shows a device to measure the movements of the intestine during surgical procedures. The mechanical transducers used in the device are called *strain gages* and they change their electrical resistance when stretched even slightly.

The energy that is converted by the input transducer may be generated by the physiological process itself, may be energy that is indirectly related to the physiological process, or may be energy produced by an external source. In the latter case, the externally generated energy interacts with, and is modified by, the physiological process, and it is this alteration that produces the measurement. For example, when externally produced x-rays are transmitted through the body, they are absorbed by the intervening tissue, and a measurement of this absorption is used to construct an image. Most medical imaging systems are based on this external energy approach.

Images can also be constructed from energy sources internal to the body as in the case of radioactive emissions from radioisotopes injected into the body. These

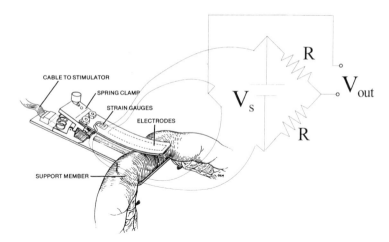

Figure 1.2 A device used to measure small movements of the intestine during surgery. These movements can be used to assess the viability of a segment of intestine. The device consists of an inflexible lower plate and a flexible upper plate. Movement of the upper plate is detected by two strain gages placed on its upper and lower surfaces. Strain gage transducers change their resistance in response to small changes in length. Subsequent electronics detect these resistance changes.

techniques make use of the fact that selected, or tagged, molecules will collect in specific tissue. The areas where these radioisotopes collect can be mapped using a gamma camera or, with certain short-lived isotopes, better localized using positron emission tomography (PET).

Many physiological processes produce energy that can be detected directly. For example, cardiac internal pressures are usually measured using a pressure transducer placed on the tip of a catheter introduced into the appropriate chamber of the heart. The measurement of electrical activity in the heart, muscles, or brain provides other examples of the direct measurement of physiological energy. For these measurements, the energy is already electrical and only needs to be converted from ionic to electronic current using an *electrode*. These sources are usually given the term *ExG*, where the *x* represents the physiological process that produces the electrical energy: ECG, electrocardiogram; EEG, electroencephalogram; EMG, electromyogram; EOG, electrooculogram; ERG, electroretinogram; and EGG, electrogastrogram. An exception to this terminology is the galvanic skin response, GSR, the electrical activity generated by the skin. Typical physiological measurements that involve the conversion of other energy forms to electrical energy are shown in Table 1.1. Figure 1.3 shows the early ECG machine where the interface between the body and the electrical monitoring equipment was buckets filled with saline (Figure 1.3E).

The *biotransducer* is often the most critical element in the system because it constitutes the interface between the subject or life process and the rest of the system. The transducer establishes the risk, or *invasiveness*, of the overall system. For

Figure 1.3 An early electrocardiogram machine.

example, an imaging system based on differential absorption of x-rays, such as a CT (computed tomography) scanner, is considered more invasive than an imaging system based on ultrasonic reflection, because CT uses ionizing radiation that may have an associated risk. (The actual risk of ionizing radiation is still an open question, and imaging systems based on x-ray absorption are considered *minimally invasive*.) Ultrasound and radiographic imaging would be considered less invasive than, for example, monitoring internal cardiac pressures through cardiac catheterization in which a small catheter is threaded into the heart chamber. Indeed, many of the outstanding problems in biomedical measurement, such as noninvasive measurement of internal cardiac pressures or intracranial pressure, await an appropriate (and undoubtedly clever) transducer mechanism.

1.2.1 Signal Encoding

Given the importance of electrical signals in biomedical engineering, much of the discussion in this text is based on electrical or electronic signals. Nonetheless, many of the principles described are general and could be applied to signals carried by any energy form. Regardless of the energy form or specific variable used to carry information, some type of encoding scheme is necessary. Encoding schemes vary in complexity: human speech is so complex that automated decoding is still a challenge for voice-recognition computer programs. Yet, the exact same information could be encoded into the relatively simple series of long and short pulses known as Morse code.

Most encoding strategies can be divided into two broad categories or domains: continuous and discrete. The discrete domain is used almost exclusively in computer-based technology, because such signals are easier to manipulate electronically. Discrete signals are usually transmitted as a series of pulses at even (synchronous transmission) or uneven (asynchronous transmission) intervals. These pulses may be of equal duration, or the information can be encoded into the pulse length. Within the digital domain, many different encoding schemes can be used. For encoding alphanumeric characters, those featured on the keys of a computer keyboard, the ASCII (American Standard Code for Information Exchange) code is used. Here each letter, the numbers 0 through 9, and many characters are encoded into an 8-bit binary number. For example, the letters *a* though *z* are encoded as 97 (for *a*) through 122 (for *z*) whereas the capital letters *A* through *Z* are encoded by numbers 65 (*A*) through 90 (*Z*). The complete ASCII code can be found in some computer texts or on the Internet.

In the continuous domain, information is encoded in terms of signal amplitude, usually the intensity of the signal at any given time. For an electronic signal, this could be the value of the voltage or current at a given time. Note that all signals are by nature *time varying*, because a single constant value contains no information. (Modern information theory makes explicit the difference between information and meaning. The latter depends upon the receiver; that is, the device or person for which the information is intended. Many students have attended lectures with a considerable amount of information that, for them, had little meaning. This text strives valiantly for both information and meaning.) If the information is linearly encoded into signal amplitude, the signal is referred to as an *analog signal*. For example, the temperature in a room can be encoded so that 0 V represents 0.0°C, 5 V represents 10°C, 10 V represents 20°C, and so on, so that the encoding equation for temperature would be as follows:

$$\text{Temperature} = 2 \times \text{Voltage amplitude}$$

Analog encoding is common in consumer electronics such as high-fidelity amplifiers and television receivers, although many applications that traditionally used analog encoding, such as sound and video recording, now use discrete or digital encoding. Nonetheless, analog encoding is likely to remain important to the biomedical engineer, if only because many physiological systems use analog encoding, and most bio-transducers generate analog-encoded signals.

The typical analog signal is one whose amplitude varies in time as follows:

$$x(t) = f(t) \qquad \text{[Eq. 1.1]}$$

When a continuous analog signal is converted to the digital domain, it is represented by a series of numbers that are discrete samples of the analog signals at a specific point in time:

$$X[n] = x[1], x[2], x[3], \ldots x[n] \qquad \text{[Eq. 1.2]}$$

Usually this series of numbers would be stored in sequential memory locations with x_1 followed by x_2, then x_3, and so forth. {It is common to use brackets to identify a discrete variable (i.e., $x[n]$); but note that the MATLAB® (MathWorks, Natick, MA) programming language used throughout this text also uses brackets in a different context.} Because digital numbers can only represent discrete or specific amplitudes, the analog signal must also be sliced up in amplitude. Hence, to *digitize* an analog signal requires slicing the signal in two ways: in time and in amplitude.

Slicing the signal into discrete points in time is termed *time sampling* or simply *sampling*. Time slicing *samples* the continuous waveform, $x(t)$, at discrete prints in time, nT_s, where T_s is the sample interval. The consequences of time slicing are discussed in Chapter 3. Slicing the signal amplitude in discrete levels is termed *quantization* (Figure 1.4). The equivalent number can only approximate the level of the analog signal, and the degree of approximation will depend on the range of binary numbers and the amplitude of the analog signal. For example, if the signal is converted into an 8-bit binary number, this number is capable of 2^8 or 256 discrete values. If the analog signal amplitude ranges between 0.0 and 5.0 V, the quantization interval in volts will be 5/256 or 0.019 V. If, as is usually the case, the analog signal is time varying in a continuous manner, it must be approximated by a series

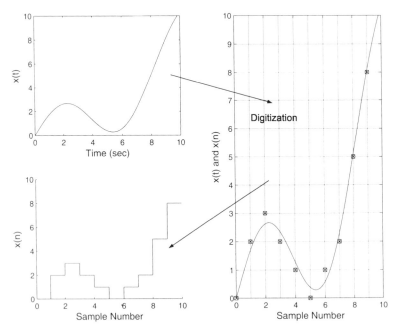

Figure 1.4 Digitizing a continuous signal (*upper left*) requires slicing the signal in time and amplitude (*right side*). The result is a series of discrete numbers (x's) that approximate the original signal, and the resultant digitized signal (*lower left*) consists of a series of discrete steps in time and value.

of binary numbers representing the approximate analog signal level at discrete points in time (Figure 1.4). The errors associated with amplitude slicing, or quantization, are also described in Chapter 3.

Example 1.1:   A 12-bit analog-to-digital converter (ADC) advertises an accuracy of ± the least significant bit (LSB). If the input range of the ADC is 0 to 10 V, what is the resolution of the ADC in analog volts?

Solution: If the input range were 10 V, the analog voltage represented by the LSB would be as follows:

$$V_{LSB} = \frac{V_{max}}{2^{Nu\ bits}} = \frac{10}{2^{12}} = \frac{10}{4,096} = 0.0024 \text{ volts}$$

Hence, the resolution would be ± 0.0024 V.

It is relatively easy, and common, to convert between the analog and digital domains using electronic circuits specially designed for this purpose. Many medical devices acquire the physiological information as an analog signal but convert it to digital format using an ADC so that it can be processed using a computer. For example, the electrical activity produced by the heart can be detected using properly placed electrodes, and the resulting signal, the ECG, is an analog-encoded signal. This signal might undergo some *preprocessing* or *conditioning* using analog electronics before being converted to a digital signal using an ADC. The converted digital signal would be sent to a computer for more complex processing and storage. (In fact, conversion to digital format is usually done even if the data are only to be stored for later use.) Conversion from the digital to the analog domain is possible using a *digital-to-analog* converter (DAC). Most personal computers include both ADCs and DACs as part of a sound card. This circuitry is specifically designed for the conversion of audio signals, but can be used for other analog signals. Data transformation cards designed as general-purpose ADCs and DACs are readily available and offer greater flexibility in sampling rates and conversion gains. These cards provide multichannel ADCs (usually eight to 16 channels) and several channels of DAC.

Basic concepts that involve signals are often introduced or discussed in terms of analog signals, but most of these concepts apply equally well to the digital domain. In this text, the equivalent digital domain equation is often presented alongside the analog equation to emphasize the equivalence. Many of the problems and examples use a computer, so they obviously are being implemented in the digital domain even if they are presented as analog-domain problems.

1.3   LINEAR SIGNAL ANALYSIS: OVERVIEW

From a mechanistic point of view, all living systems are composed of processes. These processes act, or interact, through manipulation of molecular mechanisms, chemical concentrations, ionic electrical current, and/or mechanical forces and displacements. A physiological process performs some operation(s) or manipulation(s) in response to a specific input (or inputs), which gives rise to a specific output (or outputs). In this regard, a process is the same as a system and would be systematically represented as shown in Figure 1.1. Sometimes the term *system* is reserved for larger structures composed of several processes, but the two terms are often used interchangeably, as they will be throughout this text. To study and quantify complex processes, we often impose rather severe simplifying constraints. The most common assumption is that the process and its components or subprocesses behave in a linear manner, and that their basic characteristics do not change over time. This assumption is referred to as the 'linear time-invariant' model (LTI). Such an assumption allows us to apply a powerful array of mathematical tools that are known collectively as linear systems analysis. Of course, most living systems change over time, are adaptive, and are often nonlinear. Nonetheless, the power of linear systems analysis is sufficiently seductive that assumptions or approximations are often made so that these tools can be used. Linearity can be approximated by using small-signal conditions where many systems behave more or less linearly. Alternatively, piecewise linear approaches can be used where the analysis is confined to operating ranges over which the system behaves linearly. One approach to dealing with a process that changes over time is to study that process within a short enough timeframe that it can be considered time-invariant.

The concept of linearity has a rigorous definition, but the basic concept is one of proportionality of response. If you double the stimulus into a linear system, you will get twice the response. One way of stating this proportionality property mathematically is the following: if the independent variables of linear function are multiplied by a constant, k, the output of the function is simply multiplied by k (Johnson et al., 1989):

$$y = f(x); \text{ where f is a linear function, then:}$$

$$ky = f(kx) \qquad \text{[Eq. 1.3]}$$

Note that:

$$ky = \frac{df(kx)}{dt} \text{ and } ky = \int f(kx)dt \qquad \text{[Eq. 1.4]}$$

Hence, differentiation and integration are linear operations. The major transforms described in this text, the Fourier transform (Chapter 3) and the Laplace transform (Chapter 8), are also linear processes.

Response proportionality, or linearity, is required for the application of an important concept known as *superposition*. Superposition states that if there are two (or more) stimuli acting on the system, the system responds to each as if it were the only stimulus present. The combined influence of the multiple stimuli is simply the addition of each stimulus acting alone. This allows complex stimuli to be broken down so that the problem of determining a system's response to such stimuli is greatly reduced. Approaches that rely on superposition will be found throughout this text.

1.3.1 Analysis of Linear Systems

A linear system is usually viewed as acting on a specific input signal to produce an output as shown in Figure 1.1. This is a very general concept: inputs can take many different energy forms (chemical, electrical, mechanical, or thermal), and outputs can be of the same or different energy forms. There are several ways to study a linear system. In this text, two different approaches are developed and explored: *analog analysis* using *analog models*, and *systems analysis* using *systems models*. There is potential confusion in this terminology. Although analog analysis and systems analysis are two different approaches, both are included as tools of *linear systems analysis*. Hence, linear systems analysis includes both analog and systems analysis.

The primary difference between analog and systems analysis is the way the underlying physiological processes are represented. In analog analysis, individual components are represented by analogous elements. Often these elements show detailed structures and provide some insight into the way in which a given process is implemented, although they may also represent processes more globally. In systems analysis, a whole process can be represented by a single mathematical equation. The advantage of using analog analysis is that the model is often closer to the underlying physiological processes. Conversely, analyzing at the process level, as in systems analysis, provides a more succinct description and offers a better overall view of the system under study. In addition, the more abstract representation provided by systems models emphasizes behavioral characteristics and may aid in identifying behavioral similarities between processes that contain quite different elements.

1.3.2 Analog Analysis and Analog Models

In analog analysis, there is a direct relationship between the physiological mechanism and the analog elements used in the model, although the elements may not necessarily be in the same energy modality as the physiological mechanism. For example, Figure 1.5 shows what appears to be an electric circuit. Of course it is an electric circuit, but it is also an early analog model of the cardiovascular system known as the *windkessel model*. In this circuit, voltage represents blood pressure, current represents blood flow, R_P and C_P are the resistance and compliance of the systemic arterial tree, and Z_o is the characteristic impedance of the proximal aorta. Later we will find that using an electrical network to represent what is a mechanical system is mathematically appropriate. In this analog model, the elements are not very elemental, because they represent processes distributed throughout various segments of the cardiovascular system; however, the model can be expanded to represent the system at a more detailed level.

Figure 1.6 shows an analog model of the muscle skeletal muscle that uses mechanical elements. The muscle's force originates at the *contractile element,* but this force, F_o, is modified by the muscle mechanical processes before it appears at the output, F. The internal mechanical processes include the tissue viscosity, a sort of internal friction, the parallel elastic element which represents the elastic properties of the

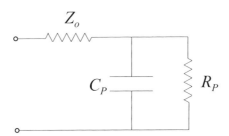

Figure 1.5 An early analog model of the cardiovascular system that used electrical elements to represent mechanical processes. In this model, voltage is equivalent to blood pressure and current to blood flow.

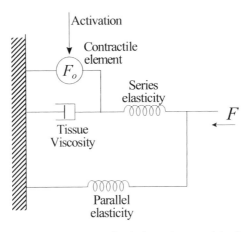

Figure 1.6 A mechanical analog model of skeletal muscle. The various elements correspond to specific properties of real muscle.

sarcolemma, and the series elastic element that reflects the elastic behavior of muscle tendons. In real muscle, these elements are nonlinear, but are often approximated as linear providing a linearized skeletal muscle model. (For a detailed look at skeletal muscles, see Devasahayam, 2000.)

This basic model of skeletal muscle shown in Figure 1.6 has been used with additional mechanical elements to construct a mechanical model of the eye movement system including a pair of extraocular muscles, the lateral and medial rectus (Figure 1.7). These muscles are the mechanical elements involved in controlling the horizontal position of the eye. Each of the two extraocular muscles shown, the lateral

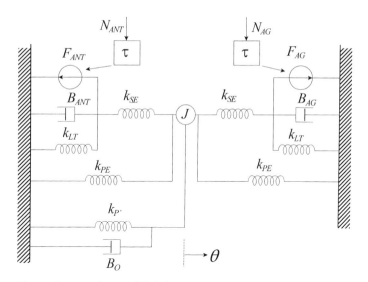

Figure 1.7 Analog model of the lateral and medial rectus muscles and associated mechanical involved in directing horizontal eye position. The neural signals, N_{ANT} and N_{AG}, are the inputs and the angular position, θ, is the output. The function of the various analog components is discussed in the text. (Adapted from Bahill and Stark, 1979.)

and medial rectus, is represented by a force generator F_{ANT} (or F_{AG}), a viscous element B_{ANT} (or B_{AG}), a series elastic element, k_{SE}, and a parallel elastic element, k_{PE}. The two muscle representations also include an additional elastic element k_{LT}. Three other elements represent the mechanical properties of the eyeball and the orbit: an inertial component, J, representing the moment of inertia of the eyeball; a viscous element Bo, representing the friction between the eye and orbit; and a parallel elastic element, $k_{P'}$, representing the elastic properties of the eye in the orbit. The neural signals, N_{ANT} and N_{AG}, are the inputs and the angular position, θ, is the output.

With the aid of a computer, this model, and all quantitative models, can be tested to see if they predict reasonable results. This is one of the primary motivations for construction of any model, the ability to 'try out' the model to see if it behaves in a manner similar to the process it represents. Programming a model into a computer to see how it behaves is known as *simulation*. Simulations of the model in Figure 1.7 have produced highly accurate predictions of the behavior of real eye movements and have also provided insight into the nature of the neural signals that activate the two muscles.

An appealing aspect of the analog-modeling approach is the relative simplicity of the mathematical description of the elements. All linear analog elements can be represented by scaling, integration, or differentiation operations between the associated variables:

$$v_1 = Av_2 \qquad \text{Scaling} \qquad \text{(Dissipative elements)}$$

$$v_1 = A\frac{dv_2}{dt} \qquad \text{Time differentiation} \quad \text{(Inertial elements)}$$

$$v_1 = A\int v_2 dt \quad \text{Time integration} \qquad \text{(Capacitive elements)} \qquad \text{[Eq. 1.5]}$$

where v_1 and v_2 are the variables associated with the analog element. The specific variables depend on the type of elements. For electrical elements, they are voltage and current, whereas for mechanical elements they are force and velocity (Table 1.2). The reason the various relationships are associated with dissipative, inertial, and capacitive elements is explained in Chapter 4.

TABLE 1.2 Variables Associated with Analog Elements and Related Conservation Laws

Element Type	Variable	Conservation Law	Element (Type)
Electrical	Voltage, V (volts)	Charge (Kirchhoff's current law)	Resistor (dissipative)
	Current, i (amps)	Energy (Kirchhoff's voltage law)	Inductor (inertial)
			Capacitor (capacitive)
Mechanical	Force, F (newtons)	Force (Newton's law)	Friction (dissipative)
	Velocity, v (cm/sec)		Mass (inertial)
			Elasticity (capacitive)

One of the advantages of analog models is that although the systems they represent can be quite complicated, the individual analog components behave in a straightforward manner as noted in Equation 1.5. Analog models become complicated because of the number of elements involved and their configuration, but the elements themselves are simple. Another advantage of analog analysis is that given the configuration of elements, a mathematical description of the overall model follows directly. We will find that by applying the conservation laws (conservation of charge, energy, and force) to a specific configuration of elements, a mathematical description follows in algorithmic fashion. One need merely follow a set of rules to obtain a mathematically complete description of the model.

It is possible to introduce nonlinear components into an analog model, but this complicates the analysis. For example, the model shown in Figure 1.5 is actually nonlinear because the capacitance changes its value as blood pressure changes. Often, a piecewise linear approach can be used where the model is analyzed over several different operating regions within which the nonlinear elements can be taken as linear.

Example 1.2: A constant force of 4 dyne is applied to a 2-g mass. Find the velocity of the mass after 5 seconds.

Solution: Inertia is one property of mass that is defined as an integral relationship between the mechanical variables force (F) and velocity (v):

$$F = m(dv/dt)$$

a modification of Newton's equation, $F = ma$. To find the velocity of the mass, solve for v in the above equation by time integrating both sides of the equation.

$$\int F dt = mv; \quad v = \frac{1}{m}\int F dt = \frac{1}{2}\int 4 dt$$

$$v = \frac{4}{2}t = 2(5) = 10\,\text{cm}/\text{sec}$$

1.3.3 Systems Analysis and Systems Models

Systems models usually represent whole processes using so-called *black box* components. Each element of a systems model consists only of an input–output relationship defined by an equation and represented by a geometric shape, usually a rectangle. No effort is made to determine what is actually inside the box; hence the term *black box*. The modeler pays no heed to what is the inside the box, only its overall input–output (or stimulus/response) characteristics. A typical element in a systems model is shown graphically as a box or sometimes as a circle when an arithmetic process is involved (Figure 1.8). The inputs and outputs of all elements are signals with a well-defined direction of flow or influence. These signals and their direction of influence are shown by lines and arrows connecting the system elements (Figure 1.8).

The letter G in the right-hand element of Figure 1.8 represents the mathematical operation that converts the input signal into an output signal, usually expressed as a ratio of output to input:

$$G = \frac{Output}{Input}; \quad Output = Input(G) \qquad \text{[Eq. 1.6]}$$

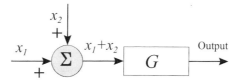

Figure 1.8 Typical elements in a system model. The left-hand element is an 'adder' whose output signal is the sum of the two inputs x_1 and x_2. The right-hand element is a general element that takes the input ($x_1 + x_2$ in this case) and operates on it with the mathematical operation G to produce the output signal.

For many elements, the mathematical operation defined by the letter *G* in Figure 1.8 can be quite complex, involving integral and differential operations, just as in an analog model. In fact, it is a straightforward task to convert from an analog model to a systems model, but not vice versa. It requires some mathematical tricks to structure calculus operations into a format that involves simply algebraic multiplication, and these techniques are described in Chapters 4 and 8.

Occasionally, the operation performed by an element is simple, such as a scaling of the input; that is, the output is the same as the input, but multiplied by a constant *gain*. In such cases, the equation for *G* would be a simple constant defining the multiplying gain. Under static or *steady-state* conditions when the inputs have a constant value, and all internal signals have also settled to a constant value, the element equations, the *G*'s, can usually be reduced to constants. If all the element functions are constants, the model can be solved, that is, the value of the output and all internal signals determined, using algebra. A steady-state solution of a systems model is given in Example 1.4 below.

One of the earliest physiological systems models, the pupil light reflex, is shown in Figure 1.9, and includes two processes. The pupil light reflex is the response of the iris to changes in light intensity falling on the retina. Increases or decreases in ambient light cause the muscles of the iris to change the size of the pupil in an effort to keep light falling on the retina constant. (This system was one of the first to be studied using engineering tools.) The two-component system receives light as the input and produces a movement of the iris muscles that changes pupil area, the aperture in the visual optics. The first box represents all of the neural processing associated with this reflex, including the light receptors in the eye. It generates a neural control signal, which is sent to the second box. The second box represents the iris musculature, including its geometric configuration. The input to this second box is the neural control signal from the first box and the output is pupil area.

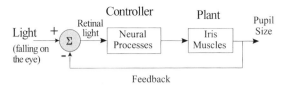

Figure 1.9 A system model of the pupil light reflex. Light falling on the retina stimulates a neural *controller* that generates a neural signal that is sent to the iris muscles, the *plant* or *effector* apparatus. The system involves *feedback* because as the pupil (the hole in the iris) reduces in size, it reduces the light falling on the retina. This is considered a negative feedback system because a positive increase in the response (in this case a reduction in pupil size), leads to a decrease in the stimulus (i.e., the light falling on the retina).

The systems model shown in Figure 1.9 demonstrates the strengths and the weaknesses of systems analysis. By compressing a number of complex processes into a single black box, and representing these processes by a single input–output equation, a systems model can provide a concise, highly simplified representation of a very complex system. You need not understand how a biological process accomplishes a given task. As long as you can document some of its behavior quantitatively (which allows you to construct the input–output equation), you can usually construct a system representation. This will allow you to analyze the system's behavior over a large (perhaps nonphysiological) stimulus range or incorporate that process into the analysis of a larger system. However, this ability to reduce complex processes to a few elements, each represented by a single equation, means that these models do not provide much insight into how the process or processes are implemented by the underlying physiological mechanisms.

The effector apparatus in Figure 1.9, the iris musculature, is often termed the *plant* in systems models. This curious terminology comes from early applications of linear systems analysis to study the control of large chemical plants. These large systems were divided into control processes under the heading *controller*, and effector processes grouped under the heading *plant*. This terminology has been transferred to physiological control models, especially those involving motor control systems.

To complete this system description of the light reflex, note that changes in pupil area, the output of the system, will alter the light falling on the retina, the input to the system. Hence, the output feeds back to the input, creating a classic *feedback control system* (Figure 1.9). The feedback is negative because an increase in light stimulation will generate an increase in the response—in this case a decrease in pupil area—decreasing the light falling on the retina and offsetting, to some extent, the increase in light stimulus. In the pupil light reflex, the decrease in retinal light produced by the decrease in pupil size does not fully compensate for the increase in stimulus, so the feedback gain is less than one (unity).

Systems models often provide more detail than that given in the very basic structure of the model in Figure 1.9. Figure 1.10 shows a more detailed model of the neural pathways that mediate the vergence eye movement response, the processes used to turn the eyes inward to track visual targets at different depths. The model shows three neural paths converging on the elements representing the oculomotor plant (the two right-most system elements). Neural processes in the upper two pathways provide a velocity-dependent signal to move the eyes quickly to approximately the right position. The lower pathway represents the processes that use visual feedback to more slowly fine-tune the position of the eyes and attain a very accurate final position. The error between the angle required to precisely image a stimulus in the two eyes and that actually attained by this neural controller is generally less than a tenth of a degree.

As with analog models, systems models can be evaluated by simulating their behavior on a computer. This not only provides a reality check—do they produce a response similar to that of the real system—but also permits evaluation of internal components and signals not available to the experimentalist. For example, what

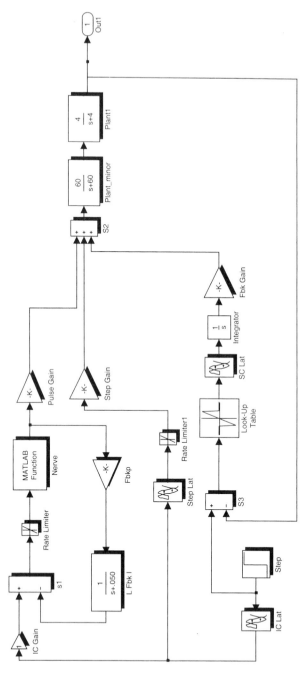

Figure 1.10 A model of the vergence eye movement neural control system showing more of the control details. This model can be simulated using MATLAB's Simulink program.

would happen in the vergence model if the neural components responsible for the pulse signal (upper pathway) were not functioning, or functioning erratically perhaps due to a brain tumor? Systems-type models are even easier to simulate than analog models because MATLAB provides a software package known as Simulink that uses graphics to convert a system model into computer code. Simulink simulations of the vergence model produce responses that are very close to actual vergence eye movements. Although Simulink is only developed for systems models, it is easy to convert analog models into system format so this software can be used to simulate analog models as well. Alternatively, there are programs such as pSpice specifically designed to simulate electronic circuits that can also be used for analog models.

Figures 1.9 and 1.10 show another important property of systems models. The influence of one process on another is explicitly stated and indicated by the line connecting two processes. This line has a direction usually indicated by an arrow, which implies that the influence or information flow travels only in that direction. If there is also a reverse flow of information, such as in the case in feedback systems, this must be explicitly stated in the form of an additional connecting line showing information flow in the reverse direction.

The next example has some of the flavor of the simulation approach, but will not require the use of Simulink.

Example 1.3: There is a MATLAB function on the disk that simulates some unknown process. The function is called $process\_x$ and takes an input variable, x, and generates a variable signal, y. (The Courier typeface is used to indicate a MATLAB variable, function, or code.) The function expects the input to be a signal represented by an array of numbers (as if x were a digitized signal), and produces an output signal that will be an array of number the same length as the input. We are to determine if process_x is a linear process over an input stimulus range of 0 to ±100. We can input to the process any signal we desire and examine the output.

Solution: Our basic strategy will be to input different signals having values within the desired range and see if the outputs are proportional. However, what is the best signal to use? The easiest might be to input two or three signals that have a constant value; for example, $x(t) = 1$, then $x(t) = 10$, then $x(t) = 100$, along with the negative values. The output should be proportional. However, what if the process contains a derivative operation? Although the derivative is a linear operation, the derivative of a constant is zero, and so we would get zero out for all three signals. Similarly, if the process contains integrations, the output to a constant could be difficult to interpret. Going back to basic calculus, recall that the derivative of a sine is a cosine, and the integral of a sine is a negative cosine. Thus, if the input signal were a sine, the output would still be sinusoidal even if the process contained integrations and/or differentiations. If the process contained derivative or integral operations, the sinusoidal output would be scaled (by the frequency), but this scaling would apply to all sine inputs.

Our strategy will be to input different sines having different amplitudes. If the output signals are proportional to the input amplitudes, we would guess that process_x is a linear process over the values tested. Because the work will be done on a computer, we can use any number of sine inputs, so let us try 100 different input signals ranging in amplitude from ±1 to ±100. If we plot the amplitude of the sinusoidal output, it should plot as straight line if the system is linear, and some other curve if it is not. The MATLAB code for this evaluation is as follows:

```
% Example 1.3 Example to evaluate an unknown process
% called 'process_x' to determine if it is linear.
%
t = 0:2*pi/500:2*pi;          % Sine wave argument
for k = 1:100                 % Amplitudes will vary from 1
                              % to 100
 x = k*sin(t);               % Generate a 1 cycle sine wave
 y = process_x(x);           % Input sine to process
 output(k) = max(y);         % Save max value of output
end
plot(output);                 % Plot and label output
xlabel('Input Amplitude');
ylabel('Output Amplitude');
```

Analysis: Within the for-loop, the program generates a one-cycle sine wave having the desired amplitude. The amplitudes are incremented from 1 to 100 as the loop progresses. The sine wave, stored as variable x, becomes the input signal to process_x. The function produces an output signal, y. The maximum value of the output signal it found using MATLAB's *max* routine, and save in variable array, output. When the loop completes, the 100 values of output are plotted.

The figure below shows one of the input signals (solid curve) and the corresponding output (dashed curve) produced by process_x. The input signal looks like a sine function as expected, whereas the output looks like a cosine function. This suggests that the process contains a derivative operation. The output signal is also slightly larger than the input signal.

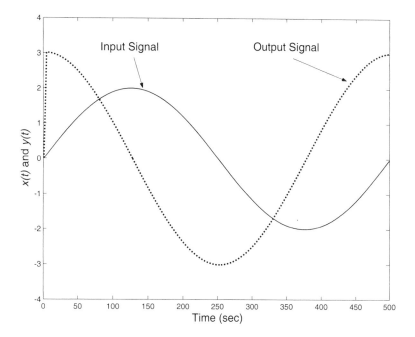

The plot of maximum output values (i.e., variable `output`) is a straight line indicating a linear relationship between the amplitude of output and input signal. Whatever the process really is, it appears to be linear.

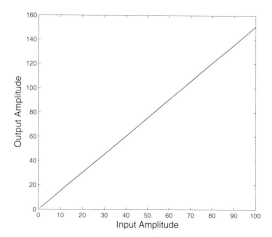

Figuring out exactly what the process is solely by testing it with external signals can be a major challenge. The field of *system identification* deals with approaches to obtain a mathematical representation for an unknown system by probing it with external signals. A few examples and problems later in this text show some of the techniques used to evaluate linear systems in this manner. Comparing the input and output sinusoids it looks like `process_x` contains a derivative and a multiplying factor that increases signal amplitude, but more input–output combinations would have to be evaluated to confirm this guess.

Example 1.4:   Find the overall input–output relationship for the systems model below. Assume that the system is in steady-state condition so that all the signals have constant values and the two elements, represented by the equations G and H, are simply gain constants.

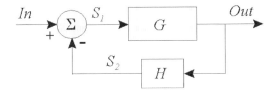

Solution: Generate an algebraic equation based on the configuration of the system and the fact that the output of each process is just the input multiplied by the associated gain term as stated in Eq. 1.6.

By definition: $G = Out/S1$; $H = S_3/Out$;   and $S_1 = In - S_3$

where G and H are constants

Hence $S_1 = Out/G$; $S_3 = Out(H)$

Since $S_1 = In - S_3$ substituting in the above: $Out/G = In - Out(H)$

Rearranging: $Out = In(G) - Out(GH)$; $Out(1 + GH) = In(G)$

$$\frac{Out}{In} = \frac{G}{1 + GH} \qquad\qquad \text{[Eq. 1.7]}$$

The solution is the classic *feedback equation*. Since the two elements, G and H, could be represented by simple gain constants, algebra alone can be used to work out the input–output equations. What of more complicated situations where the model is not in steady state and/or the processes must be defined using differential and integral operations? In Chapters 4 and 8 we develop techniques that encode calculus operations into algebraic manipulations so that this equation, along with the other algebraic operations used in this example, still applies.

1.3.4   Systems and Analog Analysis: Summary

The basic differences and relative strengths and weaknesses of systems analysis versus analog analysis have already been described. Systems analysis only tries to represent the behavior of a process whereas analog analysis makes some effort to mimic the way in which the process produces that behavior. This is done in analog models by representing the process using elements that are, to some degree, analogous to those in the actual process. Analogous elements have the same general behavior as the physiological elements they represent; hence, analog models usually represent the system at a lower level, and in greater detail, than do systems models. However, not all analog models offer this detail. This can be seen in the windkessel cardiovascular model of Figure 1.5. The single capacitor, Cp, represents the combined elastic behavior, or springlike characteristics, of the entire arterial tree. Analog models often provide better representation of secondary features such as energy use, which is usually similar between analog elements and the actual components they represent.

System models provide better clarity, particularly with regard to information flow or influences. In analog models, all components may interact to some extent and this interaction may not be obvious from inspection of the model. Referring to the eye muscle model of Figure 1.7, a change in just one parallel elastic element, k_P, will modify the force on every element in the model. In systems models, all influences are explicitly shown and their interactions are immediately apparent from an inspection of the model. For example, in the model of Figure 1.9, the fact that the iris also influences the neural controller is explicitly shown by the feedback pathway. This can be of great benefit in clarifying the control structure of a complex system. Perhaps the most significant advantage of the systems approach is that it allows processes to be rigorously represented without requiring the modeler to know the details of the underlying physiological mechanism.

1.4 NOISE AND VARIABILITY

Where there is signal, there is noise. Occasionally, the noise will be at such a low level that it is of little concern, but usually the noise limits the usefulness of the signal. This is especially true for physiological signals because they experience many potential sources of noise or variability. In most usages, *noise* is a general and relative term: noise is what you do not want, and signal is what you do want. This leads to a definition of noise as any form of unwanted variability. Noise is inherent in most measurement systems and is often the limiting factor in the performance of a medical instrument. Indeed, many medical instruments go to great lengths in terms of signal-conditioning circuitry and signal-processing algorithms to compensate for unwanted variability.

In biomedical measurements, noise or variability has four possible origins: (a) physiological variability; (b) environmental noise or interference; (c) measurement or transducer artifact; and (d) electronic noise. Physiological variability comes about because the information you desire is based on measurements subject to biological influences other than those of interest. For example, assessment of respiratory function based on the measurement of blood P_{O_2} could be confounded by other physiological mechanisms that alter blood P_{O_2}. Measurement errors due to physiological variability can be a very difficult to resolve, sometimes requiring a total redesign (or rethinking) of the approach.

Environmental noise can come from sources external or internal to the body. A classic example is the measurement of the fetal ECG signal where the desired signal is corrupted by the mother's ECG. Because it is not possible to describe the specific characteristics of environmental noise, typical noise reduction approaches such as filtering (described in Chapter 4) are not usually successful. Sometimes environmental noise can be reduced using *adaptive filters* or *noise cancellation* techniques that adjust their filtering properties on the basis of the current environment.

Measurement artifact is produced when the measurement device, or *transducer,* responds to energy modalities other than those desired. For example, recordings of electrical signals of the heart, the ECG, are made using electrodes placed on the skin. These electrodes are also sensitive to movement, so-called *motion artifact,* where the electrodes respond to mechanical movement as well as to the desired electrical activity of the heart. This artifact is not usually a problem when the ECG is recorded in the physician's office, but it can be if the recording is made during a patient's normal daily living, as in a Holter recording. Measurement artifacts can sometimes be successfully addressed by modifications in transducer design. Aerospace research has led to the development of electrodes that are relatively insensitive to motion artifact.

Unlike the other sources of variability, electronic noise has well-known sources and characteristics. Electronic noise falls into two broad classes: *thermal* or *Johnson* noise, and *shot* noise. The former is produced primarily in resistor or resistance materials whereas the latter is related to voltage barriers associated with semiconductors. Both sources produce noise that contains energy over a broad range of frequencies, often extending from DC to 10^{12} to 10^{13} Hz. Such broad-spectrum noise is referred to as *white noise* because it contains energy at all frequencies (or at least all the frequencies of interest to bioengineers) just as white light contains energy at all frequencies (or at least, all the frequencies we can see). Figure 1.11 shows a plot of the energy in a simulated white-noise waveform (actually, an array of random numbers) plotted against frequency. This is similar to a plot of the energy in a beam of light versus wavelength (or frequency) and, as with light, is also referred to as a *spectral plot* or *spectrum.* (A method for generating such a spectral or *frequency plot* from the noise waveform, or any other waveform, is developed in Chapter 3.) Note that the energy of the simulated noise is constant across the spectral range.

The various sources of noise or variability along with their causes and possible remedies are presented in Table 1.3. Note that in three of four instances, appropriate transducer design may aid in the reduction of the variability or noise. This demonstrates the important role of the transducer in the overall performance of the instrumentation system.

1.4.1 Electronic Noise

Johnson or thermal noise is produced by resistance sources and the amount of noise generated is related to the resistance and to the temperature:

$$V_J = \sqrt{4kTRBW} \text{ volts} \qquad \text{[Eq. 1.8]}$$

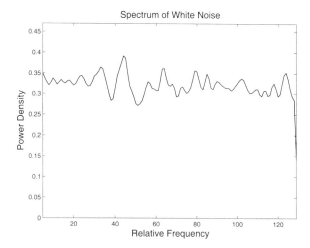

Figure 1.11   A plot of the energy in white noise as a function of frequency. The noise has a flat spectral characteristic showing similar energy levels over a wide range of all frequencies plotted. This equal-energy characteristic gives rise to the term *white noise*. Techniques for producing a signal's spectral plot are discussed in Chapter 3.

TABLE 1.3   Sources of Variability

Source	Cause	Potential Remedy
Physiological variability	Measurement only indirectly related to variable of interest	Modify overall approach
Environmental (internal or external)	Other sources of similar energy form	Noise cancellation Transducer design
Artifact	Transducer responds to other energy sources	Transducer design
Electronic	Thermal or shot noise	Transducer or electronic design

where R is the resistance in ohms, T the temperature in degrees Kelvin, and k Boltzman's constant ($k = 1.38 \times 10^{-23}$ Joules/°Kelvin). (A temperature of 310°K is often used as room temperature, in which case $4kT = 1.7 \times 10^{-20}$ J.) Here BW is the range of frequencies that is included in the signal. This range of frequencies is termed *bandwidth* and is better defined in Chapter 3. This frequency range is usually determined by the characteristics of the measurement system, often the filters used in the system. Because Johnson noise is spread over all frequencies, the greater the signal's bandwidth, the greater the noise in any given signal.

If noise current is of interest, the equation for Johnson noise current can be obtained from Eq. 1.8 in conjunction with Ohm's law:

$$I_J = \sqrt{4kT\,BW/R} \text{ amps} \qquad\qquad \text{[Eq. 1.9]}$$

In practice, there will be limits imposed on the frequencies present within any waveform (including noise waveforms), and these limits are used to determine bandwidth. In the problems given here the bandwidth is simply stated. Bandwidth is usually specified in hertz with units of inverse seconds (i.e., 1/second). Because bandwidth is not always known in advance, it is common to describe a relative noise, specifically the noise that would occur if the bandwidth were 1.0 Hz. Such relative noise specification can be identified by the unusual units required: $V/\sqrt{Hz}$ or ampere $(A)/\sqrt{Hz}$. Shot noise is defined as a current noise and is proportional to the baseline current through a semiconductor junction:

$$I_s = \sqrt{2qI_dBW} \text{ amps} \qquad [\text{Eq. 1.10}]$$

where q is the charge on an electron (1.602×10^{-19} coulomb [coul]), and I_d is the baseline semiconductor current. (In photodetectors, the baseline current that generates shot noise is termed the *dark current*, hence the letter d in the current symbol, I_d in Eq. 1.9.) Again, the noise is spread across all frequencies so the bandwidth must be specified to obtain a specific value, or a relative noise can be specified in $A/\sqrt{Hz}$.

When multiple noise sources are present, as is often the case, their voltage or current contributions add to the total noise as the square root of the sum of the squares, assuming that the individual noise sources are independent. For voltages:

$$V_T = \sqrt{V_1^2 + V_2^2 + V_3^2 + \cdots V_N^2} \qquad [\text{Eq. 1.11}]$$

A similar equation applies to current noise.

Example 1.5: A 20-mA current flows through a diode (i.e., a semiconductor) and a 200-Ω resistor. What is the net current noise, i_n? Assume a bandwidth of 1 MHz (1×10^6 Hz).

$$200\,\Omega$$

$$i = 20\,\text{mA}$$

Solution: Find the noise contributed by the diode using Eq. 1.10, the noise contributed by the resistor using Eq. 1.9, then combine them using Eq. 1.11.

$$i_{nd} = \sqrt{2qI_dBW} = \sqrt{2(1.602 \times 10^{-19})(20 \times 10^{-3})10^6} = 8.00 \times 10^{-8}\,\text{amps}$$

$$i_{nR} = \sqrt{4kT\,BW/R} = \sqrt{1.7 \times 10^{-20}(10^6/200)} = 9.22 \times 10^{-9}\,\text{amps}$$

$$i_{nT} = \sqrt{i_{nd}^2 + i_{nR}^2} = \sqrt{6.4 \times 10^{-15} + 8.5 \times 10^{-17}} = 8.1 \times 10^{-8}\,\text{amps}$$

Note that most of the current noise is coming from the diode, so the addition of the resistor's current noise does not contribute much to the diode noise current. The mathematics in this example could be simplified by calculating the square of the noise current (i.e., not taking the square roots) and using those values to get the total noise before taking the square roots.

1.4.2 Signal-to-Noise Ratio

Most waveforms consist of signal plus noise mixed together. As noted previously, signal and noise are relative terms, relative to the task: the signal is that portion of the waveform of interest whereas the noise is everything else. Often the goal of signal processing is to separate out a signal from noise, identify the presence of a signal buried in noise, or detect features of a signal buried in noise.

The relative amount of signal and noise present in a waveform is usually quantified by the *signal-to-noise ratio* (SNR). As the name implies, this is simply the ratio of signal to noise, both measured in RMS (root-mean-squared) amplitude. This measurement is rigorously defined in the next chapter. The SNR is often expressed in decibels (dB) where:

$$SNR = 20\log\left(\frac{signal}{noise}\right) \qquad \text{[Eq. 1.12]}$$

To convert from decibel scale to a linear scale:

$$SNR_{Linear} = 10^{dB/20} \qquad \text{[Eq. 1.13]}$$

For example, a SNR of 20 dB means that the RMS value of the signal is 10 times the RMS value of the noise ($10^{(20/20)} = 10$), +3 dB indicates a ratio of 1.414 ($10^{(3/20)} = 1.414$), 0 dB means the signal and noise are equal in RMS value, −3 dB means that the ratio is 1/1.414, and −20 dB means the signal is 1/10 of the noise in RMS units. Figure 1.12 shows a sinusoidal signal with various amounts of white noise. Note that is it is difficult to detect the presence of the signal visually when the SNR is −3 dB, and impossible when the SNR is −10 dB.

1.5 SUMMARY

Biological systems include a variety of physiological processes ranging from the organ level through the cellular level to the molecular level. Classic physiology is structured around large-scale biological systems such as the cardiovascular system, the endocrine system, the gastrointestinal system, and others. All biological systems communicate with one another, and with themselves, via biosignals. Such signals are be carried by electrical, chemical, mechanical, or thermal energy. All signals involve some form of coding process. For analog signals, the information is encoded into the amplitude of the signal at any given instant in time. If these analog signals are processed by digital computers, they must be converted to digital format, a process that involves slicing the signal in both amplitude and time. Amplitude slicing is known as quantization, whereas time slicing is known as sampling.

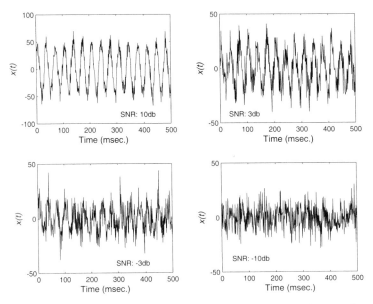

Figure 1.12   A 30-Hz sine wave with varying amounts of added noise. The sine wave is barely discernible when the signal-to-noise ratio is −3 dB and not visible when the signal-to-noise ratio is −10 dB.

The field of linear systems analysis encompasses analog and systems processes. Both representations use linear elements so that this analysis only applies to linear processes or processes that can be taken, or approximated, as linear. Electrical analog models are used to analyze electric circuits and to represent physiological processes. Mechanical analog models can also be used from either of these two perspectives. Examples were given of the early use of an electric circuit model to represent the cardiovascular system and an early mechanical model of skeletal muscle.

All signals are contaminated by noise, variability, or other artifact. Efforts to obtain meaningful information from physiological processes are often thwarted by the inability to directly measure variables of interest. All too frequently variables only loosely related to the process of interest are readily available, and these may be altered by the physiological process itself or the influence of other processes. The measurement device, the biotransducer, is often a major source of measurement errors because it responds to influences of other energy forms or environmental factors. Finally, all electrically based measurements are contaminated by thermal and/or shot noise. These two noise processes are well defined and contain noise energy over a wide range of frequencies. This broad distribution of energy means that thermal and shot noise can always be improved by limiting the frequencies contained in the signal. A variety of filters exists in analog and digital forms to limit the frequencies in a signal to only those that carry the desired information.

PROBLEMS

1. An electrical inductor has a defining equation that is the same as a mass if the variable voltage and current are substituted for force and velocity (specifically, $V_L = L\ di/dt$). A constant voltage of 10 V is placed across a 1-H inductor. How long will it take for the current through the inductor to reach 1 A? (See Example 1.2.)

2. Assume that the feedback control system presented in Example 1.4 is in steady-state or static conditions. If $G = 100$ and $H = 1$ (i.e., a unity gain feedback control system), find the output if the input equals 1. Find the output if the input is increased to 10. [Note how the output is proportional to the input, which accounts for why a system (having this configuration) is sometimes termed a *proportional control system*.] Now find the output if the input is 10 and G is increased to 1,000. Note that the difference between the input and output values depends on the value of G.

3. In the system given in Problem 1.2, the input is changed to a signal that smoothly goes from 0.0 to 5.0 in 10 seconds [i.e., $In(t) = 0.5t$ seconds]. What will the output look like? (*Note:* G and H are simple constants, so Eq. 1.7 still holds.)

4. A resistor produces 10-μV noise when the room temperature is 310°K and the bandwidth is 1 kHz. What current noise would be produced by this resistor?

5. The noise voltage out of a 1-MΩ resistor is measured using a digital volt meter as 1.5 μV at a room temperature of 310°K. What is the effective bandwidth of the voltmeter?

6. If a signal is measured as 2.5 V and the noise is 28 mV (28×10^{-3} V), what is the SNR in decibels?

7. A single sinusoidal signal is found in a large amount of noise. (If the noise is larger than the signal, the signal is sometimes said to be 'buried in noise.') The RMS value of the noise is 0.5 V and the SNR is 10 dB. What is the RMS amplitude of the sinusoid?

MATLAB Problems

8. Use the approach presented in Example 1.3 to determine if either of two processes, `process_y` or `process_z`, are linear. The two processes are found on the disk as MATLAB functions.

9. Write a MATLAB function that takes in two variables, and input variable `x` and gain variable `G`, and produces and output variable `y`. This function should implement the feedback equation Eq. 1.7, with the variable `H` set to 1.0 (i.e., a unity gain feedback system). (Name the function 'fbk_system.') Input a two-cycle sine wave as in Example 1.3 having an amplitude of 1. Plot the maximum values of the input–output relationship for this process [i.e., `max(y)/max(x)`] as a function of `G`, where `G` ranges between 1 and 1,000. (*Hint:* Put the process in a for-loop as in Example 1.3 and increment `G`. This will provide a more detailed demonstration of the relationship between the input–output ratio and the importance of the value of `G` in a feedback system.)

2 BASIC SIGNAL PROCESSING

2.1 BASIC SIGNALS: THE SINUSOIDAL WAVEFORM

Signals are the foundation of information processing, transmission, and storage. Signals also provide the interface with physiological systems and are the basis for communication between biological processes (Figure 2.1). Given the ubiquity of signals within and outside the body, it should be no surprise that understanding at least the basics of signals is fundamental to understanding, and interacting with, biological processes.

A few signals are simple and can be defined analytically, that is as mathematical functions. For example, a sinusoidal signal is defined by the equation:

$$x(t) = A\sin(\omega_p t) = A\sin(2\pi f_p t) = A\sin\left(\frac{2\pi t}{T}\right) \qquad \text{[Eq. 2.1]}$$

where A is the signal amplitude, or more accurately the *peak-to-peak amplitude*, ω_p is the frequency in radians per second, f_p is the frequency in hertz, and T is the period in seconds, and t is time in seconds. Recall that frequency can be expressed in either radians or hertz (the units formerly known as cycles per second) and are related by 2π:

$$\omega_p = 2\pi f_p \qquad \text{[Eq. 2.2]}$$

Both forms of frequency are used in the text, and the reader should be familiar with both. The frequency in Hz is also the inverse of the period, T:

$$f_p = \frac{1}{T} \qquad \text{[Eq. 2.3]}$$

The signal presented in Eq. 2.1 is completely defined by A and f_p (or ω_p, or T); once you specify these two terms, you have characterized the sine signal for all time. The sine wave signal is rather boring: if you have seen one cycle, you have seen them all. Moreover, because the signal is completely defined by A and f_p, if neither the amplitude, A, nor the frequency, f_p, changes over time, it is hard to see how this

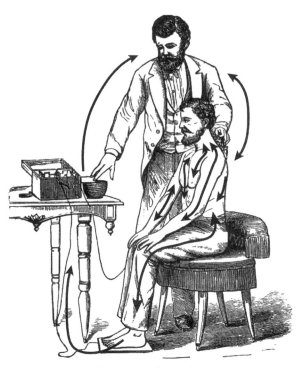

Figure 2.1 Signals continuously pass between various parts of the body. These *biosignals* are carried either by electrical energy, as in the nervous system, or by molecular signatures, as in the endocrine system and many other biological processes. Measurement of these biosignals is fundamental to diagnostic medicine and to bioengineering research.

signal could carry much information. These limitations notwithstanding, sine waves (and cosine waves) are at the foundation of many signal analysis techniques. In part, their importance stems from their simplicity and the way they are treated by linear systems. (Chapter 3 covers the 'magical' properties of sine waves in some detail.) Sine wave–like signals can also be represented by cosines, and the two are related.

$$A\cos(\omega t) = A\sin\left(\omega t + \frac{\pi}{2}\right) = A\sin(\omega t + 90 \text{ degrees})$$

$$A\sin(\omega t) = A\cos\left(\omega t - \frac{\pi}{2}\right) = A\cos(\omega t - 90 \text{ degrees}) \qquad \text{[Eq. 2.4]}$$

Note that the second representations [i.e., $A\sin(\omega t + 90$ degrees) and $A\cos(\omega t - 90$ degrees)] have conflicting units: the first part of the sine argument, ωt, is in radians, whereas the second part is in degrees. Nonetheless, this is common usage and is the form that is used throughout this text.

A general *sinusoid* (as opposed to a pure sine wave or pure cosine wave) is a sine or cosine with a general phase term as shown in Eq. 2.5:

$$x(t) = A\sin(\omega_p t + \theta) = A\sin(2\pi f_p t + \theta) = A\sin\left(\frac{2\pi t}{T} + \theta\right) \text{ or equivalently}$$

$$x(t) = A\cos(\omega_p t - \theta) = A\cos(2\pi f_p t - \theta) = A\cos\left(\frac{2\pi t}{T} - \theta\right) \qquad \text{[Eq. 2.5]}$$

where again the phase, θ, would be expressed in degrees even though the frequency descriptor ($\omega_p t$, or $2\pi f_p t$, or $2\pi t/T$) is expressed in radians or hertz. Many of the sinusoidal signals described in this text are expressed as in Eq. 2.5. Figure 2.2 shows two sinusoids that differ by 60 degrees.

To convert the difference in phase angle to a difference in time, note that the phase angle varies through 360 degrees during the course of one period, T seconds. To calculate the time difference or time delay between the two sinusoids, t_d, given the phase angle θ:

$$t_d = \frac{\theta}{360}T = \frac{\theta}{360f} \quad \text{or} \quad \theta = \frac{t_d}{T}360 = 360 t_d f \qquad \text{[Eq. 2.6]}$$

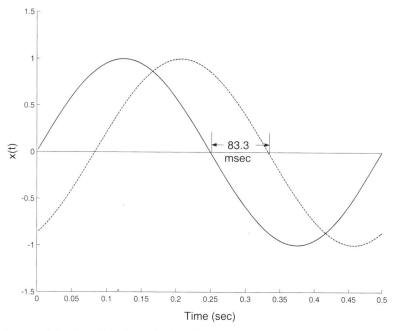

Figure 2.2 Two 2-Hz sinusoids that differ in phase by 60 degrees. This phase difference translates to a time difference or delay of 83.3 msec.

where $f = f_p = 1/T$. For the 2-Hz sinusoids in Figure 2.2, $T = 1/f = 0.5$ seconds, so:

$$t_d = \frac{\theta}{360}T = \frac{60}{360}0.5 = 0.0833 \text{ seconds} = 83.3m \text{ seconds}$$

Example 2.1:   Find the time difference or delay between two sinusoids:

$$x_1(t) = \cos(4t + 30), \text{ and } x_2(t) = -2\sin(4t)$$

Solution: Convert both to a sine or cosine (here we convert to cosines):

$$x_2(t) = -2\cos(4t - 90)$$

Thus, the angle between the two sinusoids is 120 degrees [30 − (−90)]. The period is given by:

$$T = \frac{1}{f} = \frac{1}{\omega/2\pi} = \frac{1}{4/2\pi} = 1.57 \text{ seconds}$$

and the time delay is:

$$t_d = \frac{\theta}{360}T = \frac{120}{360}1.57 = 0.523 \text{ seconds}$$

2.1.1 Sinusoidal Arithmetic

Equation 2.5 describes an intuitive way of thinking about a sinusoid, as a sine wave with a phase shift. Alternatively, Eq. 2.5 shows that a cosine could just as well be used instead of the sine to represent a general sinusoid, and in this text, we use both. Sometimes it is mathematically convenient to represent a sinusoid as a combination of a pure sine and a pure cosine. This representation can be achieved using the well-known trigonometric identity for the sum of two arguments of a cosine function:

$$\cos(x - y) = \cos(x)\cos(y) + \sin(x)\sin(y) \qquad \text{[Eq. 2.7]}$$

Based on this identity, the equation for a sinusoid can be written as:

$$C\cos(2\pi ft - \theta) = C\cos(\theta)\cos(2\pi ft) + C\sin(\theta)\sin(2\pi ft)$$

$$= a\cos(2\pi ft) + b\sin(2\pi ft)$$

where: $a = C\cos(\theta); \quad b = C\sin(\theta)$ $\qquad$ [Eq. 2.8]

To convert from a sine and cosine to a single sinusoid with angle θ, start with Eq. 2.8.

If $a = C\cos(\theta)$ and $b = C\sin(\theta)$, then to determine C:

$$a^2 + b^2 = C^2(\cos^2\theta + \sin^2\theta) = C^2$$

$$C = \sqrt{a^2 + b^2} \qquad \text{[Eq. 2.9]}$$

Equation 2.10 shows the calculation for θ given a and b:

$$\frac{b}{a} = \frac{C\sin(\theta)}{C\cos(\theta)} = \tan(\theta) \quad \text{and}\ldots$$

$$\theta = \tan^{-1}\left(\frac{b}{a}\right) \qquad\qquad \text{[Eq. 2.10]}$$

Care must be taken in evaluating Eq. 2.10 to ensure that θ is determined to be in the correct quadrant on the basis of the signs of a and b. If both a and b are positive, θ must be between 0 and 90 degrees; if b is positive and a is negative, θ must be between 90 and 180 degrees (a calculator or MATLAB will not know this and will put any negative product in the fourth quadrant); if both a and b are negative, θ must be between 180 and 270 degrees (calculators and MATLAB put positive arguments in the first quadrant even if they result from two negative numbers); and finally, if b is negative and a is positive, θ must be between 270 and 360 degrees. Again, it is common to use degrees for phase angle.

To add sine waves, simply add their amplitudes. The same applies to cosine waves:

$$a_1 \cos(\omega t) + a_2 \cos(\omega t) = (a_1 + a_2)\cos(\omega t)$$

$$a_1 \sin(\omega t) + a_2 \sin(\omega t) = (a_1 + a_2)\sin(\omega t) \qquad \text{[Eq. 2.11]}$$

To add two sinusoids [i.e., $C\sin(\omega t + \theta)$ or $C\cos(\omega t - \theta)$], convert them to sines and cosines using Eq. 2.8, add sines to sines and cosines to cosines, and convert back to a single sinusoid if desired.

Example 2.2:   Convert the sum of a sine and cosine wave, $x(t) = -5\cos(10t)$ $-3\sin(10t)$ into a single sinusoid.

Solution: Apply Eq. 2.9 and Eq. 2.10:

$$a = -5 \quad \text{and} \quad b = -3$$

$$C = \sqrt{a^2 + b^2} = \sqrt{5^2 + 3^2} = 5.83$$

$$\theta = \tan^{-1}\left(\frac{b}{a}\right) = \tan^{-1}\left(\frac{-3}{-5}\right) = 31 \text{ degrees,}$$

but θ must be in the third quadrant since both a and b are negative:

$$\theta = 31 + 180 = 211 \text{ degrees}$$

Therefore, the single sinusoid representation would be as follows:

$$x(t) = C\cos(\omega t - \theta) = 5.83\cos(10t - 211 \text{ degrees})$$

Analysis: Using Equations 2.8 through 2.11, any number of sines, cosines, or sinusoids can be combined into a single sinusoid if they are all at the same frequency. This is demonstrated in Example 2.3.

Example 2.3: Combine $x(t) = 4\cos(2t - 30$ degrees$) + 3\sin(2t + 60$ degrees$)$ into a single sinusoid.

Solution: Expand each sinusoid into a sum of cosine and sine, algebraically add the cosines and sines, and recombine them into a single sinusoid. Be sure to convert the sine into a cosine [recall Eq. 2.4: $\sin(\omega) = \cos(\omega t - 90$ degrees$)$] before expanding this term.

$$4\cos(2t - 30) = a\cos(2t) + b\sin(2t)$$

where: $a = C\cos(\theta) = 4\cos(-30) = 3.5$ and $b = C\sin(\theta) = 4\sin(-30) = -2$

$$4\cos(2t + 30) = 3.5\cos(2t) - 2\sin(2t)$$

Converting the sine to a cosine then decomposing the sine into a cosine plus a sine:

$$3\sin(2t + 60) = 3\cos(2t - 30) = 2.6\cos(2t) - 1.5\sin(2t)$$

Combining cosine and sine terms algebraically:

$$4\cos(2t - 30) + 3\sin(2t + 60) = (3.5 + 2.6)\cos(2t) + (-2 - 1.5)\sin(2t)$$

$$= 6.1\cos(2t) - 3.5\sin(2t)$$

$$= C\cos(2t + \theta) \text{ where } C = \sqrt{6.1^2 + 3.5^2} \text{ and } \theta = \tan^{-1}\left(\frac{-3.5}{6.1}\right)$$

$C = 7.0$; $\theta = 30$ (Since b is negative, θ is in fourth quadrant) so $\theta = -30$ degrees

$$x(t) = 7.0\cos(2t - 30)$$

This approach can be extended to any number of sinusoids. An example involving three sinusoids is found in Problem 4.

2.1.2 Complex Representation

An even more compact representation of a sinusoid is possible using complex notation. A *complex number* combines a *real number* and an *imaginary number*. Real numbers are commonly used, whereas imaginary numbers are the product of square roots and are represented by real numbers multiplied by the $\sqrt{-1}$. In mathematics, the $\sqrt{-1}$ is represented by the letter i, whereas engineers tend to use the letter j, the letter i being reserved for current. A *complex variable* simply combines a real and an imaginary variable: $z = x + jy$. Hence, although 5 is a real number, $j5$ is an imaginary number, and $5 + j5$ is a complex number. The arithmetic of complex numbers and some of their important properties are reviewed in Appendix E. We will use complex variables and complex arithmetic extensively in later chapters so it will be worthwhile to review these operations.

The beauty of complex numbers and complex variables is that the real and imaginary parts are *orthogonal*. One consequence of orthogonality is that the real and complex numbers (or variables) can be represented as if they are plotted on perpendicular axes (Figure 2.3).

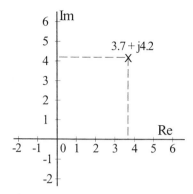

Figure 2.3 A complex number represented as an orthogonal combination of a real number on the horizontal axis and an imaginary number on the vertical axis. This graphic representation is useful for understanding complex numbers and aids in the interpretation of some arithmetic operations.

Orthogonality is discussed in more detail later, but the importance of orthogonality with respect to complex numbers is that the real number or variable does not 'interfere' with the imaginary number or variable and vice versa. Operations on one component do not affect the other. This means that a complex number behaves like two separate numbers rolled into one, and a complex variable like two variables in one. This feature comes in particularly handy when sinusoids are involved because a sinusoid at a given frequency can be uniquely defined by two variables: its magnitude and phase angle (or equivalently, using Eq. 2.8, its cosine and sine magnitudes, a and b). It follows that a sinusoid at a given frequency can be represented by a single complex number.

To find the complex representation, we will use the identity developed by the Swiss mathematician, Euler (Leonhard Euler's last name is pronounced 'oiler'. The use of the symbol e for the basis of the natural logarithmic system is a tribute to his extraordinary mathematical contributions):

$$e^{jx} = \cos x + j \sin x \qquad \text{[Eq. 2.12]}$$

The derivation for this equation is given in Appendix A. This equation links sinusoids and exponentials, providing a definition of the sine and cosine in terms of complex exponentials (Eq. 20 and Eq. 21 in Appendix C). It also provides a concise representation of a sinusoid since a complex exponential contains both a sine and a cosine, although a few extra mathematical features are required to account for the fact that the second term is an imaginary sine term. This equation will prove very useful in two sinusoidally based analysis techniques: Fourier analysis described in the next chapter and phasor analysis described in Chapter 4.

2.2   SIGNAL PROPERTIES: BASIC MEASUREMENTS

Biosignals and other information-bearing signals are often quite complicated and defy a straightforward analytical description. An archetype biomedical signal is the electrical activity of the brain as it is detected on the scalp by electrodes, the electroencephalogram (EEG) shown in Figure 2.4. Although a time display of this signal, as in Figure 2.4, constitutes a unique description, the information carried by this signal is not apparent from the time display, at least not to the untrained eye. Nonetheless, physicians and technicians are trained to extract useful diagnostic information by examining the time display of biomedical signals including the EEG. The time display of the electrocardiogram (ECG) signal is so medically useful that it is displayed continuously for patients undergoing surgery or those admitted to intensive care units (ICUs). This signal has become an indispensable image in television and movie medical dramas. Medical images, which can be thought of as two-dimensional signals, often need only visual inspection to provide information useful for diagnosis.

For some signals, a simple time display provides useful information, but many biomedical signals are not easy to interpret from their time characteristics alone. Nearly all signals will benefit from some additional signal processing. For example, the time display of the EEG signal in Figure 2.4 may have meaning for a trained

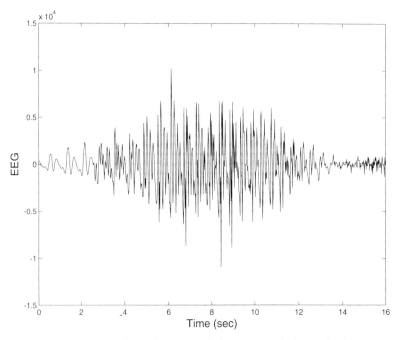

Figure 2.4   Segment of an electroencephalogram signal. (From the PhysioNet data bank, Goldberger et al., 2000.)

neurologist, but it is likely to be uninterpretable to most readers. A number of basic measurements can be applied to a signal to extract more information, while other analyses can be used to probe the signal for specific features. Transformations can be used to provide a different view of the signal. In this section, basic measurements are described followed by more involved analyses. In the next chapter a basic transformation is developed.

One of the most straightforward of signal measurements is the assessment of its average value. Averaging is most easily described in the digital domain. To determine the average of a series of numbers, simply add the numbers together and divide by the length of the series (i.e., the number of numbers in the series). This is mathematically stated as follows:

$$x_{avg} = \overline{x} = \frac{1}{N} \sum_{k=1}^{N} x_k \qquad \text{[Eq. 2.13]}$$

where k is an index number indicating a specific number in the series. The bar over the x in Eq. 2.13 stands for 'the average of . . .'. Equation 2.13 would be appropriate only for finding the average of a digital signal. An analog signal is a continuous function of time, $x(t)$, so the summation becomes an integration. The average or mean of a continuous signal, the continuous version of Eq. 2.13, is obtained by integrating the signal over time and dividing by the time length of the signal:

$$\overline{x}(t) = \frac{1}{T} \int_0^T x(t) dt \qquad \text{[Eq. 2.14]}$$

Note that the primary difference between digital and analog domain equations is the conversion of summation to integration and the use of a continuous variable, t, in place of the discrete integer, k. These conversion relationships are generally applicable, and most digital-domain equations can be transferred to continuous or analog equations in this manner. In this text, usually the reverse operation is used: the continuous domain equation is developed first, then the corresponding digital-domain equation is derived by substitution of summation for integration and an integer variable for the continuous time variable. The conditions under which a continuous analog signal and a digitized version of that signal can be considered equivalent are presented in Chapter 3.

Although the average value is a basic property of a signal, it does not provide any information about the variability of the signal. The root-mean-squared (RMS) value is a measurement that includes the signal's variability and its average. Obtaining the RMS value of a signal is just a matter of following the measurement's acronym in reverse: first squaring the signal, then taking its average, and finally taking the square root of this average:

$$x(t)_{rms} = \left[\frac{1}{T} \int_0^T x(t)^2 dt \right]^{1/2} \qquad \text{[Eq. 2.15]}$$

The discrete form of the equation can be obtained by following the simple rules described above.

$$x_{rms} = \left[\frac{1}{N} \sum_{k=1}^{N} x_k^2 \right]^{1/2}$$ [Eq. 2.16]

Example 2.4:   Find the RMS value of the sinusoidal signal:

$$x(t) = A \sin(\omega_p t) = A \sin(2\pi t / T)$$

Solution: Because this signal is periodic, with each period the same as the previous one, it is sufficient to apply the RMS equation over a single period. (This is true for most operations on sinusoids.) Neither the RMS value nor anything else about the signal will change from one period to the next. Applying Eq. 2.15:

$$\bar{x}(t)_{rms} = \left[\frac{1}{T} \int_0^T x(t)^2 dt \right]^{1/2} = \left[\frac{1}{T} \int_0^T \left(A \sin\left(\frac{2\pi t}{T_p} \right) \right)^2 dt \right]^{1/2}$$

$$= \left[\frac{1}{T} \frac{A^2}{2\pi} \left(-\cos\left(\frac{2\pi t}{T} \right) \sin\left(\frac{2\pi t}{T} \right) + \frac{\pi t}{T} \right) \Big|_0^T \right]^{1/2}$$

$$= \left[\frac{A^2}{2\pi} (-\cos(2\pi)\sin(2\pi) + \pi + \cos 0 \sin 0) \right]^{1/2}$$

$$= \left[\frac{A^2 \pi}{2\pi} \right]^{1/2} = \left[\frac{A^2}{2} \right]^{1/2} = \frac{A}{\sqrt{2}} \cong 0.707 A$$

Hence, there is a proportional relationship between the peak-to-peak amplitude of a sinusoid (*A* in this example) and its RMS value: specifically, the RMS value is $1/\sqrt{2}$ times the peak-to-peak amplitude, rounded in this text to 0.707. This relationship is only true for sinusoids. For other waveforms, the application of Eq. 2.15 or Eq. 2.16 is required.

A statistical measure related to the RMS value is the *variance*, σ^2. The variance is a measure of signal variability regardless of its average. The calculation of variance for discrete and continuous signals is as follows:

$$\sigma^2 = \frac{1}{T} \int_0^T (x(t) - \bar{x})^2 dt$$ [Eq. 2.17]

$$\sigma^2 = \frac{1}{N-1} \sum_{k=1}^{N} (x_k - \bar{x})^2$$ [Eq. 2.18]

where $\bar{x}$ is the mean or signal average. In statistics, the variance is defined in terms of an estimator known as the *expectation* operation applied to the probability distribution function of the data. Because the distribution of a signal is rarely known in advance, the equations given here are used to calculate variance in practical situations.

The *standard deviation* is another measure of a signal's variability and is simply the square root of the variance:

$$\sigma = \left[\frac{1}{T} \int_0^T (x(t) - \overline{x})^2 dt \right]^{1/2} \qquad \text{[Eq. 2.19]}$$

$$\sigma = \left[\frac{1}{N-1} \sum_{k=1}^{N} (x_k - \overline{x})^2 \right]^{1/2} \qquad \text{[Eq. 2.20]}$$

In determining the standard deviation and variance from discrete or digital data, it is common to normalize by $1/N - 1$ rather than $1/N$. This is because the former gives a better estimate of the actual standard deviation or variance when the data being used in the calculation are samples of a larger data set that has a normal distribution (rarely the case for signals). If the data have zero mean, the standard deviation is the same as the RMS value except for the normalization factor in the digital calculation. Nonetheless, they are from different traditions (statistics versus measurement) and are used to describe conceptually different aspects of a signal: signal magnitude for RMS and signal variability for standard deviation. Figure 2.5 shows the EEG data in Figure 2.4 with positive and negative values of standard deviation indicated by horizontal dotted lines.

When multiple measurements are made, multiple values or signals will be generated. If these measurements are combined or added together, the means add so that the combined value, or signal, has a mean that is the average of the individual

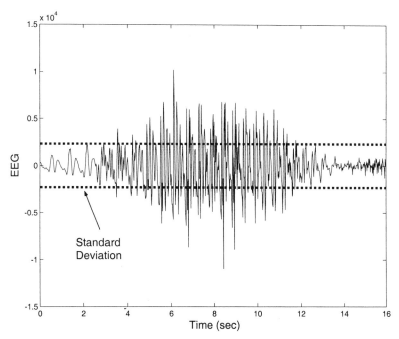

Figure 2.5 A segment of electroencephalogram signal shown in Figure 2.4 with the positive and negative standard deviation (*dotted horizontal line*).

means. The same is true for the variance: the variances add and the average variance of the combined measurement is the mean, or average, of the individual variances:

$$\overline{\sigma}^2 = \frac{1}{N}\sum_{k=1}^{N}\sigma_k^2 \qquad\qquad \text{[Eq. 2.21]}$$

where N is the number of signals being averaged. The standard deviation of the averaged signal is the square root of the variance so the standard deviations add as the $\sqrt{N}$ times the average standard deviation. Accordingly, the mean standard deviation is the average of the individual standard deviations divided by $\sqrt{N}$. Stated mathematically, from Eq. 2.21:

$$\sum_{k=1}^{N}\sigma_k^2 = N\overline{\sigma}^2;$$

Taking the square root of both sides:

$$\sum_{k=1}^{N}\sigma_k = \sqrt{N\overline{\sigma}^2} = \sqrt{N}\overline{\sigma}$$

The mean standard deviation becomes:

$$\text{Mean } \sigma = \frac{1}{N}\sum_{k=1}^{N}\sigma_k = \frac{1}{N}\sqrt{N}\overline{\sigma} = \overline{\sigma}/\sqrt{N} \qquad\qquad \text{[Eq. 2.22]}$$

In other words, averaging measurements from different sensors, or averaging multiple measurements from the same source, will reduce the standard deviation of the measurement's variability by the square root of the number of averages. For this reason, it is common to make multiple measurements whenever possible and average the results. This approach can also be applied to entire signals, a technique known as *ensemble averaging*. An example of ensemble averaging is given in the MATLAB implementation section of this chapter.

2.2.1   Decibels

It is common to compare the intensity of two signals using ratios, V_{Sig1}/V_{Sig2}, and to represent such ratios in units of *decibels*. Actually, decibels (dB) are not really units, but are simply a logarithmic scaling of ratios. The decibel has several advantageous features: (a) It provides a measurement of the effective power, or power ratio; (b) the log operation compresses the range of values (for example, a range of 1 to 1,000 becomes a range of 1 to 3 in log units); (c) when numbers or ratios are to be multiplied, they simply add if they are in log units; and (d) the logarithmic characteristic is similar to human perception. This latter feature motivated Alexander Graham Bell to develop the logarithmic unit called the *bel*. Audio power increments in logarithmic bels were perceived as equal increments by the human ear. The bel turned out to be inconveniently large, so it has been replaced by the decibel (1/10 bel). While originally defined only in terms of a ratio, decibel units are also used to express the intensity of a single signal. In this case, it has a dimension, the dimension of the signal (volts, amps, dynes, and so forth), but these units are often ignored.

When applied to a power measurement, the decibel is defined as 10 times the log of the power ratio:

$$P_{dB} = 10\log\left(\frac{P_2}{P_1}\right)dB \qquad \text{[Eq. 2.23]}$$

When applied to a voltage ratio (or simply a voltage), the decibel is defined as 10 times the log of the RMS value squared, or voltage ratio squared. Because the log is taken, this is the same as 20 times the unsquared ratio or value. If a ratio of sinusoids is involved, then peak-to-peak voltages (or whatever units the signal is in) can also be used, because they are related to RMS values by a constant (0.707), and the constants will cancel in the ratio.

$$v_{dB} = 10\log(v_2^2/v_1^2) = 20\log(v_2/v_1) \quad \text{or:}$$

$$v_{dB} = 10\log v_{RMS}^2 = 20\log v_{RMS} \qquad \text{[Eq. 2.24]}$$

The logic behind taking the *square* of the RMS voltage value before taking the log is that the RMS voltage squared is proportional to signal power. Consider the case where the signal is a time-varying voltage, $v(t)$. To draw energy from this signal, it is necessary to feed it into a resistor, or a resistor-like element that consumes energy. (Recall from basic physics that resistors convert electrical energy into thermal energy, i.e., heat.) The power (energy per unit time) transferred from the signal to the resistor is given by the following equation:

$$P = v_{RMS}^2/R \qquad \text{[Eq. 2.25]}$$

where R is the resistance. This equation shows that the power imparted to a resistor by a given voltage depends, in part, on the value of the resistor. Assuming a nominal resistor value of $1\ \Omega$, the power will be equal to the voltage squared; however, for any resistor value, the power transferred will be proportional to the voltage squared. When decibel units are used to describe a ratio of voltages, the value of the resistor is irrelevant, because the resistor values will cancel out:

$$v_{dB} = 10\log\left(\frac{v_2^2/R}{v_1^2/R}\right) = 10\log\left(\frac{v_2^2}{v_1^2}\right) = 20\log\left(\frac{v_2}{v_1}\right) \qquad \text{[Eq. 2.26]}$$

If decibel units are used to express the intensity of a single signal, the units will be proportional to the log power in the signal.

To convert a voltage from decibel to RMS, use the inverse of the defining equation (Eq. 2.26):

$$v_{RMS} = 10^{X_{dB}/20} \qquad \text{[Eq. 2.27]}$$

Decibel units are particularly useful when comparing ratios of signal and noise, the so-called *signal-to-noise ratio* (SNR) discussed in the last chapter.

Example 2.5:   A sinusoidal signal is fed into an *attenuator* that reduces the intensity of the signal. The input signal has a peak-to-peak amplitude of 2.8 V and the output signal is measured at 2 V peak-to-peak. Find the ratio of output to input voltage in decibels. Compare the power-generating capabilities of the two signals in linear units.

Solution: Convert each peak-to-peak voltage to RMS, then apply Eq. 2.26 to the given ratio. Calculate the ratio without taking the log.

$$V_{RMSdB} = 20\log(V_{out\,RMS}/V_{in\,RMS}) = 20\log\left(\frac{2.0 \times 0.707}{2.8 \times 0.707}\right)$$

$$V_{RMSdB} = -3dB$$

The power ratio is:

$$\text{Power ratio} = \frac{V^2_{out\,RMS}}{V^2_{in\,RMS}} = \frac{(2.0 \times 0.707)^2}{(2.8 \times 0.707)^2} = 0.5$$

Analysis: The ratio of the amplitude of a signal coming out of a process to that going into the process is known as the *gain*, and is often expressed in decibels. When the gain is less than 1, it means there is a loss, or reduction, in signal amplitude. In this case, the signal loss is 3 dB, so the 'gain' of the attenuator is actually −3 dB. To add to the confusion, you can reverse the logic and say that the attenuator has an attenuation (i.e., loss) of +3 dB. In this example, the power ratio was 0.5, meaning that the signal coming out of the attenuator has half the power-generating capabilities of the signal that went in. A 3-dB attenuation is equivalent to a loss of half the signal's energy. Of course, it was not necessary to convert the peak-to-peak voltages to RMS because a ratio of these voltages was taken and the conversion factor (0.707) cancels out.

2.3   ADVANCED MEASUREMENTS: CORRELATIONS AND COVARIANCES

Applying the basic measurements we have just learned to the EEG data in Figure 2.4, we find the signal has a comparatively small mean of −29.8, an RMS value of 2,309 (or 67 dB), and a standard deviation of 2,310 (Figure 2.5). (These numbers are all in relative units that relate to voltage in the brain by an unknown calibration factor.) These basic measurement numbers are not enlightening about the EEG signal or the processes that created it. More insight might be gained by comparing the EEG signal with one or more reference signals, or mathematical functions. For example, we might ask, 'How much is the EEG signal like a 10-Hz sinusoid?' Or, 'How much is it like a 12-Hz sinusoid, or a 12-Hz diamond-shaped wave, or any other function/waveform that might shed some light on the nature of the signal?' (We will find in the next chapter that comparisons to sinusoids can be surprisingly enlightening.) Such comparisons can be carried out using an operation known as *correlation*.

Another somewhat-related question we might ask of an unknown waveform such as the EEG signal is whether the EEG signal contains anything like a brief waveform such as that shown in Figure 2.6A, or other short time period waveform. This second question requires a running correlation as described in the next section.

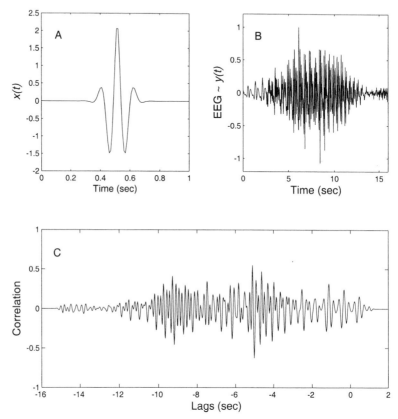

Figure 2.6 The reference waveform **(A)** is compared with the electroencephalogram signal **(B)** using a running correlation to determine to what extent the electroencephalogram signal contains this pattern. The running correlation between the two waveforms varies over time, but at maximum is only around 0.5. The running correlation operation is described and carried out in Section 2.3.2.

2.3.1 Standard Correlation and Covariance

Although it is common in everyday language to take the word *uncorrelated* as meaning *unrelated* (and thus *independent*), this is not the case in mathematical analysis, particularly if variables are related in a nonlinear manner. In the statistical sense, if two (or more) variables are independent, they are uncorrelated, but the reverse is not generally true. Moreover, signals that are very much alike can still have a mathematical correlation of zero. With these caveats in mind, correlation seeks to quantify (i.e., to assign a number to) how much one thing is like another. When comparing two mathematical functions, we use the technique of multiplying one by the other, then averaging the results. This average is often scaled by a normalizing factor. This gives us what is known as the *linear association* between two sets of variables. The same approach is used when correlation is applied to two signals. Given two functions, their average product will have the largest possible positive value when the two functions are identical. This process, since it is based on multiplication, will have the largest negative value when the two functions are exact opposites of one another (i.e., one function is the negative of the other). The average product will be zero when the two functions are, on average, completely dissimilar, again in a mathematical sense. Stated as an equation, the correlation between two signals, $x(t)$ and $y(t)$, over a time frame T is as follows:

$$Corr = \frac{1}{T}\int_0^T x(t)y(t)dt \quad \text{or in discrete form} \quad Corr = \frac{1}{N}\sum_{k=1}^N x(k)y(k) \quad \text{[Eq. 2.28]}$$

The integration (or summation in the discrete form) and scaling (dividing by T or N) simply take the average of the product. It is common to modify Eq. 2.28 by dividing by the square root of the product of the variances of the two signals. This will make the correlation value equal to 1.0 when the two signals are identical and -1 if they are exact opposites:

$$Corr_{normalized} = \frac{Corr}{\sqrt{\sigma_1^2 \sigma_2^2}} \quad \text{[Eq. 2.29]}$$

where the variances, σ^2, are defined in Eq. 2.17 and Eq. 2.18. The term *correlation* implies this normalization.

Correlation between two signals is illustrated in Figure 2.7, which shows various pairs of waveforms and the correlation between them. Note that a sine and a cosine have no (zero) correlation even though the two are alike in the sense that they are both sinusoids (upper plot). Intuitively, we see that this is because any positive correlation between them over one portion of a cycle is canceled by negative correlation over the rest of the cycle. Mathematically, this is a demonstration that a sine and a cosine of the same frequency are *orthogonal* functions, functions that, by definition, are uncorrelated. Indeed, a good way to test if two functions are orthogonal is to assess their correlation. Correlation does not necessarily measure general similarity, so a sine and a cosine of the same frequency are, by this mathematical definition, as unlike as possible, even though they have very similar oscillatory patterns.

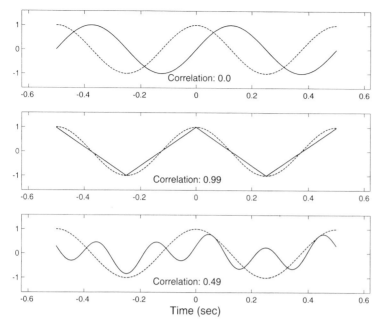

Figure 2.7 Three pairs of signals and the correlation between them as given by Equation 2.28 and normalized as in Equation 2.29. Note the high correlation between the sine and triangle wave (*center plot*) correctly expressing the general similarity between them. However, the correlation between a sine and cosine (*upper plot*) is zero, even though they are both sinusoids.

Example 2.6: Use Eq. 2.28 (continuous form) to find the correlation (unnormalized) between the sine wave and the square wave shown below. Both have an amplitude of 1.0 V (peak-to-peak) and a period of 1.0 second.

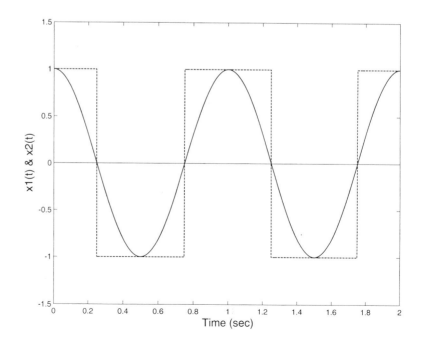

Solution: By symmetry, the correlation in the second half of the 1-second period equals the correlation in the first half, so it is only necessary to calculate the correlation period in the first half period.

$$Corr = \frac{1}{T}\int_0^T x(t)y(t)dt = \frac{2}{T}\int_0^{T/2}(1)\sin\left(\frac{2\pi t}{T}\right)dt = \frac{2}{T}\frac{T}{2\pi}\left(-\cos\left(\frac{2\pi t}{T}\right)\right)\Big|_0^{T/2}$$

$$Corr = \frac{1}{\pi}(-\cos(\pi) - -\cos(0)) = \frac{2}{\pi}$$

Covariance computes the variance that is shared between two (or more) signals. Covariance is usually defined in discrete format as follows:

$$\sigma_{xy} = \frac{1}{N-1}\sum_{k=1}^{k}(x_k - \bar{x})(y_k - \bar{y})$$ [Eq. 2.30]

The equation for covariance is similar to the discrete form of correlation except that the average values of the signals have been removed. Of course, if the signals have average values of zero, the two discrete operations (unnormalized correlation and covariance) are the same. More extensive use of correlation is presented in the section on MATLAB Implementation.

2.3.2 Autocorrelation and Cross-Correlation

The mathematical dissimilarity between a sine and a cosine is disconcerting and a real problem if you are trying to determine if a signal has general sinusoidal-like features. For example, a signal could quite similar to a cosine, but if you are correlating using a sine wave reference, you would find only a small correlation. The same would be true is you were probing a sinelike signal with a cosine reference function. You might think that these signals are not sinusoidal when in fact they were very much like a sinusoid, just not the one you selected as a reference. To circumvent this problem, you could still use only a sine (or cosine) reference, but shift this reference signal in time, performing the correlation for many different time shifts. For example, comparing a cosine with a shifted sine shows increasing correlation with greater shifts. When the sine is shifted so that its phase is modified by 90 degrees, it will be identical to a cosine and will have a correlation of 1.0. Figure 2.8 shows the correlations between a sine and a cosine as the sine is shifted relative to the cosine. Figure 2.8 (lower right) plots a cosine/sine correlation against time shift for a 2-Hz sine. When the sine is shifted by 0.125 seconds, corresponding to a phase shift of 90 degrees, the correlation reaches a maximum value of 1.0, after which it begins to decrease to a minimum of −1.0 at 0.375 seconds, corresponding to a shift of 270 degrees.

The effect of shifting the reference waveform shown in Figure 2.8 suggests an approach for using correlation to search for general signal properties such as oscillatory behavior. Rather than correlate the signal with either a sine or a cosine, correlate the signal using a sine time-shifted by different amounts, performing the correlation operation (Eq. 2.28) at each time shift. The maximum correlation will

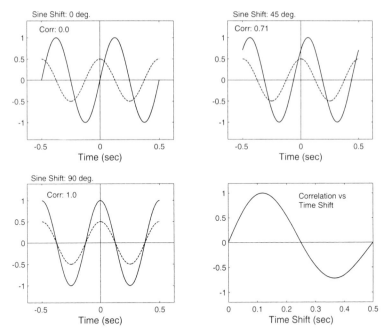

Figure 2.8 **Upper left:** The correlation between a 2-Hz cosine reference (*dashed line*) and an unshifted 2-Hz sine is 0.0. **Upper right:** When the sine time is shifted by the equivalent of 45 degrees, the correlation is 0.71. **Lower left:** When the sine is shifted by 90 degrees, the two functions are identical and the correlation is 1.0. **Lower right:** A plot of the correlation between cosine and sine as a function of the sine shift shows a peak value of 1.0 for a shift of 0.125 seconds corresponding to a shift of 90 degrees, a zero at 0.25 seconds corresponding to a shift of 180 degrees, and a correlation of −1.0 at a time shift of 0.375 seconds, corresponding to a shift of 270 degrees.

describe how much the signal is like a sinusoid. (Alternatively, a cosine could be used as the reference with similar results, although the shift required for maximum correlation would be different.) This approach also provides information on how much time shifting is required to achieve the maximum correlation, which may be of interest in some applications. This approach is demonstrated in Example 2.11 at the end of this chapter and a faster technique for probing the sinusoidal content using simple correlation (as opposed to cross-correlation) is presented in the next chapter.

When correlation is performed by time-shifting one waveform with respect to another, it is termed *cross-correlation*. This shifting correlation can be achieved by introducing a variable time delay, or time lag, or simply *lag*, into one of the two waveforms in the correlation. It does not matter which function is shifted with respect to the other, although shifting the reference waveform is more common. The correlation operations of Eq. 2.28 then become a series of correlations over different time shifts or lags. For continuous signals, the time shifting can be continuous and the correlation becomes a continuous function of the time shift. This leads to an equation for cross-correlation that is an extension of Eq. 2.28 that adds a time shift variable, τ:

$$Cross\text{-}correlation \equiv r_{xy(\tau)} = \frac{1}{T}\int_0^T y(t)x(t+\tau)dt \qquad \text{[Eq. 2.31]}$$

The variable τ is a continuous variable of time used to shift $x(t)$ with respect to $y(t)$. The variable τ is *a* time variable, but not *the* time variable (which is t). To emphasize this τ is sometimes curiously referred to as a *dummy time variable*. The correlation is now a function of the time shift, τ, also known as *lag*. Cross-correlation is often abbreviated as r_{xy}, where x and y are the two functions being correlated. Again, this equation can be converted to a discrete form by substituting summation for integration and the integers i and k for the continuous variables t and τ:

$$r_{xy}(m) = \frac{1}{N}\sum_{k=1}^{N} y(k)x(k+m) \qquad \text{[Eq. 2.32]}$$

Figure 2.9A (lower plot) shows the cross-correlation function for a sinusoid and a triangle waveform. The cross-correlation shows that they are most similar (i.e., have the highest correlation) when one signal is shifted 0.18 seconds with respect to the other. This is demonstrated by shifting one of the functions by that amount in Figure 2.9B to provide a visual demonstration of this similarity. This also suggests a useful application of cross-correlation–alignment of similar waveforms that are shifted with respect to each other.

It is also possible to shift one function with respect to itself, a process called *autocorrelation*. The autocorrelation function describes how the value of the variable at one time depends on the values at other times. This will show how well a signal correlates with various shifted versions of itself. Another way of looking at autocorrelation is that it shows how the signal correlates with neighboring portions of itself. As the shift variable τ increases, the signal is compared with more distant neighbors. A signal's autocorrelation function provides some insight into how the signal was generated or altered by intervening processes. For example, a signal that remains highly correlated with itself over a long time shift must have been produced, or modified, by a process that took into account past values of the signal. Such a process can be described as having 'memory' (Bruce, 2001), because it must remember past values of the signal (or input) and use this information to shape the signal's current values. The longer the memory, the more the signal will remain partially correlated with shifted versions of itself. Just as memory tends to fade over time, the autocorrelation function usually goes to zero for large enough time shifts.

To perform an autocorrelation, simply substitute the same variable for x and y in Eq. 2.31 or Eq. 2.32:

$$Autocorrelation \equiv r_{xx(\tau)} = \frac{1}{T}\int_0^T x(t)x(t+\tau)dt \qquad \text{[Eq. 2.33]}$$

$$r_{xx}(m) = \frac{1}{N}\sum_{k=1}^{N} x(k)x(k+m) \qquad \text{[Eq. 2.34]}$$

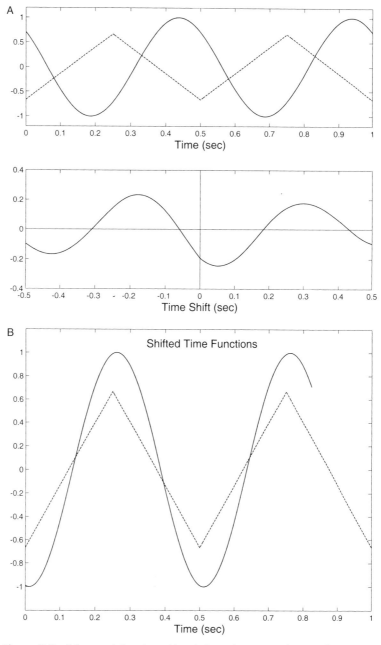

Figure 2.9   **A** (*upper plot*): a sinusoid and triangular wave at the same frequency, but not the same phase. *Lower plot:* The cross-correlation function for these two waveforms shows a peak at around –0.18 seconds when the functions are most alike. **B**: The two functions in **A** (*upper plot*) after shifting the sinusoid by an amount corresponding to the maximum cross-correlation given in **A** (*lower plot*).

Figure 2.10 shows the autocorrelation of several different waveforms. In all cases, the correlation has a maximum value of 1 at zero lag (i.e., no time shift) because when the lag (τ or n) is zero, this signal is being correlated with itself. The auto-correlation of a sine wave is another sinusoid (Figure 2.10A) because the correla-tion varies sinusoidally with the lag, or phase shift.

In Figure 2.10A, the sinusoidal pattern produced by autocorrelation falls off with increasing lags because this sinusoid had finite length. If the sinusoid were infinite in length, the autocorrelation function would be a constant amplitude cosine. A rapidly varying signal (Figure 2.10C) *decorrelates* quickly; that is, the self-correlation falls off rapidly for even small shifts of the signal with respect to itself. One could say that this signal has a very poor memory of its past values and was

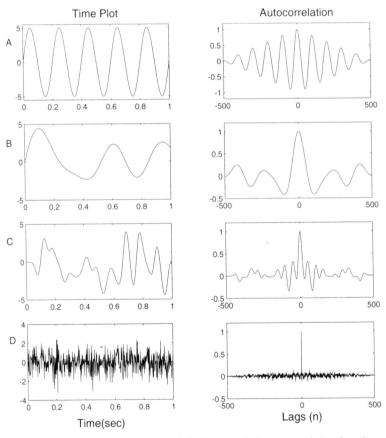

Figure 2.10   Four different signals (*left side*) and their autocorrelation functions (*right side*). **A:** A truncated sinusoid. The reduction in amplitude is due to the finite length of the signal. A true (i.e., infinite) sinusoid would have a nondiminishing cosine wave as its autocorrelation function. **B:** A slowly varying signal. **C:** A rapidly varying signal. **D:** A random signal.

probably the product of a process with a short memory. For slowly varying signals, the correlation falls slowly (Figure 2.10B). Nonetheless, for all of these signals, there is some time shift for which the signal becomes completely decorrelated with itself. For a random signal, the correlation falls to zero instantly for all positive and negative lags (Figure 2.10D). This indicates that each instant of the random signal (each instantaneous time point) is completely uncorrelated with the next instant. A random signal has no memory of its past and could not be the product of, or altered by, a process with memory.

Because shifting the waveform with respect to itself produces the same results regardless of which way the function is shifted, the autocorrelation function will be symmetrical about lag zero. Mathematically, the autocorrelation function is an even function:

$$r_{xx}(-\tau) = r_{xx}(\tau) \qquad [\text{Eq. 2.35}]$$

In addition, the value of the function at lag zero, where the waveform is correlated with itself, will be as large, or larger than, any other value. If the autocorrelation is normalized by the variance, the value will be one. (Because only one function is involved in autocorrelation, the normalization equation given in Eq. 2.29 reduces to $1/\sigma^2$.)

Figure 2.11 shows the autocorrelation function of the EEG signal shown previously. The signal decorrelates quickly, reaching a value of zero correlation after a

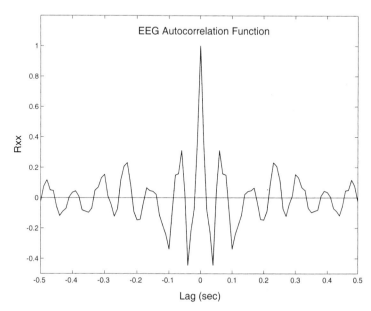

Figure 2.11 Autocorrelation function of the electroencephalogram signal in Figure 2.4. The autocorrelation function decorrelates rapidly probably due to the noise in the signal. Some correlation is seen out to 0.5 seconds.

time shift of approximately 0.03 seconds. However, the EEG signal is likely to be contaminated with noise and the autocorrelation function of a signal plus noise is the sum of the autocorrelation function of the signal plus the autocorrelation of the noise. Because noise decorrelates instantly (Figure 2.10D), some of the rapid decorrelation seen in Figure 2.11 is due to the noise. A common approach to estimating the autocorrelation of the signal without the noise is to draw a smooth curve across the peaks and use that curve as the estimated autocorrelation function of signal without noise. From Figure 2.11, we see that such an estimated function would decorrelate at a longer time shift of 0.5 to 0.6 seconds.

Two operations closely related to autocorrelation and cross-correlation are auto-covariance and cross-covariance. The relationship between correlation and covariance *functions* is similar to the relationship between standard correlation and covariance given in the last section. Covariance and correlation functions are the same except that in covariance, the means have been removed from the input signals, $x(t)$ and $y(t)$ [or just $x(t)$ in the case of autocovariance]:

$$Auto\,cov\,ariance \equiv C_{xx}(\tau) = \frac{1}{T}\int_0^T \left[x(t) - \overline{x(t)}\right]\left[x(t+\tau) - \overline{x(t)}\right]dt$$

$$C_{xx}[i] = \frac{1}{N}\sum_{k=1}^N [x(k) - \overline{x}][x(k+i) - \overline{x}] \qquad \text{[Eq. 2.36]}$$

$$Cross\,cov\,ariance \equiv C_{xy}(\tau) = \frac{1}{T}\int_0^T \left[y(t) - \overline{y(t)}\right]\left[x(t+\tau) - \overline{x(t)}\right]dt$$

$$Cross\,cov\,ariance \equiv C_{xy[i]} = \frac{1}{N}\sum_{k=1}^N \left[y(k) - \overline{y(k)}\right]\left[x(k+i) - \overline{x(k)}\right] \quad \text{[Eq. 2.37]}$$

The autocovariance function can be thought of as measuring the memory or self-similarity of the *deviation* of a signal about its mean level. Similarly, the cross-covariance is a measure of the similarity of the deviation of two signals about their respective means. An example of the application of the autocovariance to the analysis of heart rate variability is given in the next section on MATLAB Implementation.

2.4   MATLAB IMPLEMENTATION

All of the analyses described thus far are relatively easy to implement in MATLAB. In most cases, MATLAB has function that will perform these operations.

2.4.1   Mean, Variance, and Standard Deviation

Many of the techniques described in this chapter can be expeditiously, and conveniently, implemented in MATLAB. For example, the mean, variance, and standard deviations are implemented as shown in the three code lines below.

```
xm = mean(x);          % Evaluate mean of x
xvar = var(x);         % Variance of x normalizing by N-1
xnorm = var(x,1);      % Variance of x normalizing by N
xstd = std(x);         % Evaluate the standard deviation of x
```

If x is an array or series of numbers (also termed a *vector* for reasons given later) the output of these routines is a scalar representing the mean, variance, or standard deviation. If x is a matrix, the output is a row vector resulting from applying the appropriate calculation (mean, variance, or standard deviation) to each column of the matrix.

Example 2.7: Figure 2.12 shows heart rate variability for one subject under normal conditions (left side) and during a meditative state. Find the mean and standard deviation for the two conditions.

Solution: Apply the MATLAB routines `mean` and `std` (standard deviation) to the data. The program below loads the heart rate data from the .mat files `HR_pre` and `HR_med`. These files are assumed to be in workspace in this example, but are found on the accompanying CD. These files were originally obtained from the PhysioNet data base (Goldberger et al., 2000 or http://www.physionet.org) and contain approximately 500 seconds of heart rate data from a subject in a normal (`Hr_pre.mat`) and meditative state (`Hr_med.mat`). Each file contains a time variable (`t_pre` or `t_med`) and a heart rate variable (`hr_pre` or `hr_med`). The mean

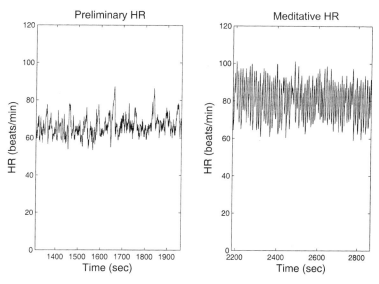

Figure 2.12 Heart rate over time during normal conditions (*left*) and during meditation (*right*). (From the PhysioNet database, Goldberger et al., 2000.)

and standard deviation of the two heart rate variables will be determined using the appropriate MATLAB routines and the two variables plotted as functions of time.

Example 2.7:   Plot the mean and standard deviation of the heart rate before and after meditation.

```
% Plots Figure 2.12.
%
load Hr_pre                        % Pre-meditative HR
load Hr_med                        % Meditative HR
%
% Calculate the averages and standard deviations
Avg_pre = mean(hr_pre)             % Average HR, normal
SD_pre = std(hr_pre)               % Standard deviation, normal
Avg_med = mean(hr_med)             % Average and std
SD_med = std(hr_med)               % HR meditative
%
% Plot the heart rate data. Label axes
subplot(1,2,1);
 plot(t_pre,hr_pre,'k');           % Plot normal HR data
 xlabel('Time (sec)'); ylabel('HR (beats/min)');
 axis([t_pre(1) t_pre(end) 0 120]);
 title('Preliminary HR');
subplot(1,2,2);
 plot(t_med,hr_med,'k');           % Plot meditative HR data
 xlabel('Time (sec)'); ylabel('HR (beats/min)');
 axis([t_med(1) t_med(end) 0 120]);
 title('Meditative HR')
```

Analysis: The program is a straightforward application of routines mean and std. The var routine could have been used if the variance was desired. In the plotting section, the axis routine was used to scale the vertical axis to be between 0.0 and 120 beats per minute. Because the time variables had different beginning and end times, the time limits were specified using the time array (t_pre or t_med) endpoints. (Recall that MATLAB is case sensitive.) The MATLAB files Hr_pre and Hr_med contain variables: hr_pre, t_pre, hr_med, and t_med.

Results:

	Premeditative	Meditative
Average heart rate (beats/min.)	66.5	81.33
Standard deviation (beats/min.)	5.36	9.35

In this subject, meditation increased the heart rate by about 22% and the standard deviation by almost 75%, not a result that might be anticipated by this Yoga-based meditation. (See the PhysioNet database for more details on the meditative conditions.)

2.4.2 Ensemble Averaging

Equation 2.22 indicates that averaging can be a simple yet powerful signal-processing technique for reducing noise when multiple observations of the signal are possible. Such multiple observations could come from multiple sensors, but in many biomedical applications, the multiple observations come from repeated responses to the same stimulus. In *ensemble averaging,* a group, or ensemble, of time responses is averaged together on a point-by-point basis; that is, an *average signal* is constructed by taking the average, for each point in time, over all signals in the ensemble. A classic biomedical engineering example of the application of ensemble averaging is the visual evoked response (VER) in which a visual stimulus produces a small neural signal embedded in the EEG. Usually this signal cannot be detected in the EEG signal, but by averaging hundreds of observations of the EEG, time-locked to the visual stimulus, the visually evoked signal emerges.

There are two essential requirements for the application of ensemble averaging for noise reduction: the ability to obtain multiple observations and a reference closely time-linked to the response. The reference shows how the multiple observations are to be aligned for averaging. Usually a time signal linked to the stimulus is used. An example of ensemble averaging is given in Example 2.8.

Example 2.8: Find the average response given a number of individual responses from the vergence eye movement system. The vergence eye movement system is responsible for turning the eye inward to view a near target. These responses are stored in MATLAB file `vergence.mat`.

Solution: Use the MATLAB averaging routine `mean`. If this routine is given a matrix variable, it averages each column. Hence, if the various signals are arranged as rows in the matrix, the mean routine will produce the ensemble average.

Example 2.8: Load eye movement data, plot the data, then construct and plot the ensemble average.

```
close all; clear all;
load vergence;             % Get vergence eye movement data
Ts = .005;                 % Sample interval = 5 msec
[nu,N] = size(data_out);   % Get data length (N)
t = (1:N)*Ts;              % Generate time vector (t = N Ts)
%
% Plot ensemble data superimposed
plot(t,data_out,'k'); hold on;
%
% Construct and plot the ensemble average
```

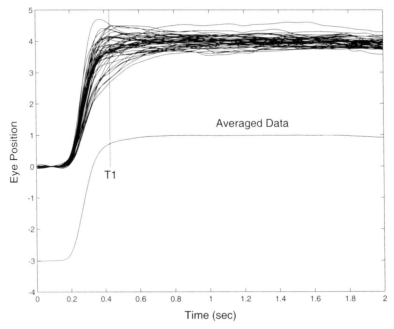

Figure 2.13   **Upper traces:** An ensemble of individual (vergence) eye movement responses to a step change in stimulus. **Lower trace:** The ensemble average, displaced downward for clarity. The ensemble average is constructed by averaging the individual responses at each point in time. Hence, the value of the average response at time T1 (*vertical line*) is the average of the individual responses at that time.

```
avg = mean(data_out);       % Calculate ensemble average and
plot(t,avg-3,'k');          % plot, separate from the other
                            % data
  xlabel('Time (sec)');     % Label axes
  ylabel('Eye Position');
plot([.43 .43],[0 5]);      % Plot horizontal line
  '-k');
text(1,1.2,'Averaged...     % Label data average
  Data');
```

The results are shown in Figure. 2.13.

2.4.3 Covariance and Correlation

MATLAB has specific functions for determining the correlation and/or covariance between two or more signals. Correlation or covariance matrices are calculated using the corrcoef or cov functions, respectively. Again, the calls are similar for both functions:

```
Rxx = corrcoef(x);        % Signal correlation
S = cov(x);               % Signal covariance
```

where x is a matrix that contains the various signals to be compared in columns. Some options are available as explained in the associated MATLAB help file. The output, Rxx, of the corrcoef routine will be an *n*-by-*n* matrix where *n* is the number of signals (i.e., columns of *x*). The diagonals of this matrix represent the correlation of the signals with themselves, r_{xx} (and, hence, will be 1), and the off diagonals represent the correlations of the various combinations. For example, r_{12} is the correlation between signals 1 and 2. Because the correlation of signal 1 with signal 2 is the same as signal 2 with signal 1, $r_{12} = r_{21}$, and in general $r_{m,n} = r_{n,m}$, so the matrix will be symmetrical about the diagonals:

$$r_{xx} = \begin{bmatrix} r_{1,1} & r_{1,2} & \cdots & r_{1,N} \\ r_{2,1} & r_{2,2} & \cdots & r_{2,N} \\ \vdots & \vdots & \ddots & \vdots \\ r_{N,1} & r_{N,2} & \cdots & r_{N,N} \end{bmatrix} \qquad \text{[Eq. 2.38]}$$

The cov routine produces a similar output, except the diagonals are the variances of the various signals and the off-diagonals are the covariances as shown in Eq. 2.39 below.

$$S = \begin{bmatrix} \sigma_{1,1}^2 & \sigma_{1,2}^2 & \cdots & \sigma_{1,N}^2 \\ \sigma_{2,1}^2 & \sigma_{2,2}^2 & \cdots & \sigma_{2,N}^2 \\ \vdots & \vdots & \ddots & \vdots \\ \sigma_{N,1}^2 & \sigma_{N,2}^2 & \cdots & \sigma_{N,N}^2 \end{bmatrix} \qquad \text{[Eq. 2.39]}$$

Example 2.8 uses covariance and correlation analysis to determine if sines and cosines of the same frequency and sine waves at multiple frequencies are orthogonal. Recall that two orthogonal signals will have zero correlation. Either covariance or correlation could be used to determine if signals are orthogonal. Example 2.9 uses both.

Example 2.9: Determine if a sine wave and a cosine wave at the same frequency are orthogonal and if sine waves at harmonically related frequencies are orthogonal. Include one sinusoid at a nonharmonic frequency.

Solution: Generate a data matrix where the columns consist of a 1.0-Hz sine and cosine, a 2.0-Hz sine and cosine, and a 3.0-Hz sine and a 3.5-Hz (i.e., nonharmonic) cosine. The six sinusoids should all be at different amplitudes. Apply the covariance (cov) and correlation (corrcoef) MATLAB functions. All of the sinusoids except the 3.5-Hz cosine are orthogonal and should show negligible correlation and covariance.

```
% Example 2.9:  Application of the correlation and
% covariance matrices to sinusoids that are orthogonal and
% nonorthogonal
```

```
%
clear all; close all;
N = 256;                        % Number of points in waveform
fs = 256;                       % Assumed sample frequency
n = (1:N)/fs;                   % Time vector: 1 sec of data
%
% Generate the sinusoids as columns of the matrix
x(:,1) = sin(2*pi*n)';          % Generate a 1 Hz sin
x(:,2) = 2*cos(2*pi*n)';        % Generate a 1 Hx cos
x(:,3) = 1.5*sin(4*pi*n)';      % Generate a 2 Hz sin
x(:,4) = 3*cos(4*pi*n)';        % Generate a 2 Hx cos
x(:,5) = 2.5*sin(6*pi*n)';      % Generate a 3 Hx sin
x(:,6) = 1.75*cos(7*pi*n)';     % Generate a 3.5 Hz cos
%
S = cov(x)                      % Print covariance matrix
Rxx = corrcoef(x)               % and correlation matrix
```

Analysis: The program defines a time vector n that is 256 points long and achieves the proper time interval by dividing by the sampling frequency, fs (also 256). (Because MATLAB is case sensitive, n and N are different variables.) The program then generates the six sinusoids using this time vector in conjunction with sin and cos functions, arranging the signals as columns of x. The program then determines the covariance (cov) and correlation (corrcoef) matrices of x.

Results: The output from this program is a covariance and correlation matrix. The covariance matrix is as follows:

```
Covariance Matrix s =
 0.5020    0.0000    0.0000    0.0000    0.0000   -0.0497
 0.0000    2.0078   -0.0000   -0.0000   -0.0000   -0.0137
 0.0000   -0.0000    1.1294    0.0000   -0.0000   -0.2034
 0.0000   -0.0000    0.0000    4.5176   -0.0000   -0.0206
 0.0000   -0.0000   -0.0000   -0.0000    3.1373   -1.2907
-0.0497   -0.0137   -0.2034   -0.0206   -1.2907    1.5372
```

The diagonals of the covariance matrix give the variance of the six signals and these differ since the amplitudes of the signals are different. The correlation matrix shows similar results except that the diagonals are now 1.0 because these reflect the correlation of the signal with itself.

```
Correlation Matrix Rxx =
 1.0000    0.0000    0.0000    0.0000    0.0000   -0.0566
 0.0000    1.0000   -0.0000   -0.0000   -0.0000   -0.0078
 0.0000   -0.0000    1.0000    0.0000   -0.0000   -0.1544
 0.0000   -0.0000    0.0000    1.0000   -0.0000   -0.0078
 0.0000   -0.0000   -0.0000   -0.0000    1.0000   -0.5878
-0.0566   -0.0078   -0.1544   -0.0078   -0.5878    1.0000
```

The covariance and correlation between the various signals are given by the off-diagonals and are zero for all combinations between signals 1 and 5, demonstrating the orthogonality of all of these harmonic signals. Conversely, nonzero covariances and correlations are found between signals 1 through 5 and signal 6, the 3.5-Hz cosine. This shows that the nonharmonically related cosine is not orthogonal to any of the other sines or cosines. Note that the bottom row is the same as the last column, reflecting the symmetry of these matrices.

It may seem a little surprising that a 1-Hz sine wave and a 2-Hz sine wave are orthogonal, but is easily demonstrated by sketching the two waveforms. Consider the product of two sine waves seen in Figure 2.14. The first half of the 1-Hz sine wave will be multiplied by a full cycle of the 2-Hz sine wave and the result will be 0.0. This would be true for any higher harmonic signal: if the 2-Hz sine wave were a 4-Hz sine wave, for example. The orthogonality of harmonically related sinusoids is a feature that will be used in the Fourier Transform described in the next chapter. It means that operations (such as correlation) involving a sinusoid do not interfere with operations that involve sinusoids at harmonically related frequencies.

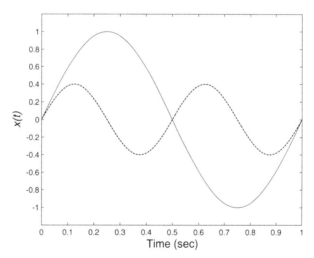

Figure 2.14 A 1-Hz sine wave plotted with a 2-Hz sine wave. The product of the two will clearly average to 0.0.

2.4.4 **Autocorrelation and Cross-Correlation**

The cross-correlation and autocorrelation operations are performed with the same MATLAB routine, with autocorrelation being treated as a special case. The program, axcor, is supplied on the accompanying CD:

$$[r, lags] = axcor(x, y);$$

Only the first input argument, x, is required. If no y variable is specified, autocorrelation is performed and the output is normalized to be 1.0 at zero lag. If both variables are given, the cross-correlation is normalized as in Eq. 2.29. The time shift extends over the entire range of the longer variable. If the MATLAB signal-processing toolbox is available, a MATLAB routine called xcorr is available that features a wider range of options. The axcor function produces an output argument, r, that is an array that is twice the length of the shortest input array. The optional output argument, lags, is simply an array containing the lag values, which is helpful in plotting the function.

Recall that auto- and cross-covariance are the same as auto- and cross-correlation if the data have zero means. Hence, autocovariance or cross-covariance can be determined using axcor simply by subtracting the variable means before calling the function.

$$[c, lags] = axcor(x-mean(x), y-mean(x));$$

The autocorrelation and autocovariance functions describe how one segment of data is correlated, *on average*, with adjacent segments. Such correlations could be due to memory-like properties in the process that generated the data. Many physiological processes are repetitive, such as respiration and heart rate, yet vary somewhat on a cycle-to-cycle basis. Autocorrelation and cross-correlation can be used to explore this variation. For example, considerable interest revolves around the heart rate and its beat-to-beat variations. Autocovariance can be used to tell us if these variations are completely random or if there is (again, on average) some correlation between beats or over several beats. In this instance, we use autocovariance, not autocorrelation, because we are interested in correlation of heart rate *variability*, not the correlation of heart rate per se. (Recall that autocovariance will remove the mean value of the heart rate from the data and analyze only the variation.) Example 2.10 analyzes the normal heart rate data presented in Figure 2.12 (Preliminary Heart Rate) to determine the correlation over successive beats.

Example 2.10: Determine if there is any correlation in the variation between the timing of successive heart beats under normal resting conditions.

Solution: Load the heart rate data taken during normal resting conditions (file Hr_pre.mat). Isolate the heart rate variable (the second column) and then take the autocovariance function. Plot this function to show potential correlation over approximately 30 successive beats.

```
% EXAMPLE 2.10 and Figure 2.15
% Use of autocovariance to determine the correlation
% of heart rate variation between heart beats
%
clear all; close all;
figure;
load Hr_pre                        % Load data
[c,lags]=axcor(hr_pre-mean(hr_pre));  % Autocovariance
                                      (mean subtracted)
```

```
plot(lags,c,'k'); hold on;           % Plot autocovariance
plot([lags(1) lags(end)], [0 0],'k') % Plot zero line for
                                     % reference
xlabel('Lags (N)'); ylabel('Autocovariance'); grid on;
axis([-30 30 -.2 1.2]);              % Limit plot range
                                     % to ± 30 beats
```

Analysis: After loading the data file, the program calculates the autocovariance using routine `axcor`. The mean is subtracted from the data variable so that auto-covariance will be performed. The data are then plotted along with a zero line and the axis is rescaled to show only the first ± 30 lags. The plotting grid is enabled.

Results: The results in Figure 2.15 show there is high correlation for heart beats that are within a few seconds of each other (approximately 0.5 within ± 4 sec). This correlation falls off rapidly so there is little or no correlation between beats that are more than about 15 seconds apart. If the variability were completely random, the autocovariance function would be 1.0 for zero lag and 0.0 everywhere else (Figure 2.10D). Problem 15 applies this analysis to the heart rate data taken during the meditative state.

One of the most popular reference signals is the sinusoid. It is common to compare the signal of interest not just with one sinusoid but with a range of sinu-soids having different frequencies. To ensure that we correlate with a sinusoid

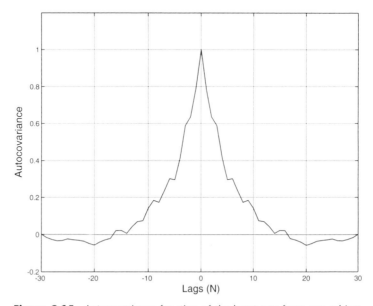

Figure 2.15 Autocovariance function of the heart rate from one subject under normal resting conditions. Some correlation is observed over approx-imately 10 successive heart beats.

having the most appropriate phase shift, we use cross-correlation and take the maximum cross-correlation values as related to the amount of sinusoid 'in' the signal at a given frequency. This strategy is demonstrated in the next example.

Example 2.11: Find the sinusoidal content in the EEG signal over a range of frequencies from 0.5 to 50 Hz. The frequency resolution of the comparison should be 0.5 Hz.

Solution: Generate a series of sine waves from 0.5 to 50 Hz in 0.5-Hz increments. (Cosine waves would work just as well.) Cross-correlate these sine waves with the EEG signal and find the maximum cross-correlation. Plot this maximum correlation as a function of the sine wave frequency. This procedure is remarkably easy to program in MATLAB.

```
% Example 2.11 and Figure 2.16
% Comparison of sinusoids at different frequencies with
   the EEG signal using cross-correlation.
%
clear all; close all;
load eeg_data;              % Get EEG data
eeg = eeg/max(eeg);         % Normalize eeg data
fs = 50;                    % Sampling frequency
t = (1:length(eeg))/fs;     % Time vector
% Cross-correlate over a range of frequencies.
for i = 1:100
  f(i) = i/2;               % Frequency range: 0.5 - 50 Hz
  x = sin(2*pi*f(i)*t);     % Generate sin
  r = axcor(eeg,x);         % Perform cross-correlation
  rmax(i) = max(r);         % Store max value
end
......labels and plot.......
```

Results: The result of the cross-correlations is seen in Figure 2.16 and an interesting structure is seen to emerge. Some frequencies show much higher correlation with sine and EEG, indicating more sine wave content at these frequencies. A particularly strong peak is seen in the region of 7 to 9 Hz, indicating the presence of an oscillatory pattern known as the *alpha wave*. A more efficient method for obtaining the same information will be given in the next chapter.

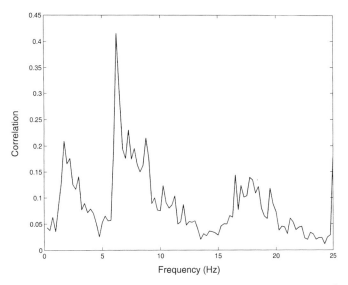

Figure 2.16 The maximum cross-correlation between sine waves and the electroencephalogram signal plotted as a function of the sine wave's frequency. A peak is seen between 7 and 9 Hz, which indicates the presence of an oscillatory pattern known as the 'alpha wave.'

2.5 SUMMARY

The sinusoidal waveform is arguably the single most important waveform in signal processing. Some of the reasons for this importance are provided in the next chapter. Because of their importance, it is essential to know the mathematics associated with sines, cosines, and general sinusoids, including complex representations.

Several basic measurements apply to any signal including mean value, RMS value, and variance or standard deviation. Although these measurements provide some essential basic information, they do not provide much information on signal content or meaning. A common approach to obtaining more information is to probe a signal by correlating it with one or more reference waveforms. One of the most popular probing signals is the sinusoid, and sinusoidal correlation is covered in detail in the next chapter.

Sometimes a signal will be correlated with another signal in its entirety, a process known as correlation, or the closely related covariance. If the correlation between the signal of interest and the reference is zero, it does not necessarily mean the two signals have nothing in common, but it does mean the signals are mathematically orthogonal.

If the probing signal is short, a running correlation known as cross-correlation may be appropriate. Cross-correlation not only shows the match between the probing signal and the signal of interest, but also where that match is greatest. A signal can also be correlated with shifted versions of itself, a process known as auto-correlation. The autocorrelation function describes the period for which a signal remains partially correlated with itself and this relates to the structure of the signal. For example, a signal consisting of random noise decorrelates immediately, whereas a slowly varying signal will remain correlated over for long period. Correlation, cross-correlation, autocorrelation, and the related covariances are all easy to implement in MATLAB.

PROBLEMS

1. Two 10-Hz sine waves have a relative phase shift of 30 degrees. What is the time difference between them? If the frequency of these sine waves doubles, but the time difference stays the same, what is the phase difference between them?
2. Convert $x(t) = 6 \sin(5t) - 5 \cos(5t)$ into a single sinusoid [i.e., M $\sin(5t + \theta)$].
3. Convert $x(t) = 30 \sin(2t + 50)$ into sine and cosine components.
4. Convert $x(t) = 5 \cos(10t + 30) + 2 \sin(10t - 20) + 6 \cos(10t + 80)$ into a single sinusoid as in Problem 2.
5. Find the delay between $x_1(t) = \cos(10t + 20)$ and $x_2(t) = \sin(10t - 20)$.
6. Equations 2.8, 2.9, and 2.10 were developed to convert a sinusoid such as $\cos(\omega t - \theta)$ into a sine and cosine wave and vice versa. Derive the equations to convert between a sinusoid based on the sine, $\sin(\omega t + \theta)$ and a sine and cosine wave. (*Hint:* Use the appropriate identity from Appendix C.)
7. Find the RMS value of the square wave with amplitude of 1.0 V and a period 0.2 seconds.
8. If a signal is measured as 2.5 V peak-to-peak and the noise is measured as 28 mV RMS, what is the SNR in decibels?
9. Use Eq. 2.28 to find the correlation (unnormalized) between $\sin(2\pi t)$ and $\cos(2\pi t)$.
10. Use Eq. 2.28 to find the correlation between a cosine and a square wave as shown below. This is the same as Example 2.6 except that the sine has been replaced by a cosine.

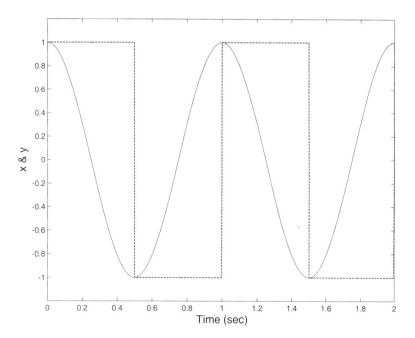

MATLAB Problems

11. Load the data in ensemble_data.mat found on the CD. This file contains a data matrix labeled 'data.' The data matrix contains 100 responses of a signal in noise. Determine whether the responses are stored in the rows or columns of the matrix. Plot several randomly selected samples of these responses. Is it possible to identify the signal from any single record? Construct and plot the ensemble average for these data. (Be sure that the matrix is in the correct orientation.) Also construct and plot the ensemble standard deviation.

12. Demonstrate that 1-Hz and 4-Hz cosine waves are orthogonal by multiplying them together and averaging the product. (Note when you multiply be sure to use the point-by-point multiplication, '.*'.)

13. The file two_var.mat contains two variables x and y. Are either of these variables random? Are they orthogonal to each other?

14. The file nerve.mat contains two signals, x and y, along with a time vector, t. The two signals represent nerve action potentials taken simultaneously at two different sections of a nerve axon along with inevitable noise. Use cross-correlation to determine the average time delay between the two signals. Assume the sampling frequency of the nerve signals was 2 kHz.

15. Develop a program along the lines of Example 2.10 to determine the correlation in heart rate variability during meditation. Load file Hr_med.mat, which contains the heart rate (beats/min as a function of time) in variable hr_med

and the time in variable t_med, then determine and plot the autocovariance. The result will show that the heart rate under meditative conditions contains some periodic elements. Can you determine the frequency of these periodic elements? The answer is presented in the next chapter, which shows how to identify the frequency characteristics of any waveform.

3 FREQUENCY TRANSFORMATIONS

The electroencephalogram (EEG) signal originally shown in Figure 2.4 looks, and is, quite complicated. Yet, in Example 2.11 this complicated waveform was probed with sinusoids at different frequencies and some semblance of internal structure was revealed: sinusoids at some frequencies showed much higher correlation with the EEG signal than sinusoids at other frequencies. This cross-correlation approach could be applied to examine the structure of any waveform using sinusoids or any other probing signals. When a high correlation is found between the signal and probe (or reference) waveform, we might say the signal is made up of, or contains, some of that probing waveform. The higher the correlation, the more the probing signal is contained in the signal being analyzed.

There are two problems with using this approach to analyze the structure of a signal. The first is that it is computationally intensive, but this may not be so serious with modern high-speed computers. The second is that you have to know what you are looking for, at least in terms of general shape. If you are probing with sinusoids, you have to know the frequencies to use in your sinusoidal probe (but not the exact phase since the cross-correlation operation evaluates correlation at all possible time shifts). For example, suppose you are examining a signal that is made up of three sinusoids, each at a different frequency. If your cross-correlation search includes the three frequencies, you should get an accurate picture of the three sinusoids contained in the signal. However, what if your search does not include the exact frequencies, but examines frequencies close to but not identical to the ones contained in the signal? As will be demonstrated in Example 3.1, you would likely get an inaccurate picture of the sinusoidal composition of the signal.

Example 3.1: Use cross-correlation to probe the two signals each containing a mixture of three sinusoids. One signal should contain sine waves at 100, 200, and 300 Hz, while the other should contain sine waves at 100, 204, and 306 Hz. Begin your search at 10.0 Hz and continue up to 500 Hz, cross-correlating every 10 Hz. Generate the signals to be analyzed assuming a sample frequency of 1.0 kHz. Use three different sine wave amplitudes to test the quantitative ability of the cross-correlation analysis.

Solution: Modify the MATLAB code in Example 2.11 to include the generation of the two sine wave mixtures. Plot out the maximum correlation for the two mixtures side-by-side.

```
% Example 3.1 and Figure 3.1
% Correlation analysis of two signals each containing
  three sinusoids
%
clear all; close all;
fs = 1000;                      % Sample frequency
N = 2000;                       % Number of points in the test
                                % signal
t = (1:N)/fs;                   % Time vector
f = [100 200 300];              % Test signal frequencies
%
% Generate the test signal as a mixture of three sinusoids
  at different freq.
x = 1.5*sin(2*pi*f(1)*t) + 1.75*sin(2*pi*f(2)*t) +...
  2.0*sin(2*pi*f(3)*t);
%
for i = 1:50                    % Analysis loop
  f(i) = i*10;                  % Frequency range: 10 - 500 Hz
  y = cos(2*pi*f(i)*t);         % Generate sinusoid
  [r,lags] = axcor(x,y);        % Cross-correlate
  [rmax(i),ix(i)] = max(r);     % Find maximum value
end
subplot(1,2,1);                 % Plot and label cross-
  plot(f,rmax,'k');                correlation results
  xlabel('Frequency (Hz)'); ylabel('Correlation');
  title('100, 200, 300 Hz'); axis([0 400 0 1]);

% Now redo for a test signal having slightly different
  frequencies.
f = [100 204 306];              % Next test signal frequencies
x = 1.5*sin(2*pi*f(1)*t) + 1.75*sin(2*pi*f(2)*t) +
  2.0*sin(2*pi*f(3)*t);
%
  for i = 1:50
  f(i) = i*10;                  % Frequency range: 10 - 500 Hz
  y = cos(2*pi*f(i)*t);         % Generate sinusoid
  [r,lags] = axcor(x,y);        % Cross-correlate
  [rmax(i),ix(i)] = max(r);     % Find maximum values
end
subplot(1,2,2);                 % Plot and label cross-
  plot(f,rmax,'k');                correlation results
```

```
xlabel('Frequency (Hz)',); ylabel('Correlation');
title('100, 204, 306 Hz'); axis([0 400 0 1]);
```

Results: As shown Figure 3.1A, the cross-correlation analysis correctly identifies the three sinusoids found in the first mixture. The analysis even shows the relative strengths of the three sine waves. However, this analysis fails to find the two sine waves in the second mixture where the frequency of these sine waves is slightly different. This is because the program is searching in increments of 10 Hz and does not compare the signal with sinusoids at exactly their frequencies of 204 Hz and 306 Hz. Moreover, this program takes over a minute and a half to run on at 1.5-GHz computer.

From Example 3.1, we can see that using cross-correlation to probe the contents of a signal works well if we know the specific pattern for which we are looking, but often we do not know what patterns a signal may contain. Of course, we could have decreased the frequency increment and used more sinusoids to probe the signals, but this would increase the computation time even further and we still may have missed some important frequencies. The question is, what probing frequencies should we use with a signal whose characteristics are unknown? If we are probing with sinusoids (a very common probing signal), and the signal we are probing is periodic, or can be taken as periodic, the answer to the question of which frequencies to use is found in an important theorem known as the *Fourier series theorem*. Before describing this theorem, it is useful to examine the sinusoid further and highlight some of its important mathematical properties.

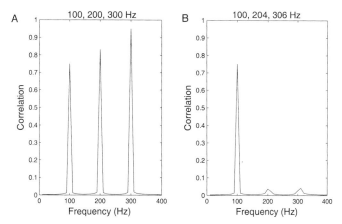

Figure 3.1 Cross-correlation analysis of two very similar signals, both containing mixtures of three sinusoids. The amplitudes of the three sine waves are 1.5, 1.75, and 2.0 (relative units) in both mixtures. **A:** The analysis of a mixture containing 100-, 200-, and 300-Hz sine waves clearly identifies the three sinusoids at the correct frequency. **B:** The analysis of the other mixture of 100, 204, and 306 Hz identifies the first sinusoid correctly, but fails to find the other two correctly.

3.1   USEFUL PROPERTIES OF THE SINUSOIDAL SIGNAL

Sinusoids have four unique properties that make them extraordinarily useful for signal and systems analysis.

1. Many signals can be broken down, or decomposed, into an equivalent representation as a series of sinusoids (Figure 3.2). The only constraint on the signal is that it be periodic, or can be taken to be periodic; that is, it repeats exactly after some period. Stated the other way, different sinusoids can be added together to reconstruct any periodic signal (or signals that can be assumed to be periodic as explained later). You may need a large number of sinusoids to represent some signals, and there are a few exceptions, but when applicable the sinusoidal representation is complete and works in both directions: Signals decomposed into a sinusoidal series can be accurately reconstructed from that series (Figure 3.2). (Because it works both ways, decomposing a signal into a sinusoidal series is known mathematically as a *bilateral transform*.) This property of sinusoids is at the heart of Fourier series analysis as described below. Why would you want to represent a waveform by a series of sinusoids? This is explained by the next two unique sinusoidal properties.

2. Sinusoids are the only functions to have energy at only one frequency, the frequency of the sinusoid. Because of this property, sinusoids are sometimes referred

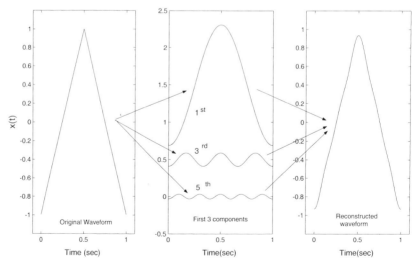

Figure 3.2   Any periodic signal (including those that can be assumed to be periodic) can be decomposed into a series of sinusoids which, in turn can be used to reconstruct the original signal. Here only three sinusoidal components are used to reconstruct the original triangle waveform. Even with only three components, the reconstruction is quite good. More components would lead to a more accurate representation and a better reconstruction (see Figure 3.6).

to as *pure* and the tones they produce as sound waves sound somehow pure or basic. Because they have energy at only one frequency, sinusoids can be easily converted to an alternate representation known as the *frequency representation* or *frequency domain*. Specifically, any specific sinusoid can be represented as two points on a frequency plot, one specifying the amplitude the other the phase, at the frequency of the sinusoid (Figure 3.3). Combining this characteristic with the decomposition properties described above provides us with a technique for converting any periodic waveform into a frequency representation. If a waveform can be decomposed into equivalent sinusoids, and each sinusoid plots directly into the frequency domain, sinusoids can be used as an intermediary between any periodic function and its frequency representation. The frequency representation of a signal is also referred to as its *spectrum*. Thus, any periodic signal can be represented by a magnitude and phase spectrum (Figure 3.4), a feature that is exploited in the rest of this chapter.

3. If the input to any linear system, no matter how complex, is a sinusoid, the output is a similar sinusoid. The only differences between the input and output are the magnitude and phase of the sinusoid: the frequency and sinusoidal wave shape will be the same (Figure 3.5). Combining this feature with the above two sinusoidal characteristics, we develop a powerful paradigm for analyzing the behavior of linear systems, at least for periodic input signals. If any periodic function can be broken down into sinusoids, and the magnitude and phase change produced by the linear system can be determined for each sinusoid, then the output

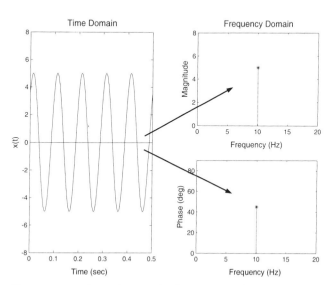

Figure 3.3   A sinusoid is completely represented by its magnitude and phase at a given frequency. In general, the frequency characteristics of a signal are presented as plots of magnitude and phase against frequency.

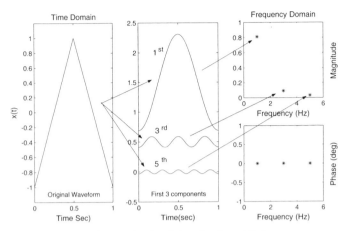

Figure 3.4   Combining the simplicity of a sinusoid's frequency representation (a single point on the magnitude and phase frequency plot) with the decomposition feature stated above, sinusoids can be used to convert any periodic signal from a time representation to a frequency representation: if the sinusoidal components of a signal can be determined its frequency characteristics can also be determined.

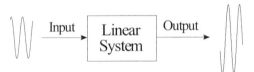

Figure 3.5   The output of any linear system driven by a sinusoid is a sinusoid at the same frequency. Only the magnitude and phase of the output sinusoid differ from the input.

of the system to any periodic function can be determined. Simply decompose the signal into its sinusoidal constituents, finding the output for each sinusoid, then reconstruct the actual output signal from the sinusoidal components. This sounds involved but is actually quite easy to implement with computer processing. This *decompose and conquer* approach is developed in Chapter 6.

4. If the sinusoids are harmonically related, they are orthogonal. This means that if we can restrict our decomposition of periodic waveforms to harmonically related sinusoids, each of the components in the decomposition will be independent of the others. For example, suppose we decided to decompose a waveform into 10 harmonically related sinusoids, but later decided to use more components to attain a more accurate decomposition. Adding more components will not change the value of the components we already have because all components are orthogonal.

3.2 FOURIER SERIES ANALYSIS

The Fourier series theorem states that any periodic signal, no matter how complicated, can be represented by a sum of sinusoids; specifically, a series of sinusoids that are the same or multiples of the signal frequency. That is, the signal can be equivalently represented by sinusoids that are *harmonically* related to the base frequency of the signal. Harmonically related means that sinusoids are related by frequencies that are multiples of a base frequency. Thus, the signals sin(2*t*), sin(4*t*), and sin(6*t*) are harmonically related. Stated another way, sin(4*t*) and sin(6*t*) are both *harmonics* of the base frequency sin(2*t*). For example, if a nonsinusoidal, periodic signal repeats every 10 seconds (or can be taken to repeat every 10 seconds for the sake of analysis), that signal can be completely represented by sinusoids having frequencies of 0.1 Hz (base frequency = 1/10 Hz), 0.2 Hz, 0.3 Hz, 0.4 Hz, and so on. In theory, we may need a large, possibly infinite, number of sinusoids to achieve an accurate breakdown of a given periodic signal, but for most real signals, the magnitude of the sinusoidal components becomes negligibly small as frequency increases so these components contribute little to the signal representation.

To put the Fourier series theorem in mathematical terms, note that if the period of a periodic function $x_T(t)$ is T, the base or *fundamental frequency* is:

$$f_1 = \frac{1}{T} \qquad\qquad \text{[Eq. 3.1]}$$

then the base cosine wave, the cosine at the fundamental frequency, becomes:

$$\text{Fundamental} = \cos(2\pi f_1 t) = \cos\left(\frac{2\pi t}{T}\right) \qquad\qquad \text{[Eq. 3.2]}$$

and the series of harmonically related cosine waves becomes:

$$Series(n) = \cos(2\pi n f_1 t) = \cos\left(\frac{2\pi n t}{T}\right) \quad n = 1, 2, 3, \ldots \qquad \text{[Eq. 3.3]}$$

The Fourier series theorem states that a signal can be represented by a series of sinusoids (not necessarily only a sine wave, or only a cosine wave), so the cosine may contain a phase term, θ. The amplitude and phase of these cosines must be allowed to vary for different values of components in the series (i.e., different value of n). Allowing the amplitude and phase to vary, the harmonically related series would be stated mathematically as:

$$Series(n) = C_n \cos\left(\frac{2\pi n t}{T} + \theta_n\right) \quad n = 1, 2, 3, \ldots \qquad \text{[Eq. 3.4]}$$

The Fourier series theorem simply states that any periodic function can be *completely and equivalently* represented by a summation of this series, as specifically stated in Eq. 3.5.

$$x_T(t) = \frac{C_0}{2} + \sum_{n=1}^{\infty} C_n \cos\left(\frac{2\pi nt}{T} + \theta_n\right) \quad \text{or in terms of } f_1$$

$$= \frac{C_0}{2} + \sum_{n=1}^{\infty} C_n \cos(2\pi nf_1 t + \theta_n) \qquad \text{[Eq. 3.5]}$$

where $x_T(t)$ is a periodic function of period T, and the first term, C_0, accounts for any nonzero average value of the signal. The term is also known as the *direct current* (DC) term. (The term DC is used broadly to mean any signal or signal component that is constant over time, even though most signals have nothing to do with current.) If the signal has zero mean, as is often the case, then this term will be zero. Often $2\pi nf_1$ is represented in terms of radians, where $2\pi nf_1 = n\omega_1$. Using frequency in radians makes the equations look cleaner, but in practice frequency is usually measured in hertz. Both will be used here.

As mentioned previously, and as Eq. 3.5 states, the number of sinusoids in the sum required to represent $v_T(t)$ is theoretically infinite, but in practice, the number of sine and cosine components that have significant amplitudes is limited. Often, only a few sinusoids are required to represent the signal. Figure 3.6 shows the reconstruction of a square wave using Eq. 3.5 and a series consisting of three, six, 12,

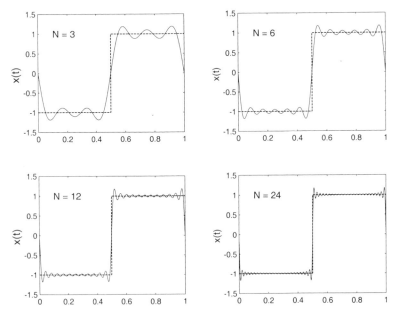

Figure 3.6   Reconstruction of a square wave using three, six, 12, and 24 sinusoids. The square wave is one of the most difficult signals to represent with a sinusoidal series. The oscillations seen in the sinusoidal approximations are known as *Gibbs oscillations*, and they increase in frequency, but do not diminish in amplitude as more sinusoids are added to the summations.

and 24 sinusoids. The square wave is one of the most difficult waveforms to represent using a sinusoidal series because of the sharp transitions. The reconstruction produces a fair representation of a square wave using only six sinusoids.

The cosine wave series of Eq. 3.5 can also be represented in terms of a sine and cosine series. Substituting the sine/cosine representation in Eq. 2.8 for the sinusoidal function in Eq. 3.5, the Fourier series equation becomes:

$$x_T(t) = \frac{a_0}{2} + \sum_{n=1}^{\infty} a_n \cos(2\pi n f_1 t) + \sum_{n=1}^{\infty} b_n \sin(2\pi n f_1 t) \qquad \text{[Eq. 3.6]}$$

The fact that periodic functions can be completely represented by sinusoids at the same or multiple frequencies tells us that if we are looking for sinusoids in a signal, we need only cross-correlate the periodic signal with sinusoids at those specific frequencies. When cross-correlating a signal with

$$r_{xy}(\theta) = \frac{1}{T} \int_0^T x(t) \cos(2\pi n f_1 t + \theta_n) dt \quad n = 1, 2, 3, \ldots$$

sinusoids, it is easier mathematically to vary the phase of the cosine rather than shift the time. Time shifting and phase shifting are proportionally related by the period $[\tau = \theta \ (T/360); \ \text{Eq. 2.6}]$. This leads to an extension of the cross-correlation equation (Eq.2.31):

$$r_{xy}(\theta) = \frac{1}{T} \int_0^T x(t) \cos(2\pi n f_1 t + \theta_n) dt \quad n = 1, 2, 3, \ldots \qquad \text{[Eq. 3.7]}$$

where, again, $f_1 = 1/T$.

The Fourier series theorem tells us that we need only one sinusoid per frequency component (i.e., each value of n). This sinusoidal component will have the specific phase shift θ that maximizes the cross correlation $r_{xy}(\theta)$. To find this component analytically, it would be necessary to set $dr_{xy}(\theta)/d\theta = 0$, solve for θ_{max}, then find $r_{xy}(\theta_{max})$. This is not too difficult to do using a computer program such as MATLAB and was the approach used to find the sinusoidal components in Example 2.11 and Figure 3.1. However, following this procedure analytically can be tedious, especially if $x(t)$ is a complicated function.

There are two ways to simplify the calculations to find the sinusoidal components: Convert the sinusoid in Eq. 3.7 to a sine and cosine or use complex notation to represent the sinusoid. Both allow the direct calculation of the sinusoidal component values. Under the first approach, the sinusoid is divided into a sine and cosine and each is correlated with $x(t)$ separately. Because the sine and cosine do not have a phase, there is no time shift required and the cross-correlation becomes simple correlation as in Eq. 2.28:

$$a_n = \frac{2}{T} \int_0^T x(t) \cos(2\pi n f_1 t) dt \quad n = 0, 1, 2, \ldots$$

$$b_n = \frac{2}{T} \int_0^T x(t) \sin(2\pi n f_1 t) dt \quad n = 0, 1, 2, \ldots \qquad \text{[Eq. 3.8]}$$

These equations solve for the series coefficients in Eq. 3.6. It is not difficult to derive these equations directly from Eq. 3.6 as shown in Appendix A. The factor of 2 is required because the coefficients a_n and b_n are defined in Eq. 3.5 and Eq. 3.6 as amplitudes, not correlations (other scalings may be used). The digital versions of these equations are given in the discussion of the complex representation.

The a_0 term comes directly from Eq. 3.8 when $n = 0$:

$$a_0 = \frac{2}{T} \int_0^T x(t)\,dt \qquad\qquad \text{[Eq. 3.9]}$$

This equation explains why a_0 is divided by two in Eq. 3.5 and Eq. 3.6. It is calculated as twice the average value in order to be compatible with the Fourier transform equations of Eq. 3.8.

To carry out the correlation and integration in Eq. 3.8, there are a few constraints on $x(t)$. First, $x(t)$ must be capable of being integrated; that is:

$$\int_0^T |x(t)|\,dt < \infty \qquad\qquad \text{[Eq. 3.10]}$$

Second, while $x(t)$ can have discontinuities, they must be finite in number and have finite amplitudes. Finally, the number of maxima and minima must also be finite. These three criteria are sometimes referred to as the *Dirichlet conditions* and are generally met by real-world signals.

If we are decomposing the signal into sinusoidal components to plot the signal's frequency components (Figure 3.4), it is desirable to represent the signal by a series of single sinusoids with varying magnitude and phase. This allows us to plot the magnitude and phase frequency characteristics directly as shown in Figure 3.4. This single sinusoidal representation can be obtained by first calculating the cosine and sine coefficients using Eq. 3.8, then converting to a single sinusoid using Eq. 2.9 and Eq. 2.10, and repeated here:

$$x_T(t) = \frac{C_0}{2} + \sum_{n=1}^{\infty} C_n \cos(2\pi n f_1 t + \theta_n) \quad n = 1, 2, 3, \ldots$$

$$\text{where } C_n = \sqrt{a_n^2 + b_n^2} \text{ and } \theta_n = -\tan^{-1}(b_n/a_n) \qquad \text{[Eq. 3.11]}$$

Because θ_n is negative in the equation that led to Eq. 2.10 [i.e., $C_n\cos(2\pi n f_1 t - \theta_n)$], but positive in the equation above, θ_n equals the *negative* of $\tan^{-1}(b_n/a_n)$.

The fact that periodic signals can be completely represented by a sinusoidal series has implications well beyond simply limiting the appropriate sinusoids for probing a periodic signal. It implies that any periodic signal, $x_T(t)$, is as well represented by the sinusoidal series as by the signal itself. In other words, the series of sine and cosine coefficients, a_n and b_n in Eq. 3.8, or the equivalent single sinusoid magnitudes and phases, C_n and θ_n in Eq. 3.11, are as good a representation of $x_T(t)$ as $x_T(t)$ itself. This is evident from the fact that, given only a_n and b_n (or, equivalently C_n and θ_n) and the frequencies at which they occur (or simply the fundamental frequency, f_1), you could reconstruct the signal, $x_T(t)$ using Eq. 3.6.

Converting a signal into its sinusoidal series equivalent using Eq. 3.8 is known as a *transformation* since it transforms $x_T(t)$ into an alternative representation. This transformation based on sinusoids is often referred to as the *Fourier transform*, but this term should be reserved for *aperiodic* (sometimes referred to as *transient*) signals as described below. It is important to use technical terms carefully, so here the term *Fourier series analysis* will be used to describe the transformation between the time and frequency representation of a *periodic* signal. In general, a transformation can be thought of as a remapping of the original signal into an alternative form in the hope that the new form will be more useful or have more meaning. Because it is possible to reverse the transformation and go from the sinusoidal series back to the original function using Eq. 3.5 or Eq. 3.6, the Fourier series transformation would be referred to as a *bilateral transformation*.

3.2.1 Symmetry

Some waveforms are symmetrical or antisymmetrical about $t = 0$, so that one or the other of the components, a_n or b_n in Eq. 3.7, will be zero. Specifically, if the waveform has mirror symmetry about $t = 0$; that is, $x(t) = x(-t)$ (Figure 3.7, upper plot), multiplications with sine functions will be zero leading to b_n terms that are zero. Such mirror symmetry functions are termed *even* functions. Similarly, if the function has antisymmetry, $x_T(t) = -x(t)$ (Figure 3.7, middle plot), termed an *odd* function, all multiplications with cosines will be zero resulting in a_n coefficients that are zero. Finally, functions that have *half-wave* symmetry will have no even coefficients, and both a_n and b_n will be zero for $n =$ even. These are functions where the second

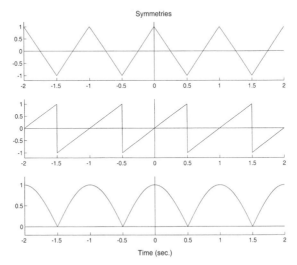

Figure 3.7 Waveform symmetries. Each waveform has a period of 1 second. **Upper plot:** even symmetry; **middle plot:** odd symmetry; **lower plot:** half-wave symmetry. Note that the upper plot is also half-wave symmetric and the lower waveform has even symmetry.

TABLE 3.1   Function Symmetries

Function Name	Symmetry	Coefficient Values
Even	$x(t) = x(-t)$	$b_n = 0$
Odd	$x(t) = -x(-t)$	$a_n = 0$
Half-wave	$x(t) = x(T - t)$	$a_n = b_n = 0$; for n even

half of the period looks like the first half flipped left to right [i.e., $x(t) = x(T - t)$ (Figure 3.7, lower plot)]. Functions having half-wave symmetry can also be either odd or even functions. These symmetries were useful for reducing the complexity of solving for the coefficients when such computations were done manually. Even when the Fourier series is calculated on a computer (usually the case), these properties can help verify the computer solution. Table 3.1 and Figure 3.7 summarize these properties.

3.3   FREQUENCY REPRESENTATION

Functions other than sinusoids can be, and are, used to perform transformations, but sinusoids are especially useful because of their unique frequency characteristics mentioned earlier: a sinusoid contains energy at only one frequency (Figure 3.3). Hence, if we know the sinusoidal makeup of a signal, it is easy to determine its frequency characteristics, or *spectrum*. A complete description of a waveform's frequency characteristics consists of two plots (or two sets of data points): a plot of the magnitude versus frequency and a plot of the phase versus frequency. Although both magnitude and phase plots are necessary to completely represent the signal and to convert the frequency representation back into the time representation, often only the magnitude spectrum is of interest. Each sinusoidal component gives us a single point on the two frequency curves (magnitude and phase) at a frequency related to the component number n; specifically:

$$f = \frac{n}{T} = nf_1 \qquad \text{[Eq. 3.12]}$$

For example, each C_n (from Eq. 3.11) would appear as a single point on the *magnitude* (upper) plot while each θ_n would show as a single point on the phase (lower) curve (Figure 3.3). These frequency plots (or equivalent set of data values) are known as the 'frequency domain' representation, just as the untransformed waveform, $x_T(t)$, is known as the *time domain* representation. Hence, the Fourier series analysis in Eq. 3.8 can be used to convert a periodic signal into its equivalent frequency representation. Indeed, this time-to-frequency conversion is the primary *raison d'être* for Fourier analysis. The Fourier series (and related Fourier transform) is not the only path to a signal's frequency or spectral characteristics, but it is the most general approach; it makes the fewest assumptions about the signal, and its digital version can be calculated with great speed using an algorithm known as the *fast Fourier transform* (FFT). To go in the reverse direction, from the frequency to the time domain, the Fourier series equations, Eq. 3.5 or Eq. 3.6, can be used.

The resolution of a spectrum can be loosely defined as the frequency difference that a spectrum can resolve; that is, how close two frequencies can get and still be identified as two frequencies. This resolution clearly depends on the frequency spacing between harmonic numbers which in turn is equal to $1/T$ (Eq. 3.12). Hence, the longer the data period being analyzed, the higher the spectral resolution. This holds for discrete data as well.

Example 3.2: Find the Fourier series of the triangle waveform defined below. Find the first four Fourier series components.

$$x(t) = \begin{cases} t & 0 < t \le 0.5 \\ 0 & 0.5 < t \le 1.0 \end{cases}$$

Solution: Use Eq. 3.8 to find the cosine (a_n) and sine (b_n) coefficients. Then convert to magnitude (C_n) and phase (θ_n) if desired. The magnitude and phase representation is more useful for plotting.

To find the sine coefficients use the lower equation in Eq. 3.8:

$$b_n = \frac{2}{T}\int_0^T x(t)\sin(2\pi n f_1 t)dt = \frac{2}{T}\int_0^{0.5} t\sin(2\pi n t)dt$$

$$= \frac{1}{2\pi^2 n^2}[\sin(2\pi n t) - 2\pi n t \cos(2\pi n t)]_0^{0.5}$$

$$= \frac{1}{2\pi^2 n^2}[\sin(\pi n) - \pi n \cos(\pi n)] = \frac{-1}{2\pi n}(\cos(\pi n))$$

$$= \frac{1}{2\pi}, \frac{-1}{4\pi}, \frac{1}{6\pi}, \frac{-1}{8\pi} = 0.159, -0.080, 0.053, -0.040$$

To find the cosine coefficients, apply the upper equation of Eq. 3.8:

$$a_n = \frac{2}{T}\int_0^T x(t)\cos(2\pi n f_1 t)dt = \frac{2}{T}\int_0^{0.5} t\cos(2\pi n t)dt$$

$$= \frac{1}{2\pi^2 n^2}[\cos(2\pi n t) + 2\pi n t \sin(2\pi n t)]_0^{0.5}$$

$$= \frac{1}{2\pi^2 n^2}[\cos(\pi n) + \pi n \sin(\pi n) - 1] = \frac{1}{2\pi^2 n^2}(\cos(\pi n) - 1)$$

$$= -\frac{1}{\pi^2}, 0, -\frac{1}{9\pi^2}, 0 = -0.101, 0, -0.011, 0$$

These two coefficient terms can then be combined into the single sinusoidal representation as magnitude (C_n) and phases (θ_n) using Eq. 3.11. Care must be taken in computing the angle to ensure that it represents the proper quadrant:

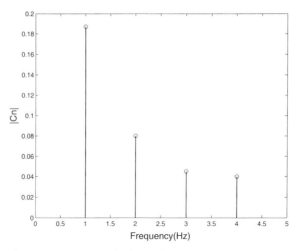

Figure 3.8 Magnitude plot of the frequency characteristics of the waveform given in Example 3.2. Only the first four components are shown. These components were determined by taking the square root of the sum of squares of the first four cosine and sine terms.

$$C_n = \sqrt{a_n^2 + b_n^2} = 0.187, 0.080, 0.054, 0.040$$

$$\theta_n = -\tan^{-1}(b_n/a_n) = -58\deg(2^{\text{nd}}), -90\deg(3^{\text{rd}}), -77\deg(2^{\text{nd}}), -90(3^{\text{rd}})\deg$$

$$\theta_n = -(180-58) = -122\deg, -90\deg, -(180-77) = -103\deg, -90\deg$$

For a complete description of $x(t)$ we would need more components and the a_0 term. The $a_0/2$ term is twice the average value of $x(t)$, which can be obtained using Eq. 3.9:

$$\frac{a_0}{2} = \overline{x}(t) = \frac{1}{T}\int_0^T x(t)dt = \frac{1}{1}\int_0^{0.5} tdt = \frac{t^2}{2}\Big|_0^{0.5} = 0.125$$

The most common way of presenting the frequency information determined in Example 3.2 is as a plot of the single sinusoid coefficients, C_n and θ_n, against frequency. Since the period of the function is 1.0 seconds, the fundamental frequency f_1 is 1 Hz. Hence, the first four values of n represent 1 Hz, 2 Hz, 3 Hz, and 4 Hz. The magnitude plot is shown in Figure 3.8.

3.4 COMPLEX REPRESENTATION

Euler's identity, Eq. 2.12, allows us to describe the sine and cosine functions in terms of imaginary exponentials as shown below and derived in Appendix A:

$$\cos(2\pi n f_1 t) = \frac{1}{2}\left(e^{+j2\pi n f_1 t} + e^{-j2\pi n f_1 t}\right)$$

$$\text{and}\quad \sin(2\pi n f_1 t) = \frac{1}{j2}\left(e^{+j2\pi n f_1 t} - e^{-j2\pi n f_1 t}\right) \qquad \text{[Eq. 3.13]}$$

Substituting these definitions into the Fourier series equation, Eq. 3.6:

$$x(t) = \frac{a_0}{2} + \sum_{n=1}^{\infty}\left(\frac{a_n}{2}e^{j2\pi n f_1 t} + \frac{a_n}{2}e^{-j2\pi n f_1 t} + \frac{b_n}{2j}e^{j2\pi n f_1 t} - \frac{b_n}{2j}e^{-j2\pi n f_1 t}\right)$$

Removing the j terms from the denominator (multiplying numerator and denominator by j), and collecting terms:

$$x(t) = \frac{a_0}{2} + \sum_{n=1}^{\infty}\left[\left(\frac{a_n - jb_n}{2}\right)e^{j2\pi n f_1 t} + \left(\frac{a_n + jb_n}{2}\right)e^{-j2\pi n f_1 t}\right]$$

Now defining a new coefficient, C_n:

$$C_n = \frac{a_n - jb_n}{2} \qquad \text{[Eq. 3.14]}$$

Then substituting for a_n and b_n in Eq. 3.14, the integral term from the Fourier series equations given in Eq. 3.8 becomes:

$$C_n = \frac{2}{2T}\int_0^T x(t)\cos(2\pi n f_1 t)dt - \frac{2j}{2T}\int_0^T x(t)\sin(2\pi n f_1 t)dt \quad n = 0, 1, 2, \ldots$$

Combining:

$$C_n = \frac{1}{T}\int_0^T x(t)[\cos(2\pi n f_1 t) - j\sin(2\pi n f_1 t)]dt \quad n = 0, 1, 2, \ldots$$

Using Euler's identity (Eq. 2.12), the term in the brackets can be replaced by a single imaginary exponential giving the complex form of the Fourier series analysis as:

$$C_n = \frac{1}{T}\int_0^T x(t)e^{-2j\pi n f_1 t}dt \quad n = 0, \pm 1, \pm 2, \pm 3, \ldots \qquad \text{[Eq. 3.15]}$$

where $f_1 = 1/T$. In this complex representation, the $a_0/2$ term (the average value or DC term) does not require a separate equation since when $n = 0$ the exponential reduces to 1, and the integral computes the average value. From the definition of C_n (Eq. 3.14), it can be seen that this single complex variable contains both the cosine and sine coefficients. The cosine coefficients (a_n) can be obtained by taking twice the real part of C_n, whereas the sine coefficients (b_n) can be obtained as twice the imaginary part of C_n. Note that from the complex arithmetic:

$$|C_n| = \sqrt{\mathrm{Re}^2 + \mathrm{Im}^2} = \sqrt{\frac{a_n^2 + b_n^2}{2}} = 0.707\sqrt{a_n^2 + b_n^2} \qquad \text{[Eq. 3.16]}$$

$$\theta_{C_n} = -\tan^{-1}\left(\frac{\mathrm{Im}}{\mathrm{Re}}\right) = -\tan^{-1}\left(\frac{b_n/2}{a_n/2}\right) = -\tan^{-1}\left(\frac{b_n}{a_n}\right) \qquad \text{[Eq. 3.17]}$$

Hence the magnitude of C_n is equal to 0.707 times the magnitude of the sinusoidal components and the angle of C_n is equal to the phase of the sinusoidal component. The complex variable C_n not only contains both sine and cosine coefficients, but these components can be readily obtained in either the single sinusoidal magnitude and phase form (most useful for plotting) or the sine/cosine form.

The complex version of the inverse Fourier series requires the summation be done for $n = \pm\infty$:

$$x(t) = \sum_{n=-\infty}^{\infty} C_n e^{+j2\pi n f_1 t} \qquad \text{[Eq. 3.18]}$$

Example 3.3:   Find the Fourier series of the pulse waveform shown below with a period T and a pulse width of W. Use the complex equation.

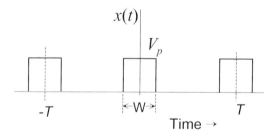

Solution: Apply Eq. 3.15 directly except since the signal is and even function [$x(t) = x(-t)$], it is easier to integrate from $t = -T/2$ to $+T/2$.

$$C_n = \frac{1}{T}\int_{-T/2}^{T/2} x(t)e^{-j2\pi n f_1 t}\,dt = \frac{1}{T}\int_{-W/2}^{W/2} V_p e^{-j2\pi n f_1 t}\,dt$$

$$C_n = \frac{V_p}{T(-j2\pi n f_1 t)}\left[e^{-j2\pi n f_1 W/2} - e^{j2\pi n f_1 W/2}\right]$$

$$C_n = \frac{V_p W}{T}\frac{\sin(2\pi n f_1 W/2)}{2\pi n f_1 W/2}; \quad n = 0, \pm1, \pm2, \pm3, \dots$$

Given specific values for T and W it is possible to solve for the complex values of C_n. The magnitude spectrum would be equal to $|C_n|/0.707$, whereas the phase spectrum would be equal to the angle of C_n (i.e., $\angle C_n$, see Appendix A). In general, it is easier to solve for the Fourier series analytically using the sine and cosine equations, Eq. 3.8, but the complex representation is more often used when discussing the digital implications of Fourier series analysis because of its succinct presentation. It is also the approach used in computer algorithms, and is the format cited in journal articles.

3.5   THE CONTINUOUS FOURIER TRANSFORM

The Fourier series analysis is a good approach to determining the frequency or spectral characteristics of a periodic waveform, but what if the signal is not periodic? Most real signals are not periodic, and for many physiological signals, such as the EEG signal in Figure 2.4, only a part of the signal is available. The segment of EEG signal shown in Figure 2.4 is much less than the amount of data actually recorded, but no matter how long the recording, the EEG signal existed before the recording and will continue after the recording session ends (unless the EEG recording session was so traumatic as to cause untimely death!). If the signal is *aperiodic*, an extension of the Fourier series analysis can be used. An aperiodic signal is one that exists for a finite period, but is zero at all other times (Figure 3.9). Aperiodic signals are sometimes referred to as *transient* signals.

To extend the Fourier series analysis to aperiodic signals, such signals could be treated as a periodic, but where the period goes to infinity, (i.e., $T \rightarrow \infty$). Note that if $T \rightarrow \infty$, then $f_1 = 1/T \rightarrow 0$; however, nf_1 does not necessarily go to zero because n may still go to infinity. Rather as f_1 becomes smaller and smaller, so does the

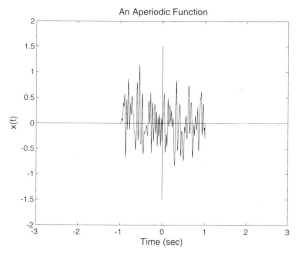

Figure 3.9 An aperiodic function exists for a finite length of time and is zero everywhere else.

increment between harmonics and nf_1 becomes a continuous variable f. Thus, in the Fourier series equations (Eq. 3.8 and 3.15), the $2\pi nf_1$ becomes $2\pi f$ for aperiodic functions. If radians are used instead of Hz, then $n\omega_1$ goes to ω in all these equations. This leads to the equation for the continuous Fourier transform:

$$\lim_{\substack{T\to\infty \\ f_1\to 0}} C_n = \int_0^T x(t)e^{-j2\pi nf_1 t}dt = \int_{-\infty}^{\infty} x(t)e^{-j2\pi ft}dt$$

$$\left.\begin{array}{l} X(f) \equiv C(f) \\ X(\omega) \equiv C(\omega) \end{array}\right\} = \int_{-\infty}^{\infty} x(t)e^{-j2\pi ft}dt = \int_{-\infty}^{\infty} x(t)e^{-j\omega t}dt \qquad \text{[Eq. 3.19]}$$

or, in terms of the sine and cosine equations:

$$\left.\begin{array}{l} A(f) \\ A(\omega) \end{array}\right\} = \int_{-\infty}^{\infty} x(t)\cos(2\pi ft)dt = \int_{-\infty}^{\infty} x(t)\cos(\omega t)dt$$

$$\left.\begin{array}{l} B(f) \\ B(\omega) \end{array}\right\} = \int_{-\infty}^{\infty} x(t)\sin(2\pi ft)dt = \int_{-\infty}^{\infty} x(t)\sin(\omega t)dt \qquad \text{[Eq. 3.20]}$$

These transforms now produce a continuous function as an output and it is common to denote these Fourier transform variables with capital letters. In addition, the transform equation is no longer normalized by the period because $1/T \to 0$ since $T \to \infty$. Although the transform equation has limits between $\pm\infty$, the actual limits will be over the length of $x(t)$, which must be finite or the transform does not exist.

Although it is rarely used in analytical computations, the inverse continuous Fourier transform is given as:

$$x(t) = \frac{1}{2\pi} \int_{-\infty}^{\infty} X(\omega)e^{j\omega t}d\omega \qquad \text{[Eq. 3.21]}$$

where again, ω is frequency in radians.

Example 3.4:   Find the Fourier transform of the pulse in Example 3.3 assuming the period, T, goes to infinity.

Solution: We could apply Eq. 3.19, but because $x(t)$ is an even function, we could also use cosine portion of the Fourier transform in Eq. 3.20.

$$X(f) = A(f) = \int_{-\infty}^{\infty} x(t)\cos(2\pi ft)dt = \int_{-W/2}^{W/2} V_p \cos(2\pi ft)dt = \frac{V_p}{2\pi f}\left|\sin(2\pi ft)\right|_{t=-W/2}^{t=W/2}$$

$$X(f) = \frac{V_p}{2\pi f}(\sin(\pi fW) - \sin(-\pi fW)) = \frac{V_p}{\pi f}\sin(\pi fW)$$

A plot of $X(f)$ is shown in Figure 3.10. Note that this is a complex function so the magnitude is just the absolute value of the function shown in Figure 3.10, and the phase would alternate between 0 degrees when the function is positive and 180 degrees when the function is negative.

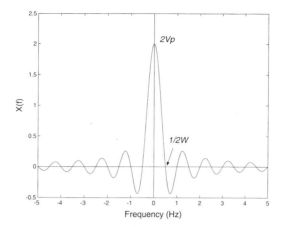

Figure 3.10   The complex spectrum of an aperiodic pulse as determined in Example 3.4.

3.6   DISCRETE DATA: THE DISCRETE FOURIER TRANSFORM

Most Fourier analysis is done using a digital computer and is applied to discrete data. Discrete data differ from continuous, periodic data in two fundamental ways: They are time- and amplitude-sampled, as discussed in Section 1.2.1, and they are finite. The finite data can be considered as an aperiodic signal or as one period of a periodic signal. Either way, the analysis is the same: a discrete version of the Fourier analysis equations:

$$X[n] = \sum_{m=1}^{M} x[m] e^{-j2\pi mn/M} \qquad \text{[Eq. 3.22]}$$

where m is the index of the data array of length M, and n is the harmonic number. In the most common implementation of this equation, the length of n and m are equal. Equation 3.22 produces a series of numbers that describe the amplitude and phase of a harmonic series of sinusoids. The fundamental period of this series is still $1/T$, which can also be given in terms of the data length, M, and the sampling frequency, f_s:

$$f_1 = \frac{1}{T}; \quad T = \frac{M}{f_s}; \quad f_1 = \frac{f_s}{M} \qquad \text{[Eq. 3.23]}$$

If M is finite, as must always be the case, Eq. 3.23 is actually the discrete version of the Fourier series; if the data are actually periodic, this equation is performing a *discrete Fourier series* analysis. On the other hand, if the assumption is that the data represent an aperiodic function (commonly made, even if untrue), and the data at hand are a truncated version of infinitely long data, this equation can be taken as performing the *discrete Fourier transform* (DFT). Regardless of how this equation

TABLE 3.2 Terms and Abbreviations Used in Fourier Analysis

Type of Signal	Type of Analysis	Method
Periodic and continuous (analog)	Fourier series	Analytical Eq. 3.8 or Eq. 3.18
Periodic and discrete	Discrete Fourier series	Eq. 3.21, DFT or FFT
Aperiodic and continuous	Fourier transform	Analytical Eq. 3.19
Aperiodic (assumed) and discrete	DFT	Eq. 3.21, DFT or FFT

DFT, discrete Fourier transform; FFT, fast Fourier transform.

is interpreted, it is usually implemented using a high-speed algorithm know as the FFT.

It should be apparent that terminology associated with Fourier analysis is complicated by the use of several different terms and abbreviations. There is the temptation to simply call all these operations *Fourier transforms*, or even worse, *FFTs*. Using appropriate terminology will reduce confusion about what you are actually doing and may even impress less informed bioengineers. To aid in being *linguistically correct*, Table 3.2 summarizes the terms and abbreviations used to describe the various aspects of Fourier analyses.

The discrete form of the inverse Fourier transform is similar to the inverse Fourier series equation (Eq. 3.18):

$$x[n] = \sum_{m=1}^{M} X[m]e^{j2\pi mn/M}$$ [Eq. 3.24]

Again, discrete-time data differ from continuous data in two important ways: The data are time- and amplitude-sampled and the data are shortened to fit within the finite constraints of computer memory. Each of these operations has an effect on the discrete Fourier transform, particularly time sampling and data truncation. The influence of each of these operations on the digitized signal is described in the next section.

3.6.1 Data Sampling: Sampling Theorem

To import a continuous time domain (i.e., analog) signal into a digital computer requires, among other things, slicing that signal into discrete time intervals (usually evenly spaced), a process known as *sampling* (Figure 3.11).

This sampling process has some very peculiar effects on the discrete Fourier transform (or discrete Fourier series). Figure 3.12A shows an example of the magnitude frequency spectrum of a 1-second periodic, continuous signal that might be found using continuous Fourier series analysis (Eq. 3.8 or Eq. 3.15). The fundamental frequency is $1/T = 1$ Hz and the first 10 harmonics are plotted out to 10 Hz. For this particular signal, there is little energy above 7.0 Hz. Figure 3.12B shows the theoretical magnitude spectrum that would be obtained by applying the DFT to the same

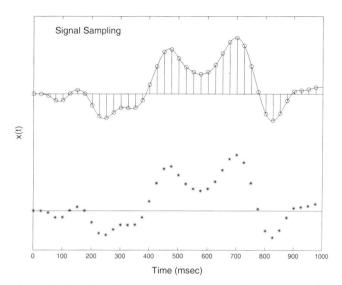

Figure 3.11 A continuous signal (*upper trace*) is sampled at discrete points in time and stored in memory as an array of numbers proportional to the continuous signal's amplitude at the time it was sampled (*lower trace*).

signal that produced the spectrum in Figure 3.12A, but after it is sampled at 20 Hz. Due to the sampling process, many more frequencies exist than those in the original. The output of the DFT is itself periodic, with the period equal to the *sample* frequency, f_s (in this case 20 Hz). The spectrum even contains negative frequencies and has even symmetry about $f = 0.0$ Hz as well as about positive and negative multiples of f_s. Because of the even symmetry, the portion between 0 and 20 Hz has mirror symmetry about $f_s/2$ (in this case 10 Hz).

The spectrum of the sampled signal presents a problem because if the sampled signal spectrum is different from the original, unsampled signal spectrum (which it is), it stands to reason that the sampled signal is different from the original. If the sampled signal is different from the original and if we cannot recover the sampled signal from the original, digital signal processing is a lost cause. We will be processing something unrelated to the original signal. The critical question then becomes, Given that the sampled signal is different from the original, can we at least reconstruct the original signal from the sampled signal? The frequency domain version of that question is, Can we reconstruct the unsampled spectrum from the sampled spectrum? The answer is, Well, maybe. It depends on the sampling frequency—as shown below.

Figure 3.13 shows just one period of the DFT shown in Figure 3.12B, the period between 0 and f_s Hz. In fact, this is the only portion of the DFT spectrum that is produced by the DFT algorithm, the rest of the frequencies are theoretical (but not inconsequential). Comparing this spectrum to the spectrum of the original signal

A

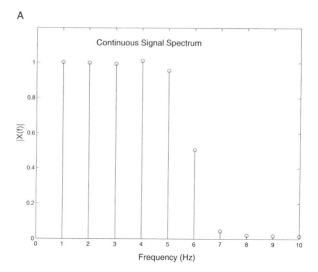

B

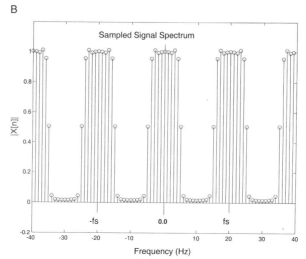

Figure 3.12   **A:** The spectrum of a continuous signal. **B:** The spectrum of this signal after being sampled at $f_s = 20$ Hz. Sampling produces a larger number of frequency components not in the original spectrum, even components having negative frequency. The sampled signal has a spectrum that is periodic at the sampling frequency (20 Hz) and has even symmetry about 0.0 Hz and positive and negative multiples of the sampling frequency, f_s. Since the sampled spectrum is periodic, it goes on forever and only a portion of can be shown.

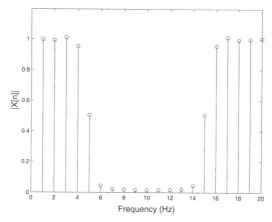

Figure 3.13 One period of the discrete Fourier transform shown in Figure 3.12B, the period between 0 and f_s Hz.

(Figure 3.12A), we see that the two are the same for the first half of the spectrum up to $f_s/2$, and the second half is just the mirror image. The spectrum above $f_s/2$ is a result of the periodicity produced by the sampling process combined with the even symmetry. (Because of the periodicity, it is in the frequency range above $f_s/2$ where the theoretical negative frequencies make their impact.) It would appear that in this case at least, we could obtain a frequency spectrum that was identical to the original if we simply disregarded all frequencies above $f_s/2$. Under this strategy, these frequencies are ignored because they are generated by the sampling process and do not belong to the original, continuous signal. The frequency $f_s/2$ is often referred to as the *Nyquist frequency*.

This strategy of ignoring all frequencies above the Nyquist frequency ($f_s/2$), works well and is the approach that is commonly adopted. But, it only works if the original signal did not have components at, or above $f_s/2$. For example, consider a situation where four sinusoids having frequencies of 100, 200, 300, and 400 HZ are sampled at a sampling frequency of 1,000 Hz. The spectrum produced by the DFT actually contains eight frequencies (Figure 3.14A): the four original frequencies plus the four mirror image frequencies reflected about $f_s/2$ (500 Hz). As long as we know, in advance, that the sampled signal does not contain any frequencies above the Nyquist frequency (500 Hz), we will not have a problem. We know that the first four frequencies are those of the signal and the second four, above the Nyquist frequency, are the reflections that can be ignored. However, a problem occurs if the signal contains frequencies higher than the Nyquist frequency. The reflections of these frequency components will carry back into the lower half of the spectrum. This is shown in Figure 3.14B where the signal now contains two additional frequencies at 650 and 850 Hz. These frequency components have their reflections in the lower half of the spectrum—at 350 and 150 Hz, respectively. It is now no longer possible to determine if the 350-Hz and 150-Hz signals are part of the true

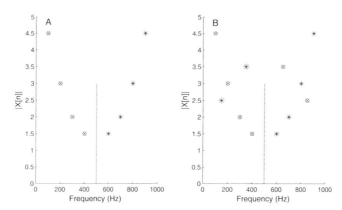

Figure 3.14 **A:** Four sine waves between 100 and 400 Hz are sampled at 1 kHz. Only one period of the sampled spectrum is shown, the period between 0 and f_s Hz. Sampling essentially produces new frequencies not in the original signal. Because of the periodicity and even symmetry of the sampled spectrum, the additional frequencies are a mirror image reflection around $f_s/2$, the Nyquist frequency. If all the frequency components of the sampled signal are below the Nyquist frequency, the upper frequencies can be ignored. **B:** If the sampled signal contains frequencies above half the sampling frequency, these are reflected in the lower half of the spectrum (*large asterisks*). It is no longer possible to determine which frequencies belong where, a problem known as *aliasing*.

spectrum of the signal (i.e., the spectrum of the signal before it was sampled) or whether these are the reflections of signals with frequency components greater than $f_s/2$. Both halves of the spectrum now contain mixtures of frequencies above and below the Nyquist frequency, and it is impossible to know where they really belong. This confusing condition is known as *aliasing*. One obvious way—in fact the only way—of resolving this ambiguity is to insure that all frequencies in the original signal are less than the Nyquist frequency, and this is exactly what is done.

Because the Fourier transform is bilateral, if you cannot determine the original spectrum from the one in the computer, you cannot reconstruct the original signal from the one stored in the computer. The frequencies above the Nyquist frequency have hopelessly corrupted the signal stored in the computer. Fortunately, the converse is also true. If there are no corrupting frequency components in the original signal (i.e., the signal contains no frequencies above half the sampling frequency), the spectrum in the computer will be a true reflection of the signal's spectrum, if we eliminate or disregard the frequencies above the Nyquist frequency. Elimination of frequencies above the Nyquist frequency can be achieved by lowpass filtering, so the original signal can be reconstructed from the one in the computer by lowpass filtering. This leads to the famous *sampling theorem* of Shannon: The original signal can be recovered from a sampled signal if the sampling frequency is more than twice the *maximum* frequency contained in the original:

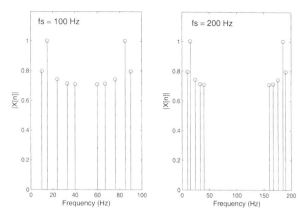

Figure 3.15 The same signal is sampled at two different sampling frequencies. The higher sampling frequency provides much greater separation between the signals of interest and those produced by the sampling process.

$$f_s > 2^* f_{max}$$ [Eq. 3.25]

Usually the sampling frequency is under software control, and it is up to the biomedical engineer doing the sampling to ensure that f_s is high enough. To make elimination of the unwanted higher frequencies easier, it is common to sample at three to five times f_{max}. This increases the spacing between the frequencies in the original signal and those generated by the sampling process (Figure 3.15). The temptation to oversample, setting f_s to be much higher than is really necessary, is strong, and it is a strategy often pursued. However, excessive sampling frequencies leads to large data storage and processing requirements that could overtax the computer system.

3.6.2 Amplitude Slicing: Quantization (Optional)

By selecting an appropriate sampling frequency, it is possible to circumvent problems associated with time slicing, but the digitization process also requires the data be sliced in amplitude because the signal value must be represented by a discrete number (usually in binary format). Amplitude resolution is given in terms of the number of bits in the binary output with the assumption that the least significant bit (LSB) in the output is accurate (which may not always be true). Typical analog-to-digital converters (ADCs) feature 8-, 12-, and 16-bit output with 12 bits presenting a good compromise between conversion resolution and cost. In fact, most signals do not have sufficient signal-to-noise ratio to justify a higher resolution; you are simply obtaining a more accurate conversion of the noise.

The number of bits used for conversion sets an upper limit on the resolution, and determines the quantization error (Figure 3.16). The more bits that are used to represent the signal, the finer the resolution of the digitized signal and the smaller the

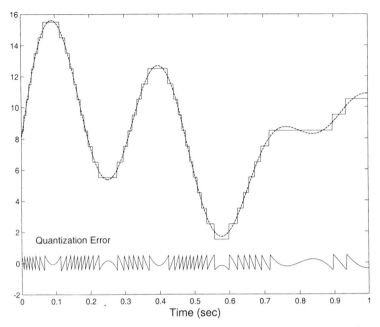

Figure 3.16 Quantization (amplitude slicing) of a continuous waveform. The lower trace shows the error between the quantized signal and the input.

quantization error. The quantization error is the difference between the original, continuous signal value and the digital representation of that value after it is sampled. This error can be thought of as a noise process added to the digitized signal. If a sufficient number of quantization levels exist (say n above 64, equivalent to 7 bits), the distortion produced by quantization error may be modeled as additive independent white noise with zero mean and a variance determined by the quantization step size, δ. The quantization step size is just the maximum voltage the ADC can convert divided by the number of steps which is $2^n - 1$; hence, $\delta = V_{MAX}/2^{n-1}$. Assuming that the error, η, is uniformly distributed between $-\delta/2$ to $+\delta/2$, the variance of the noise would be:

$$\sigma^2 = \int_{-\delta/2}^{\delta/2} \frac{\eta^2}{\delta} dn = \frac{\delta^2}{12} = \frac{V_{MAX}^2}{12(2^n - 1)^2} \qquad \text{[Eq. 3.26]}$$

where V_{MAX} is the maximum voltage the ADC can convert and n is the number of bits out of the ADC. Assuming a uniform distribution for the quantization noise, the RMS (root mean square) value of the noise would be approximately equal to the standard deviation, σ, as shown in Chapter 2.

Example 3.5: What is the equivalent quantization noise (RMS) of a 12-bit ADC that has an input voltage range of ±5.0 V.

Solution: Apply Eq. 3.26

$$\sigma^2 = \frac{V_{MAX}^2}{12(2^n - 1)^2} = \frac{5^2}{12(2^{12} - 1)^2} = 1.24 \times 10^{-7}$$

$$V_{RMS} = \sigma = \sqrt{1.24 \times 10^{-7}} = 0.35\text{mV}$$

3.6.3 Data Length: Truncation

The digitized waveform must necessarily be truncated to the length of the memory storage array, a process described as *windowing*. The windowing process can be thought of as multiplying the data by some window shape. If the waveform is simply truncated and no further shaping is performed on the resultant digitized waveform (as is often the case), then the window shape is rectangular by default (Figure 3.17). Other shapes can be imposed on the data by multiplying the digitized waveform by the desired shape. The influence of windowing is generally quite subtle and, except for very short data segments, a rectangular window (i.e., simple truncation) is usually acceptable. (See Semmlow, 2004, for a thorough discussion of the influence of various nonrectangular window functions.)

The length of a truncated data segment is determined during the data acquisition process and is usually a compromise between the desire to acquire enough of the signal and the need to minimize computer memory usage.

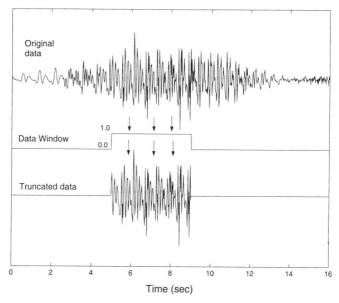

Figure 3.17 Data truncation or shortening can be thought of as mathematically as multiplying the original data by a window function that is one over the length of the truncated data and zero everywhere else.

The length of the data segment determines the apparent period of the data and hence the frequency range of the DFT. Recall that the frequency components obtained by the Fourier series analysis depend on the period, specifically:

$$f = \frac{n}{T} = nf_1 \qquad\qquad \text{[Eq. 3.27]}$$

where n is the harmonic number and f_1 the fundamental frequency which equals $1/T$. For a discrete signal where N is the total number of points in the digitized signal, the equivalent period is the total number of points, N, divided by the sample frequency, f_s:

$$T_{effective} = \frac{N}{f_s} \qquad\qquad \text{[Eq. 3.28]}$$

and the equivalent frequency, f, can be written as:

$$f = \frac{n}{T} = \frac{n}{N/f_s} = \frac{nf_s}{N} \qquad\qquad \text{[Eq. 3.29]}$$

This equation is often used in the MATLAB routines to generate the horizontal axis of a frequency plot (i.e., `f = (1:N)*fs/N`, where N is the length of the signal. See Example 3.6). The last data point from the DFT would have a frequency value of f_s because:

$$f_{last} = f|_{n=N} = \frac{Nf_s}{N} = f_s \qquad\qquad \text{[Eq. 3.30]}$$

In the continuous Fourier series, the frequency increases for each harmonic number by $1/T$ (Eq. 3.12). Accordingly, the resolution of the continuous Fourier series depends on the period; that is, the time length of the data. For the DFT, the same relationship between data length and frequency resolution holds; specifically, the resolution is proportional to f_s/N as given in Eq. 3.29. Hence, for a given sampling frequency, the larger N, the smaller the frequency increment between successive DFT data points. In other words, the more points sampled, the higher the spectral resolution.

Once the data has been acquired, it would seem that the number of points representing the data, N, is fixed, but there is a method that can be used to increase the data length post hoc. We can extend the data simply by tacking on constant values, usually zeros. This may sound like cheating, but it is justified by the underlying assumption that the signal is zero outside of the actual data segment. In signal processing, adding zero data points to extend the period of a signal is known as *zero padding*. Other, more complicated, *padding* techniques can be used, such as padding with the end points, but zero padding is by far the most common strategy for extending data.) This gives the *appearance* of a spectrum with higher resolution. Figure 3.18 shows what appear to be advantages of extending the period by adding zeros to the data segment. The signal is a triangle wave with an actual period of 1 second but which has been extended to 2 and 8 seconds with zero padding. Zero

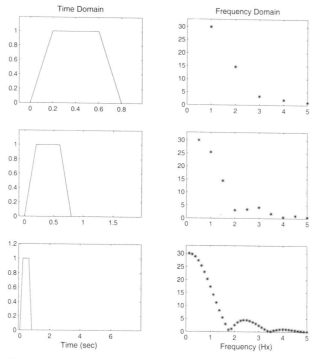

Figure 3.18 A waveform having an original period of one second (*upper left*) and its associated frequency spectrum (*upper right*). Extending the period to 2 and 8 seconds (*middle and lower plots*) by adding zeros decreases the spacing between frequency points, producing a smoother-looking frequency curve.

padding the data seems to improve resolution because the more closely spaced frequency points show more of the spectrum's details. In fact, extending the period with zeros does *not* increase the information in the signal, and the resolution of the signal is not any better. Zero padding provides an interpolation of the points in the unpadded signal; that is, it fills in the gaps of the original spectrum using an estimation process. Overstating the value of zero padding is a common mistake of practicing engineers. Zero padding does not increase the resolution of the spectrum, only the apparent resolution. However, the interpolated spectrum will certainly *look* better when plotted.

3.7 POWER SPECTRUM

The *power spectrum* is commonly defined as the Fourier transform of the autocorrelation function. In continuous notation and discrete notation, the power spectrum equation becomes the following:

$$PS(f) = \int_0^T r_{xx}(t)e^{-2pft}dt \quad P = 1, 2, 3 \ldots$$

$$PS(f) = \sum_{n=0}^{N-1} r_{xx}[n]e^{-j2pnfT_s} \quad P = 1, 2, 3 \ldots \qquad \text{[Eq. 3.31]}$$

where $r_{xx}[n]$ is the autocorrelation function described in Chapter 2. Because the autocorrelation function has even symmetry, the sine terms, b_n, will all be zero (Table 3.1), and Eq. 3.31 can be simplified to include only real cosine terms:

$$PS(f) = \int_0^T r_{xx}(t)\cos(2pmft)dt \quad P = 1, 2, 3 \ldots$$

$$PS(f) = \sum_{n=0}^{N-1} r_{xx}[n]\cos(2pnfT_s) \quad P = 1, 2, 3 \ldots \qquad \text{[Eq. 3.32]}$$

Equation 3.32 is sometimes referred to as the *cosine transform*. A more popular method for evaluating the power spectrum is the so-called direct approach. The direct approach is motivated by the fact that the energy contained in an analog signal, $x(t)$, is related to the magnitude of the signal squared integrated over time:

$$E = \int_{-\infty}^{\infty} |x(t)|^2 dt \qquad \text{[Eq. 3.33]}$$

By an extension of a theorem attributed to Parseval (Stearns and David, 1996) it is easy to show that:

$$\int_{-\infty}^{\infty} |x(t)|^2 dt = \int_{-\infty}^{\infty} |X(f)|^2 df \qquad \text{[Eq. 3.34]}$$

The term $|X(f)|^2$ is called the *periodogram* and equals the energy density function over frequency, also referred to as the energy spectral density, the *power spectral density*, or simply the power spectrum. (Traditionally, evaluation of the power spectrum involved the averaged periodogram as described in Section 3.7.1, but it is now common to refer to the unaveraged periodogram as the power spectrum; hence the two terms have become interchangeable.) In the direct approach, the power spectrum or periodogram is calculated as the magnitude squared of the Fourier transform (or Fourier series) of the waveform of interest:

$$PS(f) = |X(f)|^2 \qquad \text{[Eq. 3.35]}$$

This direct approach of Eq. 3.35 has displaced the cosine transform for determining the power spectrum because of the efficiency of the FFT. (However, a variation of this approach is still used in some advanced signal processing techniques involving time and frequency transformation.) One of the problems compares the power spectrum obtained using the direct approach of Eq. 3.35 with the traditional cosine transform method represented by Eq. 3.32.

Unlike the Fourier transform, the power spectrum does not contain phase information. Hence, the power spectrum is not a bilateral transformation—it is not possible to reconstruct the signal from the power spectrum. However, the power spectrum has a wider range of applicability and is defined for some signals that do not have a meaningful Fourier transform (such as those resulting from random processes). Because the power spectrum does not contain phase information, it is applied in situations where phase is not considered useful or to data that contain a lot of noise, since phase information is easily corrupted by noise.

3.7.1 Spectral Averaging

Although the power spectrum can be evaluated by applying the FFT to the entire waveform, it can also be applied to isolated segments of the data. The periodogram determined from each of these segments can then be averaged to produce a power spectrum that better represents the overall, or *global*, features of the spectrum. This approach is popular when the available waveform is only a sample of a longer signal. In such very common situations, spectral analysis is necessarily an estimation process, and averaging improves the statistical properties of the result.

Averaging is usually achieved by dividing the waveform into a number of possibly overlapping segments and evaluating the periodogram on each of these segments (Figure 3.19). The final power spectrum is constructed from the ensemble average of the power spectra obtained from each segments.

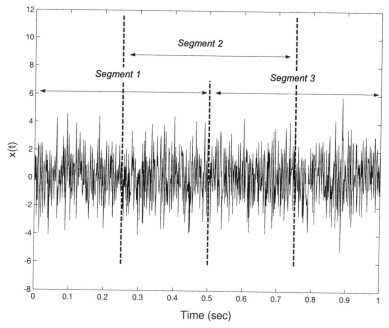

Figure 3.19 A waveform is divided into three segments with a 50% overlap between each segment. As described below, in the Welch method of spectral analysis, the power spectrum of each segment would be computed separately and an average of the three transforms would provide the result.

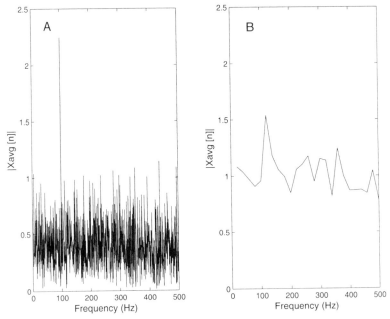

Figure 3.20 Power spectra obtained from a waveform consisting of a 100-Hz sine wave buried in white noise (signal-to-noise ratio of −16 dB) with **(B)** and without **(A)** averaging. In the unaveraged spectrum, a spike at 100 Hz is clearly seen. For the averaged spectrum, the 100-Hz component is not as obvious and could easily be missed; however, the averaging technique produces a smoother estimate of the white noise spectrum, which ideally should be a flat horizontal line.

Averaging involves tradeoffs between spectral resolution, which is reduced by averaging, and the desire for better statistical reliability. Segmentation necessarily reduces the number of data samples evaluated by the periodogram in each segment. As stated in Eq. 3.29, frequency resolution of a spectrum is proportional to f_s/N, where N is now the number of samples in a segment. Choosing a short segment length (a small N) will provide more segments for averaging and improve the reliability of the spectral estimate, but will also decrease frequency resolution. Figure 3.20 shows spectra obtained from a 1,024-data array consisting of a 100-Hz sinusoid and white noise. In Figure 3.20A, the periodogram is taken from the entire waveform, whereas in Figure 3.20B, the waveform is divided into 32 nonoverlapping segments and the power spectrum from each segment is averaged. The periodogram produced from the segmented data is much smoother, but the loss in frequency resolution is apparent because the 100-Hz sine wave is no longer visible. In practice, the selection of segment length and averaging strategy is usually based on experimentation with the actual data.

One of the most popular procedures to evaluate the average periodogram is attributed to Welch and is a modification of the segmentation scheme originally developed by Bartlett (Maple, 1987). In this approach, overlapping segments are used and a shaping window is applied to each segment. By overlapping segments, more segments can be averaged for a given segment and data length. Power spectra based on averaged periodograms obtained from noisy data traditionally use half-overlapping segments, that is, segments that overlap by 50%. Higher amounts of overlap have been recommended in other applications and, when computing time is not a factor, maximum overlap has been recommended. Maximum overlap means shifting over by just a single sample to get the new segment. Examples of this approach to estimating the power spectrum are provided in a subsequent section on MATLAB implementation.

3.8   SIGNAL BANDWIDTH

The concept of representing a signal in the frequency domain brings with it additional concepts relating to a signal's spectral characteristics. One of the most important of these new concepts is signal bandwidth. Later we will extend the concept of bandwidth to processes as well as signals, and we will find that the definitions related to bandwidth are essentially the same. The bandwidth of a signal is defined by the range of frequencies found in the signal. Figure 3.21A shows the spectrum of a hypothetical signal that contains energy at frequencies in equal measure up to

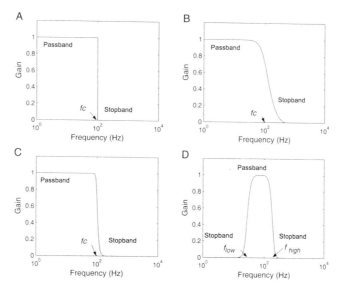

Figure 3.21   Frequency characteristics of ideal and realistic signals. The frequency plots shown here have a linear vertical axis, but often the vertical axis is plotted in decibels. The horizontal axis is in log frequency. **A:** Idealized signal having a sharp attenuation in energy at frequency f_c. **B:** Realistic signal with a gentle attenuation of energy. **C:** Realistic signal with a sharp attenuation characteristic. **D:** Signal with attenuation at both low and high frequencies.

frequency f_c, above which no energy is found. We would say that the signal has a *flat* spectrum up to f_c, the *cutoff frequency*, above which its spectrum is zero. Because the signal contains energy only between 0 and f_c Hz, the bandwidth of this signal would be 0 to f_c Hz, or simply f_c Hz.

Although some real signals can be quite flat over selected frequency ranges, they are unlikely to show such an abrupt cessation of energy above a given frequency. Figure 3.20B shows a more realistic spectral characteristic, where signal energy begins to decrease at a specific frequency but decreases gradually with frequency until there is no more energy in the signal. When the decrease in signal energy takes place gradually, as in Figure 3.20B, defining the bandwidth is problematic. If we are to define a single bandwidth for this signal, we need to define a cutoff frequency: a frequency boundary between the region of substantial energy and the region of minimal energy. This boundary has been somewhat arbitrarily defined as the frequency when the signal's RMS value has declined by 3 dB with respect to its average unattenuated value. (In Figure 3.21, the unattenuated value for all filters is 1.0.) The 3-dB boundary is not entirely arbitrary, because when the signal is attenuated 3 dB, its RMS amplitude is 0.707 of its unattenuated value and it has half the power of its unattenuated power. (Recall the previous section on decibels: $0.707^2 = 1/2$.) Accordingly, this boundary point is also known as the *half-power point*. The terms *cutoff frequency*, minus *3-dB point*, and *half-power point* are synonymous. In Figure 3.21B, the signal would have a bandwidth of 0.0 to f_c Hz, or simply f_c Hz. The signal in Figure 3.21C has a sharper decline in energy, referred to as the frequency *rolloff*, but it would still have a bandwidth given by the 3-dB point at f_c Hz.

It is possible that a signal rolls off at both the low-frequency and high-frequency ends as shown in Figure 3.21D. In this case, the signal's frequency characteristic has two cutoff frequencies, one labeled f_{low} and the other f_{high}. In this case the bandwidth is defined as the range between the two cutoff frequencies (or 3 dB points), that is, $f_{high} - f_{low}$ Hz.

3.9   MATLAB IMPLEMENTATION

MATLAB provides a variety of methods for calculating spectra, particularly if the signal processing toolbox is available. The basic Fourier transform routine is implemented as:

```
Xf = fft(x,n)              % Calculate the Fourier Transform
```

where x is the input waveform and Xf is a complex vector providing the sinusoidal coefficients. (It is common to use capital letters for the Fourier transformed variable.) The argument n is optional and is used to modify the length of data analyzed: If n is less than the length of x, the analysis is performed over the first n points. If n is greater than the length of x, the signal is padded with trailing zeros to equal n. The fft routine implements Eq. 3.22 above and uses a high-speed algorithm. The FFT algorithm requires the data length to be a power of two, and although the MATLAB routine will interpolate, calculation time then becomes highly dependent on data length. The algorithm is fastest if the data length is a power of two or if the length has many prime factors. For example, on one machine a 4,096-point FFT takes 2.1 seconds, but requires 7 seconds if the sequence is 4,095 points long, and 58 seconds if the sequence is 4,097 points long. If possible, use data lengths that are powers of two.

The magnitude of the frequency spectra can be easily obtained by applying the absolute value function, abs, to the complex output Xf:

```
Magnitude = abs(Xf);        % Take the magnitude of Xf
```

This MATLAB function simply takes the square root of the sum of the real part of Xf squared plus the imaginary part of Xf squared. The phase angle of the spectra can be obtained by application of the MATLAB angle function:

```
Phase = angle(Xf)           % Find the angle of Xf
```

The angle function takes the arc tangent of the imaginary part divided by the real part of Xf. Unlike most handheld calculators (and the MATLAB atan function), the angle routine does take note of the signs of the real and imaginary parts and will generate an output in the proper quadrant. The magnitude and phase of the spectrum can then be plotted using standard MATLAB plotting routines.

Example 3.6: Construct the waveform used in Example 3.2 (repeated below) and

$$x(t) = \begin{cases} t & 0 < 5 \le 0.5 \\ 0 & 0.5 < 5 \le 1.0 \end{cases}$$

determine the Fourier transform using both the MATLAB fft routine and a direct implementation of the defining equations (Eq. 3.8).

Solution: The MATLAB fft routine does no scaling, so its output should be multiplied by 2/N, where N is the number of points to get the correct coefficients in RMS value. To get the peak-to-peak values the output will have to be further scaled by dividing by 0.707.

```
% Example 3.6 and Fig 3.22
% Find the Fourier Transform of half triangle waveform
% used in Example 3.2. Use both the MATLAB fft and a
% direct implementation of Eq. 3.8
%
clear all; close all;
N = 256;                    % Data length
t = (1:N)/N;                % Generate time vector 1 sec long
fs = N;                     % fs for 1 sec data
f = (1:N)*fs/N;             % Frequency vector for plotting
x = t;                      % Generate the signal: initial
```

```
x(129:N) = 0;                   % section is a ramp, then zeros
%
Xf = fft(x);                    % Take Fourier Transform, scale
Mag = abs(Xf(2:end))/... % remove first point (DC value)
      (N/2);
Phase = angle(Xf(2:end))*(360/(2*pi));
%
plot(f(1:20),Mag(1:20),'xb'); hold on; % Plot magnitude
xlabel('Frequency (Hz)'); ylabel('|X(f)|');
%
% Calculate the Fourier series using the basic equations
% (Eq. 3.8)
for n = 1:20
  a(n) = (2/N)*sum(x.*(cos(2*pi*n*t)));
  b(n) = (2/N)*sum(x.*(sin(2*pi*n*t)));
  % Calculate magnitude and phase
  C(n) = sqrt(a(n).^2 + b(n).^2);
  theta(n) = -(360/(2*pi)) * atan(b(n)./a(n));
end
plot(f(1:20),C(1:20),'sr');       % Plot superimposed
%
% Output numerical values
disp([a(1:4)' b(1:4)' C(1:4)' Mag(1:4)' theta(1:4)'
  Phase(1:4)'])
```

The spectrum produced by the two methods is identical as seen by the perfect overlap of the x points and square points in Figure 3.22.

The numerical values produced by this program are given below.

a_n	b_n	C_n	Mag(fft)	Theta	Phase (fft)
−0.1033	0.1591	0.1897	0.1897	−57.0182	121.5756
0.0020	−0.0796	0.0796	0.0796	−88.5938	−91.4063
−0.0132	0.0530	0.0546	0.0546	−76.0053	99.7760
0.0020	−0.0398	0.0398	0.0398	−87.1875	−92.8125

Both methods produce identical magnitude spectra as seen in both Figure 3.22 and the data above (compare C_n with Mag[fft]) The column *Theta* shows the angles calculated using Eq. 3.8 and these are incorrect because the MATLAB atan function does not determine the correct quadrant. Both magnitudes and the phase found by the fft routine closely match the values determined analytically in Example 3.2. Note that the values for a_2 and a_4 are not exactly zero as they were determined to be analytically. Because of these small errors, the phase angle for the second and

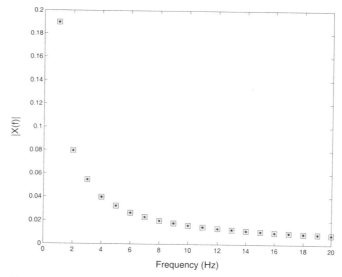

Figure 3.22   Magnitude frequency spectra produced by the MATLAB
fft routine (*points) and a direct implementation of the Fourier transform equations (Eq. 3.8) (*square boxes*).

fourth components is not exactly −90 degrees as determined analytically. It is worth
noting that analytic solutions can be more accurate than computer solutions, and
that computer solutions are prone to small errors even for a simple problem such
as given in this example.

An example applying the MATLAB fft to a waveform containing sinusoids and
white noise is provided below along with the resultant spectra in Figure 3.23. Other
applications are explored in the problem set at the end of this chapter. Example 3.7
uses a special routine, sig_noise, found on the enclosed CD. This routine generates data consisting of sinusoids and noise and can be useful in evaluating spectral analysis algorithms. The calling structure for sig_noise is:

```
[x,t] = sig_noise([f],[SNR],N);      % Generate a signal
                                      % in noise
```

where f specifies the frequency of the sinusoid(s) in hertz, SNR specifies the desired
noise associated with the sinusoid(s) in decibels, and N is the number of points. If
f is a vector, a number of sinusoids are generated, each with a signal-to-noise ratio
specified by SNR assuming it is a vector. If SNR is a scalar, its value is used for the
SNR of all the frequencies generated. The output waveform is in x and t is a time
vector useful in plotting. The routine assumes a sample frequency of 1 kHz.

Example 3.7:   Construct a waveform consisting of a single 250-Hz sine wave and
white noise with an SNR of −14 dB. Calculate the Fourier transform of this waveform and plot the magnitude spectrum.

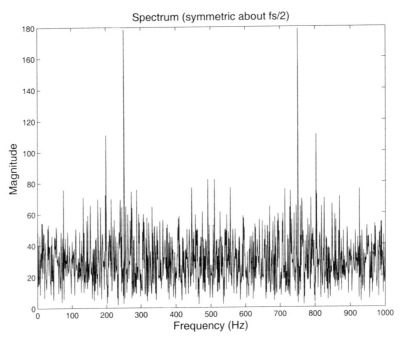

Figure 3.23   Plot produced by the MATLAB program. The peak at 250 Hz is apparent. The sampling frequency of these data is 1 kHz, hence, the spectrum is symmetric about the Nyquist frequency, $f_s/2$ (500 Hz). Normally, only the first half of this spectrum would be plotted. (Signal-to-noise ratio equals −14 dB; $N = 1,024$.)

Solution: Use `sig_noise` to generate the waveform, take the Fourier transform using `fft`, obtain the magnitude using `abs`, and plot.

```
% Example 3.7 and Figure 3.23 Determine the Power Spectrum
% of a noisy waveform
% First generate a waveform consisting of a single sine in
% noise, calculate the Power Spectrum from FFT and plot
%
clear all; close all;
N = 1024;                             % Number of data points
% Generate data using 'sig_noise'
% 250 Hz sin plus white noise; N data points; SNR = -14 dB
[x,t] = sig_noise (250,-14,N); % Generate signal and noise
%
fs = 1000;                            % Sample frequency is 1 kHz.
Xf = fft(x);                          % Calculate FFT
```

```
Mag = abs(Xf);                  % Calculate the magnitude
f = (1:N)*fs/N;                 % Frequency vector for plotting
plot(f,Mag,'k');                % Plot the magnitude spectrum
   title('Spectrum (symmetric about fs/2)');
  xlabel('Frequency (Hz)'); ylabel('Magnitude');
```

Analysis: The program is straightforward. After constructing the waveform using the routine `sig_noise`, the program takes the discrete Fourier transform with `fft` and then plots the magnitude (constructed using `abs`) versus frequency. A frequency vector the same length as the data (N points) is generated to aid in plotting. The frequency vector is based on Eq. 3.29 and increases linearly from 1.0 to f_s.

Results: The spectrum is shown in Figure 3.23 and the peak related to the 250-Hz sine wave is clearly seen. As expected, the spectrum above $f_s/2$ (i.e., 500 Hz) is a mirror image of the lower half of the spectrum. The white noise is not very well represented in this analysis because averaging is not used. Ideally, the background spectrum should be a constant value, yet the background is highly variable, with occasional peaks that could be mistaken for signals. A better way of determining the spectrum of a broadband feature such as white noise would be to use an averaging strategy as shown in Examples 3.8 and 3.9.

The heart rate data shown in Figure 2.12 show considerable differences in the mean and standard deviation of the rate between meditative and normal states. Applying the autocovariance to the meditative heart rate data (Problem 2.15 in Chapter 2) indicates a possible repetitive structure for the variation in heart rate during the meditative average rate and variance. Example 3.8 searches for possible structure in the frequency characteristics of both normal and meditative heart rates using the Fourier transform.

Example 3.8:   Determine and plot the power spectra of heart rate variability during normal and meditative states.

Solution: The frequency characteristics may be found by Fourier transform using the direct method given in Eq. 3.35. However, the heart rate data should first be converted to time data, and this is a bit tricky. The data set obtained by a download from the PhysioNet database provides the heart rate at unevenly spaced times, where the sample times are provided as a second vector. The heart rate data need to be rearranged into even time positions. This will be done through interpolation using MATLAB's `interp1` routine. This routine takes in the unevenly spaced *x-y* pairs as two vectors along with a vector containing the desired evenly spaced *x* values. The routine then uses linear interpolation (other options are possible) to approximate the *y* values that match the evenly spaced x values. Details can be found in the MATLAB help file for `interp1`. In the program below, the uneven *x-y* pairs are in vectors `t_pre` and `hr_pre`, respectively, the evenly spaced vector is `xi`, and the corresponding *y* values are placed in vector `yi`.

```
% Example 3.8 and Figure 3.24
% Frequency analysis of heart rate data in the
% normal and meditative state
% After loading the data, the program converts the data
% to evenly spaced time data using interpolation
%
clear all; close all;
%
fs = 1.0;                    % Sample interval
load Hr_pre;                 % Load normal and meditative data


%
% Convert to evenly spaced time data using interpolation
% First generate and evenly space time vectors having one
% second intervals and extending over the data time range
%
xi = (ceil(t_pre(1)):fs:floor(t_pre(end)));
yi = interp1(t_pre,hr_pre,xi'); % Interpolate
yi = yi - mean(yi);             % Remove average
f = (1:length(yi))*fs/...       % Vector for plotting
    length(yi);
%
% Now determine the Power Spectrum
YI = abs((fft(yi)).^2);    % Direct approach (Eq. 3.35)
subplot(1,2,1);
plot(f,YI,'k');                 % Plot spectrum
  xlabel('Frequency (Hz)'); ylabel('Power Spectrum');
  axis([0 .15 0 max(YI)*1.25]);
%
% Repeat for meditative data
```

Analysis: To convert the heart rate data to a sequence of evenly spaced points in time, a time vector, xi, is first created that increases in increments of 1.0 second between the lowest and highest values of time in the original data. A 1.0-second increment was chosen because this was approximately the average time spacing of the regional data. Evenly spaced time data, yi, were generated using the MATLAB interpolation routine interp1. The example requested the power spectrum of heart rate *variability*, not heart rate per se; in other words, the changes in beat-to-beat rate, not the beat-to-beat rate itself. To get this change, we simply subtract out the average heart rate before evaluating the power spectrum.

After interpolation and removal of the mean heart rate, the spectrum is determined using fft as in the last example. The power spectrum is calculated by taking the square of the magnitude component. The frequency plot is limited to an area of interest between 0.0 and 0.15 Hz by the axis routine.

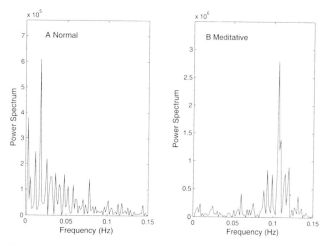

Figure 3.24 **A:** Power spectrum of heart rate variability under normal conditions. The power decreases with frequency. **B:** Power spectrum of heart rate variability during meditation. Strong peaks in power around 0.12 Hz are seen indicating that some of the variation in heart rate is organized around these frequencies. Note the larger scale of the meditative power spectrum.

Results: The power spectrum of normal heart rate variability is low and decreases with frequency, showing little energy above 0.1 Hz (Figure 3.24). The meditative state shows large peaks at around 0.1 to 0.12 Hz, indicating some repetitive process is active at these frequencies, frequencies corresponding to a time frame of around 10 seconds. Speculation as to the mechanism behind this heart rate rhythm is left to the reader, who may be more knowledgeable in Yoga meditation than the author.

As stated in Section 3.7.1, averaging can be used to improve the spectral representation of broadband processes. Example 3.9 explores this notion by applying spectral averaging to the same heart rate variability data used in Example 3.8.

Example 3.9: Determine and plot the frequency characteristics of heart rate variability during normal and meditative states using averaging.

Solution: Write a general program called `welch` to generate an average power spectrum given the data, segment size, and the number of overlapping points in adjacent segments. This routine should also take in, as an optional parameter, the sampling frequency to be used in generating a frequency vector. Output the power spectrum and the frequency vector. Output only the nonredundant points (i.e., up to $f_s/2$).

```
% Example 3.9 Influence of averaging on heart rate data.
% Loads files Hr_pre and Hr_med that contain the heart rate
```

```
% data.
% Calculates the Power Spectrum using signal averaging.
%   Uses eight segments with a 99% overlap. Assumes the
%   heart rate variable is yi with fs = 1.0 Hz
%
... Data loading and restructure as in Example 3.8 ...
%
segment_length = fix(length(yi)/8);  % Average 8 segments
[PS_avg,f] = welch(yi,segment_length,segment_length-1,fs);
subplot(1,2,1)
  plot(f,PS_avg,'k');                    % Plot averaged PS
  xlabel('Frequency (Hz)'); ylabel('Power Spectrum');
  axis([0 .2 0 max(PS_avg)*1.2]);    % Limit horizontal axis
....... Repeat for meditative data .......
```

This example uses the routine `welch` shown below to do the averaging.

```
function [PS,f] = welch(x,nfft,noverlap,fs);
%Function to calculate averaged spectrum
%[ps,f] = welch(x,nfft,noverlap,fs);
% Output arguments
%   sp spectrogram
%   f frequency vector for plotting
% Input arguments
%   x data
%   nfft window size
%   noverlap number of overlapping points in adjacent
%   segments
%   fs sample frequency
% Uses Hanning window
[N xcol] = size(x);               % Make row vector
if N < xcol
  x = x';
  N = xcol;
end
half_segment = fix(nfft/2);       % Half segment length
if isempty(noverlap) == 1
  noverlap = half_segment;        % Default overlap at 50%
end
if isempty(fs) == 0
  f = (1:half_segment)*           % Calculate freq. vector
      fs/nfft;
else
  f = (1:half_segment)*           % Default freq. vector
      pi/nfft;
end
increment = nfft - noverlap;
```

```
nu_avgs = fix((N-nfft)/...          % Find number of segments
        increment)-1;
%
for i = 1:nu_avgs                    % Calculate spectra
  first_point = 1 + (i-1)            % Isolate the correct
            * increment;            % data segment
  data = x(first_point: first_point+nfft-1);
  if i == 1
PS = abs((fft(data)).^2)/...         % Calculate PS (1st time)
     (nfft*nu_avgs);
else
PS = PS + abs((fft(data))...         % Calculate PS avg
     .^2)/(nfft*nu_avgs);
  end
end
PS = PS(1:half_segment);             % Remove redundant points
```

Analysis: The coding of routine `welch` illustrates a number of MATLAB tricks. The initial section tests the dimensions of the input to determine if it is arranged as a row or column vector. If it is a row vector, the number of rows, `N`, will be less than the number of columns, `xcol`, and the vector is transposed insuring that `x` is now a column vector. Next, the program checks if a desired overlap is specified (i.e., if `noverlap` is not an empty variable) and if so, sets the overlap to a default value of 50% (i.e., half the segment length, `nfft`). Then a frequency vector, `f`, is generated from 1 to π if `fs` is unspecified, or from 1 to `fs` if it is given. Next, the number of segments to be averaged is determined based on the segment size (`nfft`) and the overlap (`noverlap`). A loop is used to take the Fourier transform of each segment, calculate the power spectrum, and sum the individual spectra. Note that the spectra are normalized by both the number of spectra in the average and the segment length. Finally, the averaged power spectrum is shortened to eliminate redundant points.

The `welch` routine is used in Example 3.9 in a straightforward manner to calculate the average power spectrum. The heart rate data are divided into eight segments somewhat arbitrarily. In practice, the best segment length would have to be determined empirically. The segment length chosen (1/8 of the data length) seems to give a good representation of the background level without losing the peaks in the meditative state spectrum.

Results: The results in Figure 3.25 show much smoother spectra than those of Figure 3.23, but they also lose some of the detail. The power spectrum of heart rate variability taken under normal conditions now appears to decrease smoothly with frequency. In fact, replotting the spectra in Figure 3.26 in decibels, by taking 20 log (power spectrum) before plotting, shows that the normal spectrum in dB actually decreases linearly with log frequency. The decibel plot also indicates a similar linear decrease for the meditative spectrum, but with a large peak at 0.12 Hz superimposed on this decrease.

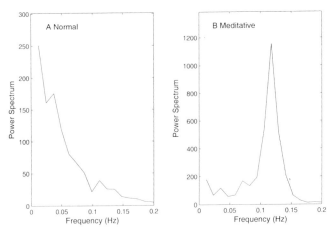

Figure 3.25 Power spectra taken from the same heart rate variability data used to determine the spectra in Figure 3.24, but constructed using an averaging process. The spectra produced by averaging are considerably smoother. **A:** Normal conditions. **B:** Meditative conditions.

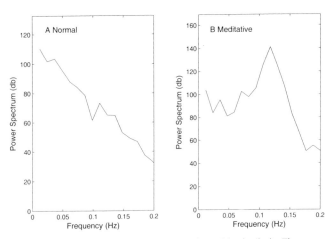

Figure 3.26 Spectra of Figure 3.25 replotted in decibels. The spectrum obtained under normal conditions (**A**) is now seen to decrease linearly with frequency. The spectrum obtained under meditative conditions (**B**) also shows a linear decrease but with a large peak at 0.12 Hz superimposed on this decrease.

Example 3.10 explores the effects of averaging on a controlled data set consisting of a mixture of signals: a broadband signal, two closely spaced sinusoidal signals, and noise. The spectrum of the broadband signal and the sinusoids without the noise are shown in Figure 3.27 and the spectrum with the noise added with and without averaging is shown in Figure 3.28.

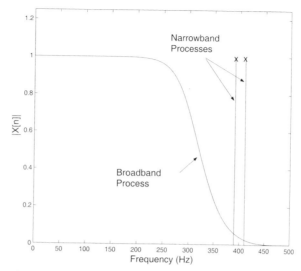

Figure 3.27 The power spectrum of a mixture of signals used to create the test signal in file `broadband1`. The mixture consists of the output of a lowpass filter and two sinusoidal signals at 390 and 410 Hz. Noise was then added to this mixture to create the test signal.

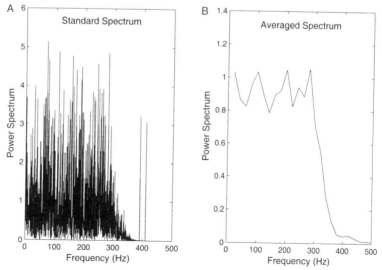

Figure 3.28 **A:** Standard spectral analysis identifies the two closely spaced sinusoids at 390 and 410 Hz, but the broadband signal between 0 and 300 Hz is noisy and poorly defined. **B:** Spectral averaging significantly improves the spectrum of the broadband signal, but the two sinusoids can barely be seen.

Example 3.10: Evaluate the influence of power spectrum averaging on a combination of broadband and narrowband processes with added noise. The data may be found in file `broadband1.mat`.

Solution: Load the test data file `broadband1` containing the narrowband and broadband processes shown in Figure 3.27. First, calculate and display the unaveraged power spectrum. Then apply power spectrum averaging using routine `welch` as in Example 3.5.

```
% Example 3.10 Investigation of the influence of averaging
% to improve broadband spectral characteristics in the
% Power Spectrum.
% Loads file broadband1 that contains broadband and
% narrowband signals (i.e., sinusoids)and noise.
% Calculates the standard Power Spectrum and one obtained
% using segment averaging. The averaging uses 80 segments
% with a 99% overlap
% The data is variable x in the .mat file and is assumed
% to be acquired at a sampling frequency of 1.0 kHz
%
close all; clear all;
load broadband1;                  % Load data (variable x)
fs = 1000;                        % Sampling frequency
PS = abs((fft(x)).^2)/length(x);  % Calculate unaveraged PS
half_length = fix(length(PS)/2);  % Find data length /2
f = (1:half_length)*...           % Frequency vector for
    fs/(2*half_length);           % plotting
subplot(1,2,1)
plot(f,PS(1:half_length),'k');    % Plot unaveraged Power
                                  % Spectrum
  xlabel('Frequency (Hz)'); ylabel('Power Spectrum');
  title('Standard Spectrum');
%
seg_length = fix(length(x)/80);   % Average 80 segments
% Use 99% overlap
[PS_avg,f] = welch(x,seg_length,seg_length-1,1000);
%
subplot(1,2,2)
  plot(f,PS_avg,'k');             % Plot averaged Power
                                  % Spectrum
  xlabel('Frequency (Hz)'); ylabel('Power Spectrum');
  title('Averaged Spectrum');
```

Analysis: The procedure is similar to that used in Example 3.9 except the number of segments was increased to 80 because the data set was longer (4,000 points) and a good estimate of the broadband process was desired.

Results: In the unaveraged power spectrum (Figure 3.28A) the two sinusoids are clearly seen at 390 and 410 Hz; however, the broadband signal is noisy and poorly defined. Averaging greatly improves the representation of the broadband signal, but now the two sinusoids are not evident although there is a very small peak where they should be. This demonstrates one of those all-so-common engineering compromises. Spectral techniques that produce good representation of *global* features such as broadband signals are not good at resolving narrowband or *local* features such as sinusoids and vice versa.

3.10 SUMMARY

The sinusoid [i.e., $C \cos(\omega t + \theta)$] is a unique signal with a number of special properties. A sinusoid can be completely defined by three values: its amplitude, C, its phase, θ, and its frequency, ω (or $2\pi f$). Any periodic signal can be broken down into a series, possibly infinite, of harmonically related sinusoids. Similarly, any periodic signal can be reconstructed from a series of sinusoids. Thus, any periodic signal can be equivalently represented by a sinusoidal series. A sinusoid is also a pure signal in that has energy at only one frequency, the only waveform to have this property. This means that the sinusoids can serve as intermediaries between the time domain representation of a signal and its frequency domain representation. When the input to a linear system is a sinusoid, the output will also be a sinusoid at the same frequency. Only the magnitude and phase of a sinusoid can be altered by a linear system. Finally, harmonically related sinusoids are orthogonal so they operate independently.

The technique for determining the sinusoidal series representation of a periodic signal is known as Fourier series analysis. To determine the equivalent sinusoidal series, the signal of interest is correlated with sinusoids at harmonically related frequencies. This correlation provides the amplitude and phase of the sinusoidal series that represents the periodic signal. The equivalent sinusoidal series can be used to construct a plot of the frequency composition of the signal: plots of the magnitude and phase compositions of the signal over a range of frequencies. Fourier series analysis is often described and implemented using the complex representation of a sinusoid.

If the signal is not periodic, but exists for finite time, Fourier decomposition is still possible by assuming this aperiodic signal is in fact periodic, but the period is infinite. This approach leads to the Fourier transform where the correlation is now between the signal and infinite number of sinusoids at continuously varying frequencies. The frequency plots then become continuous curves. The inverse Fourier transform can be also constructed from the continuous frequency representation.

Fourier decomposition can be applied to digitized data and is known as the discrete Fourier analysis or discrete Fourier transform. In fact, most all Fourier analysis is done in a computer usually using a high-speed algorithm known as the FFT (Fast Fourier Transform). The equations follow the same pattern as those developed for continuous time signals except integration becomes summation and both the sinusoidal and signal variables are discrete. The discrete Fourier transform can be used to understand the relationship between a continuous time signal and the sampled version of that signal. This frequency based analysis shows that the original, unsampled signal can be recovered if the sampling frequency was more that twice the highest frequency component in the unsampled signal, a rule known as *Shannon's sampling theorem*.

The Fourier series or Fourier transform can be used to construct the power spectrum of a signal. The power spectral curve describes how the signal's power varies with frequency. The power spectrum is particularly useful for random data were phase characteristics have little meaning. By dividing the signal into a number of possibly overlapping segments and averaging the spectrum obtained from each segment, a smoothed power spectrum can be obtained. The resulting frequency curve will emphasize the broadband or general characteristics of a signal's spectrum, but may lose some of the fine detail.

PROBLEMS

1. Find the Fourier series of the square wave below using Eq. 3.8. (*Hint:* Take advantage of symmetries to simplify the calculation.)

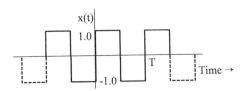

2. Find the Fourier series of the waveform below. The period, T, is 1 second.

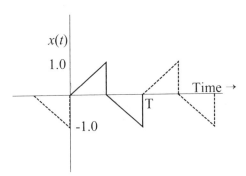

3. Find the Fourier series of the half-wave rectified sinusoidal waveform below. Use symmetry properties.

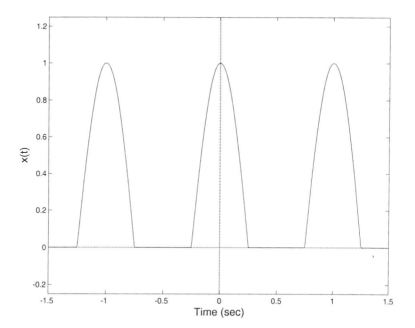

4. Find the Fourier series of the sawtooth waveform below.

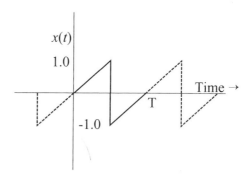

5. Find the continuous Fourier transform of an aperiodic pulse signal given in Example 3.4 using the complex equation, Eq. 3.19.

MATLAB Problems

6. Generate a 512-point waveform consisting of two sinusoids at 200 and 400 Hz. Assume a sampling frequency of 1 kHz. Also generate another waveform containing frequencies at 200 and 900 Hz. Take the Fourier transform of both waveforms and plot the magnitude of the full spectrum (i.e., 512 points). [*Hint:* To generate the sine waves, first construct a time vector:

   $t = (1:N)T = (1:N)/f_s$ using the MATLAB code:
   ```
   t = (1:512)/1000.
   ```

   Then generate the signal using the code: `x = sin(2*pi*f1*t) + sin(2*pi*f2*t)` where `f1` = 200 for both signals, while `f2` = 400 for one waveform, and `f1` = 900 for the other.]

7. Load the file `sample_rate` that contains signals x and y. Are either of these signals oversampled (i.e., $f_s/2 \geq f_{max}$)? Alternatively, could the sampling rate of either signal be safely reduced? Justify your answer.

8. Construct two arrays of white noise: one 128 points in length and the other 1,024 points in length. Take the Fourier transform of both. Does increasing the length improve the spectral estimate of white noise? (Eliminate the first point—the average or DC term—when you plot the spectra and plot only nonredundant points.)

9. Generate a 256-point waveform consisting of a 300 Hz signal in 12 dB of noise (i.e., SNR = −12 dB). (MATLAB call: `x = sig_noise(300, −12,256);`) Calculate and plot the Fourier transform using two different methods. In the first method, use `fft` applied directly to the waveform, x. In the second method, use the traditional approach of taking the Fourier transform of the autocorrelation function; that is, calculate the autocorrelation function using `axcor`, then take the `fft` of the autocorrelation function. Plot magnitude and phase for both techniques.

10. Use MATLAB routine `sig_noise` to generate two arrays, one 128 points long and the other 512 points long. Include two closely spaced sinusoids having frequencies of 320 and 340 Hz with an SNR of −12 dB. The MATLAB call should be:
    ```
    x = sig_noise([320 340],-12,N); where N = either 128 or
    512.
    ```
    Calculate and plot the (unaveraged) power spectrum.

11. Use `sig_noise` to generate a 128-point array containing 320- and 340-Hz sinusoids as in Problem 10. Calculate and plot the unaveraged power spectrum of this signal for an SNR of −12 dB, −14 dB, and −16 dB. How does the presence of noise affect the ability to detect and distinguish between the two sinusoids?

12. Load the file `broadband2` that contains variable x, a broadband signal with added noise. Assume a sample frequency of 1 kHz. Calculate the averaged power spectrum using routine `welch`. Evaluate the influence of segment length

using segment lengths of N/4 and N/16, where N is the length of the variable, x. Use the default overlap.

13. Load the file eeg_data that contains the EEG data shown in Figure 3.1 as variable eeg. Analyze these data using the unaveraged power spectral technique and an averaging technique using the welch routine. Find a segment length that smoothes the background spectrum but retains any important spectral peaks. Use a 99% overlap.

14. Load the file chirp that contains a sinusoidal signal, x, which increases its frequency linearly over time. This type of signal is called a *chirp* signal because of the sound it makes when played through an audio system. [If you have an audio system, you can listen to this signal after loading the file using the MATLAB command: sound(x,1,000).] Take the Fourier transform of this signal and plot magnitude and phase. Note that the signal frequencies are merged together and there is no information on the timing. (Actually, information on signal timing is buried in the phase plot.) Time-frequency methods covered in more advanced signal processing are necessary to recover the timing information.

4 CIRCUIT AND ANALOG ANALYSIS IN SINUSOIDAL STEADY STATE

4.1 CIRCUITS AND ANALOG SYSTEMS

The next two chapters provide the tools for analyzing systems containing analog elements. The major application of these tools is the analysis of electrical and electronic circuits. Biomedical instrumentation often contains analog circuitry that provides an interface between devices that monitor physiological variables (biotransducers) and the computer that processes the data. Figure 4.1 shows an example electronic circuit that was used in another common application: as an interface between the computer and an effector device, in this case an electrode that delivers electrical stimulation. Based on the computer's digital-to-analog converter output, this circuit delivers controlled high voltage pulses for stimulation of electrically excitable tissue. The analysis of electronic circuits is further explored in Chapter 10.

A second application of these tools is the analysis of analog models of physiological systems. Figure 4.2 shows what looks like an electric circuit but is actually an analog model of the cardiovascular system. In this representation, model currents describe blood flow in different parts of the cardiovascular system while voltages describe various cardiovascular pressures.

To analyze an electric circuit or analog model generally calls for the solution of one or more differential equations. Even the solution of a simple circuit or model that can be defined by a first-order differential equation can become tedious if the signals are complicated. Two popular techniques exist for simplifying the analysis of processes or systems that are represented by differential equations. (The terms *system* and *process* are used interchangeably in this text. The *system* term is more common when coupled with the word *analysis*, as in *systems analysis*.) Both methods transform the differential equations into algebraic equations, obviously a highly desirable modification. In addition, these techniques generate equations that provide more insight into the underlying processes than classical differential equations. Both approaches require that the underlying processes be linear. This can present severe constraints for bioengineers, because most living systems are nonlinear. Nonetheless, the power of these simplifying techniques is so seductive that

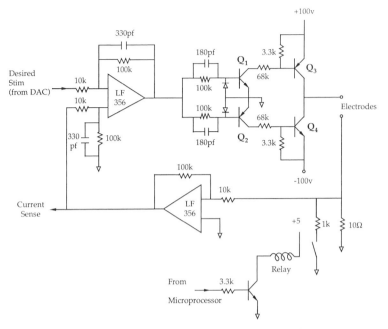

Figure 4.1   An electronic circuit that provides an interface between a computer output ('from DAC' on the schematic, the output of the computer's digital-to-analog converter) and stimulation electrodes. This circuit generates a high-voltage stimulus (up to 100 V) where the stimulus *current* is proportional to the computer control voltage.

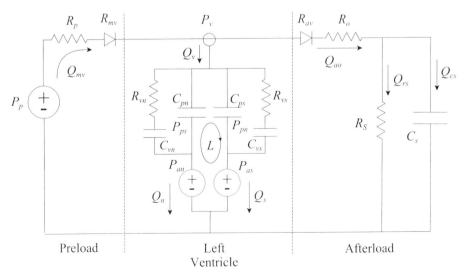

Figure 4.2   Analog model that represents elements of the cardiovascular system as electrical elements. (From Prof. J.-K. Li, Rutgers University, with permission.)

linearity is often assumed (rightly or wrongly) or imposed using the methods mentioned in Chapter 1.

Which of the two simplifying approaches is used depends on the type of signals or operating conditions encountered by the system. The *Laplace transform* is the most general approach and is the most frequently used simplification technique. It applies to systems exposed to one-time, or *transient* stimulus signals or steplike changes in operating conditions. The Laplace transform is covered in Chapter 8. The *phasor* approach described in this chapter applies to systems exposed to sinusoidal stimulation, yet as we will discover over the next three chapters, it has many additional applications. Nearly all of the techniques associated with phasor analysis apply equally well to the Laplace transform approach. Before we can adopt these tools, we need to know something about what electrical and mechanical systems are made of and what variables are used to describe their behavior.

4.2 SYSTEM VARIABLES AND ELEMENTS

Electric or electronic circuits consist of an arrangement of elements that define relationships between voltages and currents. In electric circuits, such a configuration of elements may be used to generate, detect, and/or process signals. In analog models, elements are used to represent a biological mechanism with an analogous electrical or mechanical element. For example, a voltage source in combination with a resistor might be used to represent sodium conduction across a nerve membrane as shown in the next chapter. In the cardiovascular model shown in Figure 4.2, the analogous elements do not process the same type of energy as the elements they represent: mechanical cardiovascular processes are represented by electrical elements. Regardless of the type of elements used, the mathematical representations and the tools for analyzing these representations are the same. The approaches developed over the next three chapters are of value to biomedical engineers for the design of medical instruments, but also for the analysis of physiological systems. Many of the tools developed in these chapters are used again in the chapter on basic electronics (see Chapter 10).

In both electric circuits and mechanical system, only two variables are needed to describe the system's behavior. An element, be it electrical or mechanical, can be viewed as simply defining the relationship between these two variables. Although these variables are different for electrical and mechanical systems, they share much in common: one variable is associated with potential energy, and the other variable is related to kinetic energy or movement. The potential energy variable may be viewed as the cause of an action whereas the kinetic energy variable is the effect. In mechanical systems, the two variables are force and velocity. Force applied to a movable object will cause that object to have a velocity. Force is related to potential energy because energy, or work, is the integral of force over distance:

$$E = Work = \int F ds = Fx \text{ joules} \quad \text{(if } F \text{ constant over } x\text{)} \qquad \text{[Eq. 4.1]}$$

where F is force in newtons and x is distance in meters.

The kinetic energy variable, also called the *flow* variable, is velocity and is directly related to the kinetic energy of a mass by the well-known equation:

$$E = 1/2\,mv^2 \qquad\qquad \text{[Eq. 4.2]}$$

Power, the energy per unit time, is always the product of the potential and kinetic energy variables. For a mechanical element:

$$P = dE/dt = F(dx/dt) = Fv \text{ watts} \qquad\qquad \text{[Eq. 4.3]}$$

where P is power, F is force, x is distance, and v is velocity.

The mechanical analysis described in Section 4.4 of this chapter, is sometimes applied to mechanical systems, but for biomedical engineers, the main application is in mechanical analog models of physiological processes.

For electric circuits or electric analog models, the major variables are voltage and current. Voltage is related to potential energy and is sometimes even called *potential*. Voltage applied to a circuit element will *cause* a current to flow through that element. Voltage is the potential energy with respect to a given charge:

$$v = dE/dq \left(\frac{\text{joules}}{\text{coul}} \right) \text{or V} \qquad\qquad \text{[Eq. 4.4]}$$

where v is voltage, E is the energy of an electric field, and q is charge. (Slightly different typeface will be used to represent voltage, v, and velocity, v, to minimize confusion.)

The kinetic energy or flow variable is current that is simply the flow of charge:

$$i = dq/dt \left(\frac{\text{coul}}{\text{sec}} \right) \text{or A} \qquad\qquad \text{[Eq. 4.5]}$$

where i is current and q is charge.

The relationship for power and energy is the same in the electrical domain as in the mechanical domain; specifically, power is the product of the two variables v and i. Rearranging the equation that defines voltage (Eq. 4.4):

$$dE = vdq$$

and from the definition of P in Eq. 4.3:

$$P = dE/dt = v(dq/dt) = vi \text{ watts} \qquad\qquad \text{[Eq. 4.6]}$$

Table 4.1 summarizes the variables used to describe the behavior of mechanical and electrical systems. Because the tools that will be developed in this chapter are more

TABLE 4.1   Major Variables in Mechanical and Electrical Systems

Domain	Potential Energy Variable (Units)	Kinetic Energy Variable (Units)
Mechanical	Force, F (J/m = newton)	Velocity, v (m/sec)
Electrical	Voltage, v (J/coulomb = volts)	Current, i (coulomb/sec = ampere)

often used to analyze electric circuits (or electric analog models), they will be introduced in the *electrical domain*. However, these tools apply equally well to certain mechanical systems as shown later in the chapter.

4.2.1 Electrical Variables

The two variables that describe the behavior of electric circuits and electric analog models are voltage and current. Voltage is always a relative variable: it is the difference between the voltages at two points. In fact, the proper term for voltage is *potential difference* (abbreviated p.d.), but this term is rarely used in electronics. Subscripts are sometimes used to indicate the points from which the potential difference is measured. For example, in Eq. 4.7 the notation v_{ba} means 'the voltage at point b with respect to point a':

$$v_{ba} = v_b - v_a$$ [Eq. 4.7]

The positive voltage point, point b in Eq. 4.7, is indicated by a plus sign when drawn on a circuit diagram, as shown in Figure 4.3.

It is also common to say that there is a voltage drop from point b to point a, or a voltage rise from a to b. By this convention, it is logical that v_{ab} should be the negative of v_{ba}: $v_{ab} = -v_{ba}$. Voltage always has a direction, or *polarity*, that is usually indicated by a plus sign to show the side assumed to have a greater voltage (Figure 4.3). A source of considerable confusion is that the plus sign only indicates the point that is *assumed* to have a more positive value for the purpose of analysis or discussion. In fact, it could be that the voltage polarity is the opposite of what was originally assumed or assigned. As an example, suppose there is actually a rise in voltage from b to a in Figure 4.3 even though we assumed that b was the more positive as indicated by the plus sign. When this occurs, we do *not* change our original polarity assignment. We merely state that v_{ba} has a negative value. So a negative voltage does not imply negative potential energy, it is just that the actual polarity is the reverse of that assumed or assigned.

By convention, but with some justification as shown later, the voltage of the earth is assumed to be at 0.0 V, so voltages are often measured with respect to the voltage of the earth, or some common point referred to as *ground*. A common ground point is indicated by either of the two symbols shown at the bottom of the simple circuit shown in Figure 4.4. Some sources use the symbol on the right side to mean a ground

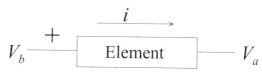

Figure 4.3 A generic electric circuit element demonstrating how voltage and current directions are specified. The straight lines on either side indicate wires connected to the element.

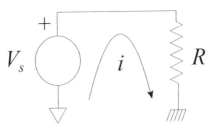

Figure 4.4 A simple circuit consisting of a voltage source, V_s, and a resistor R. Two different symbols for the grounding point are shown. Sometimes the symbol on the right is taken to mean an actual connection to earth, but this is not standardized.

that is actually connected to earth and the symbol on the left side to mean a common reference point not necessarily connected to the earth, but still assumed to be 0.0 V. However, this usage is not standardized and the only assumption that can be made with certainty is that both symbols represent a common grounding point, which may or may not be connected to earth. Hence when a voltage is given with only one subscript, v_a, it is understood that this is voltage at a with respect to ground or a common reference point.

Current is a flow so it must have a direction. This direction, or rather the *assumed* direction, is indicated by an arrow as in Figure 4.4. By convention, the direction of the arrow indicates the direction of assumed positive charge flow. In electronic circuits, charge is usually carried by electrons, which have a negative charge, so the particles that are actually flowing, the electrons, are flowing in the opposite direction of (assumed) positive charge flow. Nonetheless, the convention of defining positive charge flow was established by Benjamin Franklin before the existence of electrons was known and has not been modified because it really does not matter which direction is taken as positive as long as we are consistent. As with voltage, it may turn out that positive charge flow is actually in the direction opposite that indicated by the arrow and, again, we do not change the direction of the arrow, but rather give the current a negative value. Thus, a negative value of current flow does not mean that some strange antiparticles are flowing, but only that the actual current direction is opposite to our assumed/assigned direction.

This approach to voltage polarity and current direction may seem confusing, but it is actually quite liberating. It means that we can make our polarity and direction assignments without worrying about reality; that is, the voltage polarity or current direction that actually exists in a given circuit. We can make these assignments more or less arbitrarily (there are some rules that must be followed as described below) and allow the positive or negative values to indicate the actual direction or polarity.

4.2.2 Electrical Elements

The elements as described here are idealizations: true elements only approximate the characteristics described. However, actual electrical elements come quite close to these idealizations, so their shortcomings can often be ignored, at least with respect to that famous engineering phrase 'for all practical purposes.'

Electrical elements are divided into two categories based on their energy generation characteristics: active elements can supply energy to a circuit, whereas passive elements do not. Active elements do not always supply energy, in some cases they actually absorb energy. For example, a battery is an active element that can supply energy, but when it is being charged it is absorbing energy. Passive elements are divided into two categories: those that use up, or dissipate, energy and those that store energy.

4.2.2.1 Resistors

The only element in the first group of passive elements, elements that dissipate energy, is the resistor, which dissipates energy in the form of heat. The basic equation that describes the two-variable, voltage–current relationship for a resistor is the classic Ohm's law:

$$v_R = Ri_R \, \text{V} \tag{Eq. 4.8}$$

where R is the resistance in volts/amp, better known as ohms (Ω); i is the current in amps (A); and v is the voltage in volts (V). The resistance value of a resistor is a consequence of a natural property of the material from which it is made, known as resistivity, ρ. Specifically:

$$R = \rho \frac{l}{A} \, \Omega \tag{Eq. 4.9}$$

where ρ is the resistivity of the resistor material, l is the length of the material, and A is the area of the material. Table 4.2 shows the resistivity, ρ, of several materials commonly used in electric components.

TABLE 4.2 Resistivity

Conductors	ρ (Ohm-Meters)	Insulators	ρ (Ohm-Meters)
Aluminum	2.74×10^{-8}	Glass	10^{10}–10^{14}
Nickel	7.04×10^{-8}	Lucite	$>10^{13}$
Copper	1.70×10^{-8}	Mica	10^{11}–10^{15}
Silver	1.61×10^{-8}	Quartz	75×10^{16}
Tungsten	5.33×10^{-8}	Teflon	$>10^{13}$

The power that is dissipated by a resistor can be determined by combining Eq. 4.6 and Eq. 4.8:

$$P = vi = v(v/R) = v^2/R \text{ watt or}$$

$$P = vi = (iR)i = i^2R \text{ watt} \qquad \text{[Eq. 4.10]}$$

The voltage–current relationship expressed by Eq. 4.8 can also be stated in terms of the current:

$$i = \frac{1}{R}v = Gv \text{ A} \qquad \text{[Eq. 4.11]}$$

The inverse of resistance, R, is termed the *conductance*, G, and has the units of mhos (ohms spelled backward, a subtle example of engineering humor) and is symbolized by the upside-down omega, $\mho$ (even more amusing). Equation 4.9 can be exploited to make a device that varies in resistance, usually by varying the length l, as shown in Figure 4.5. Such a device is termed a *potentiometer* or *pot* for short.

   By convention, power is positive when it is being lost or dissipated. Hence, resistors must always have a positive value for power. In fact, one way to define a resistor is to say that it is a device for which $P > 0$ for all t. For P to be positive, the voltage and current must be in the same direction; that is, the current direction must point in the direction of the voltage drop. This polarity restriction is indicated in Figure 4.6A along with the symbol that is used for a resistor in electric circuit diagrams or electric *schematics*. Stated another way, the voltage and current polarities must be set so that current flows into the positive side of the resistor. Either the voltage polarity (the plus side) or the current direction can be chosen arbitrarily, but not both. Hence, once the voltage polarity or the current direction is selected, the other is fixed by power considerations. The same will be true for other passive elements, but not for source elements. This is because source elements can, and do, supply energy, so their associated power is usually negative. In this case, current would flow out of the positive side of the element rather than into it. Figure 4.6B shows two symbols used to denote a variable resistor such as shown in Figure 4.5.

Example 4.1:   Determine the resistance of 100 feet of no. 14 American Wire Gauge (AWG) copper wire.

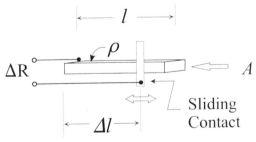

Figure 4.5   A variable resistor made by changing the effective length, Δl, of the resistive material.

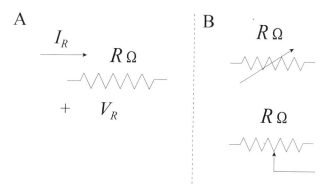

Figure 4.6 **A:** The symbol for a resistor along with its polarity conventions. For a resistor, as with all passive elements, the current direction must be such that it flows from positive to negative. In other words, current flows into the positive side of the element. **B:** Two symbols that represent a variable resistor or potentiometer.

Solution: A no. 14 AWG (also known as or B & S which stands for Brown and Sharpe gauge) wire has a diameter of 0.064 inches (see Appendix D). The value of ρ for copper is 1.70×10^{-8} Ω-m. Converting all units to the centimeters, grams, and dynes (cgs) metric unit and applying Eq. 4.9:

$$l = 100 \, \text{ft} \left(\frac{12 \, \text{in}}{1 \, \text{ft}} \right) \left(\frac{2.54 \, \text{cm}}{\text{in}} \right) = 3,048 \, \text{cm}$$

$$A = \pi r^2 = \pi \left(\frac{d}{2} \right)^2 = \pi \left(\frac{2.54(0.064)}{2} \right)^2 = \pi \left(\frac{0.1626}{2} \right)^2 = 20.76 \times 10^{-3} \, \text{cm}^2$$

$$R = \rho \left(\frac{l}{A} \right) = 1.7 \times 10^{-10} \left(\frac{3,048}{20.76 \times 10^{-3}} \right) = 2.49 \times 10^{-5} \, \Omega$$

4.2.2.2   Inductors

Energy storage devices can be divided into two classes: *inertial* elements and *capacitive* elements. The corresponding electrical elements are the inductor and capacitor, respectively, and the voltage—current equations for these elements involve differential or integral equations. Current flowing into an inductor carries energy that is stored in a magnetic field. The voltage across an inductor is the result of a self-induced electromotive force, which opposes that voltage and is proportional to the time derivative of the current:

$$v_L = L \, di_L / dt \, \text{V} \qquad\qquad\qquad \text{[Eq. 4.12]}$$

where L is the constant of proportionality termed the *inductance* measured in henrys (h). (The henry is actually Weber-turns per ampere, or volts per ampere per second, and is named for the American physicist Joseph Henry, 1797–1878.) An inductor is simply a coil of wire that uses mutual flux coupling (i.e., mutual inductance) between the wire coils. The inductance is related to the magnetic flux, Φ, carried by the inductor and by the geometry of the coil and the number of loops, or *turns*, N:

$$L = \frac{N\Phi}{i} \text{ henrys} \qquad\qquad \text{[Eq. 4.13]}$$

The energy stored can be determined from the equation for power (Eq. 4.6) and the voltage–current relationship of the inductor (Eq. 4.12):

$$P = vi = Li\,di/dt = dE/dt \text{ solving for } dE$$

$$dE = P\,dt = Li(di/dt)dt = Li\,di \qquad\qquad \text{[Eq. 4.14]}$$

The total energy stored as current increases from zero to some value i is:

$$E = \int dE = L\int_0^i i\,di = \frac{1}{2}Li^2 \text{ joules}$$

The similarity between the equation for kinetic energy of a mass (Eq. 4.2) and the energy in an inductor (Eq. 4.14) explains why an inductor is considered an inertial element. It behaves as if the energy is stored as kinetic energy associated with a mass of moving electrons, although it is actually stored in an electromagnetic field. Inductors follow the same polarity rules as resistors. Figure 4.7 shows the symbol for an inductor, a schematic representation of a coil, with the appropriate voltage–voltage directions.

If the current through an inductor is constant, so-called *direct current* (DC), then there will be no energy stored in the inductor and the voltage across the inductor will be zero, regardless of the amount of steady current flowing through the inductor. (The term *DC* stands for direct current, but it has been modified to mean *constant value* and can be applied to either current or voltage, as in *DC current* or *DC voltage*, or even non-electrical variables). The condition when voltage across an element is zero for any current is known as a *short circuit*. Hence, an inductor appears as a short circuit to a DC current.

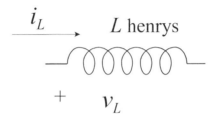

Figure 4.7 Symbol for an inductor showing the polarity conventions for this passive element.

The *v-i* relationship of an inductor can also be expressed in terms of current. Solving Eq. 4.12 for *i*:

$$v_L = L \, di_L/dt; \quad di_L = \frac{1}{L} v_L dt; \quad \int di_L = \int \frac{1}{L} v_L dt$$

$$i_L = \frac{1}{L} \int v_L dt \qquad\qquad \text{[Eq. 4.15]}$$

The integral of any function will be continuous, even if that function contains a discontinuity as long as that discontinuity is finite. A continuous function is one that does not change instantaneously. Thus, for a continuous function, $f(t)$:

$$f(t-) = f(t+) \text{ for any } t \qquad\qquad \text{[Eq. 4.16]}$$

Because the current through an inductor is the integral of the voltage across the inductor, the current will be continuous in real situations since any voltage discontinuities will surely be finite. Thus, the current though an inductor can change slowly or rapidly (depending on the voltage), but it can never change in a discontinuous (i.e., steplike) manner. In mathematical terms, for an inductor:

$$i_L(t-) = i_L(t+) \qquad\qquad \text{[Eq. 4.17]}$$

The integral relationship between current and voltage in an inductor ensures that inductor current is always continuous. Indeed, one of the popular usages of an inductor is to reduce current spikes by passing the current through an inductor. The integration property tends to *choke off* the current spikes, so an inductor used in this manner is sometimes called a *choke*.

4.2.2.3 Capacitors

A capacitor also stores energy, in this case in an electromagnetic field created by oppositely charged plates. (Capacitors are nicknamed *caps* and engineers frequently use that term. Curiously, no such nicknames exist for resistors or inductors, although in some applications inductors are called *chokes* as noted in the previous paragraph.) In the case of a capacitor, the energy stored is proportional to the charge on the capacitor, and charge is related to the time integral of current flowing through the capacitor. This gives rise to voltage–current relationships that are the opposite of the relationships for an inductor:

$$v_C = \frac{1}{C} \int i_C dt \text{ V} \qquad\qquad \text{[Eq. 4.18]}$$

or solving for i_C:

$$i_C = C \, dv_C/dt \text{ A} \qquad\qquad \text{[Eq. 4.19]}$$

where C, the *capacitance*, is the constant of proportionality and is given in units of *farads*, which are coulombs per volt. (The farad is named after Michael Faraday, an English chemist and physicist who, in 1831, discovered electromagnetic induction.) The inverse relationship between the voltage–current equations of inductors and capacitors is an example of *duality*, a property that occurs often in electric circuits. The symbol for a capacitor is two short parallel lines reflecting the parallel plates of a typical capacitor (Figure 4.8).

The capacitance describes the ability of a capacitor to store (or release) charge with respect to changes in voltage, specifically:

$$C = q/v \text{ farads} \qquad\qquad \text{[Eq. 4.20]}$$

where q is charge in coulombs and v is volts. A large capacitor can take on or release charge, q, with little change in voltage, whereas a small capacitor shows a larger change in voltage for a given in charge. The largest capacitor readily available to us, the earth, is considered a near-infinite capacitor ($C = \infty$, almost): its voltage remains constant no matter how much current flows into or out of it. This is why the earth is a popular ground point or reference voltage; it is always at the same voltage, so we just all agree that it is at 0.0 V.

Most capacitors are constructed from two approximately parallel plates, which are often rolled up to make a circular tube. The charge is held on the opposing plates. The capacitance for such a parallel plate capacitor is given as:

$$C = q/v = \varepsilon \frac{A}{d} \text{ f} \qquad\qquad \text{[Eq. 4.21]}$$

where A is the area of the plates, d is the distance separating the plates, and ε is a property of the material separating the plates termed the *dielectric constant*. Although Eq. 4.21 is only an approximation for a real capacitor, it does correctly indicate that capacitance can be increased either by increasing the plate area, A, or by decreasing the plate separation, d. However, decreasing plate separation will also decrease the voltage-handling capabilities of the capacitor and increase its leakage characteristics (described below). For this reason, real capacitors having larger capacitance values usually require larger physical volume, and capacitors with higher voltage-handling characteristics require a larger volume, because the plate separation of these capacitors must be larger. Alternatively, special dielectrics that can sustain higher voltages with smaller plate separations can be used, but they are more expensive.

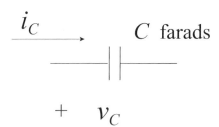

Figure 4.8 The symbol for a capacitor showing the polarity conventions.

Example 4.2: Calculate the dimensions of a 1-f capacitor. Assume a plate separation of 1.0 mm with air between the plates.

Solution: Use Eq. 4.21 and the dielectric constant for a vacuum. The dielectric constant for a vacuum is $\varepsilon_0 = 8.85 \times 10^{-12}$ C^2/mm^2 and is used for air.

$$A = \frac{Cd}{\varepsilon_0} = \frac{1f(10^{-3}\,m)}{8.85 \times 10^{-12}\,coul^2/newton\,m^2} = 1.13 \times 10^8\,m^2$$

This is an area of about 6.5 miles on a side! This large size is related to the units of farads, which are very large for practical capacitors. Typical capacitors are in the microfarads (10^{-6}) or picofarads (10^{-12}), giving rise to much smaller plate sizes. An example calculating the dimensions of a practical capacitor is given in the problems at the end of this chapter.

The energy stored in a capacitor can be determined using modified versions of Eq. 4.4 and Eq. 4.20:

$$v = q/C \text{ and from Eq. 4.4: } dE = vdq = \frac{1}{C}qdq$$

Hence, for a capacitor holding a specific charge, Q:

$$E = \int dE = \frac{1}{C}\int_0^Q qdq = \frac{1}{2}\frac{Q^2}{C}$$

$$\text{Substituting } V = Q/C \text{ joules} \quad E = \frac{1}{2}CV^2 \text{ joules} \qquad \text{[Eq. 4.22]}$$

Capacitors in parallel essentially increase the effective size of the capacitor plates so when two or more capacitors are connected in parallel, their values add. If they are connected in series, their values add as reciprocals. Such series and/or parallel combinations are discussed at length in Chapter 7.

Whereas inductors will not allow an instantaneous change in current, capacitors will not allow an instantaneous change in voltage. Because the voltage across a capacitor is the integral of the current, capacitor voltage will be continuous on the basis of the same arguments used for inductor current. Thus, for a capacitor:

$$v_C(-) = v_C(t+) \qquad \text{[Eq. 4.23]}$$

It is possible to change the voltage across a capacitor slowly or rapidly depending on the current, but never instantaneously. For this reason, capacitors are frequently used to reduce voltage spikes just as inductors are sometimes used to reduce current spikes. The fact that the behavior of voltage across a capacitor is similar to the behavior of current through an inductor is an example of a principle termed duality.

Capacitors and inductors have reciprocal responses to situations where their voltages and currents are constant. Again, such conditions are referred to as *DC conditions*. Because the current through a capacitor is proportional to the derivative of voltage (Eq. 4.19), if the voltage across a capacitor is constant, the capacitor current

TABLE 4.3 Energy Storage and Response to Discontinuous and Direct Current Variables in Inductors and Capacitors

Element	Energy Stored	Continuity Property	Direct Current Property
Inductor	$E = \dfrac{1}{2}LI^2$	Current continuous $i_L(0-) = i_L(0+)$	If i_c = constant (direct current) $v_L \rightarrow 0$ (short circuit)
Capacitor	$E = \dfrac{1}{2}CV^2$	Voltage continuous $v_C(0-) = v_C(0+)$	If v_C = constant (direct current voltage) $i_C \rightarrow 0$ (open circuit)

will be zero regardless of the value of the voltage. An *open circuit* is defined as an element having zero current for any voltage; hence, capacitors appear as open circuits to DC current. For this reason, capacitors are said to *block DC* and are sometimes used for exactly that purpose.

The continuity and DC properties of inductors and capacitors are summarized in Table 4.3. A general summary of passive and active electrical elements will be presented later in Table 4.4 at the end of Section 4.2.3.

4.2.2.4   Electrical Elements: Reality Check

The equations given above for passive electrical elements are idealizations of the actual elements. In fact, the elements do have voltage–current characteristics that are nearly linear. However, all real electric elements will contain some resistance, inductance, and capacitance, and these undesired characteristics are termed *parasitic elements*. For example, a real resistor will have some inductor- and capacitor-like characteristics, although these will generally be small and can be ignored except at very high frequencies. (Resistors made by winding resistance wire around a core, so-called *wire-wound resistors*, have a large inductance as might be expected for this coillike configuration. However, these are rarely used in electronic applications.) Similarly, real capacitors also closely approximate ideal capacitors except for some parasitic resistance. This parasitic element appears as a large resistance in parallel with the capacitor (Figure 4.9), leading to a small *leakage* current through the capacitor. Low-leakage capacitors can be obtained at additional cost with parallel resistance exceeding 10^{12} to 10^{14} Ω, resulting in very small leakage currents. Inductors are constructed by winding a wire into coil configuration. Because all wire contains some resistance, and a considerable amount of wire may be used in an inductor, real inductors generally include a fair amount of series resistance. This resistance can be reduced by using wire having a larger diameter (as suggested by Eq. 4.9), but this results in increased physical size.

In most electrical applications, the errors introduced by real elements can be ignored. It is only under extreme conditions, involving high-frequency signals or the need for very high resistances, that these parasitic contributions need be taken into account. While the inductor is the least ideal of the three passive elements, it is also

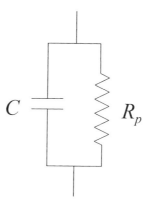

Figure 4.9 The schematic of a real capacitor showing the parasitic resistance that would lead to a leakage of current through the capacitor.

the least used in conventional electronic circuitry, so its shortcomings are not that consequential.

4.2.3 Active Elements

Active elements can supply energy to a system and in the electrical domain come in two flavors: voltage sources and current sources. These two devices are somewhat self-explanatory. Voltage sources supply a specific voltage that may be constant or time varying but is always *defined* by the element. In the ideal case, the voltage is independent of the current through the source: a voltage source is concerned only about maintaining its specified voltage; it does not care about what the current is doing. The voltage polarity is part of the voltage definition and must be given with the symbol for a voltage source as shown in Figure 4.10. The current through the source can be in either direction (again, the source does not care). If the current is flowing into the positive end of the source, the source is being 'charged' and is removing energy from the circuit. If current flows out of the positive side, the source is supplying energy. The equation for a voltage source is simply $v = V_{Source}$. The energy supplied or taken up by the source is given by Eq. 4.6 ($P = vi$). The voltage source in Figure 4.10 is shown as *grounded*; that is, one side is connected to ground. Voltage sources are commonly used in this manner, and many commercial power supplies have this grounding built in. Voltages sources that are not grounded are referred to as *floating* sources. A battery is an example of a floating voltage source.

An ideal current source supplies a specified current, which can be fixed or time varying. It cares only about the current running though it. Current sources are less intuitive because current is thought of as an affect (of voltage) not as a cause. One way to think about a current source is that it is a voltage source whose output voltage is somehow automatically adjusted to produce the desired current. A current source manipulates the cause, voltage, to produce a desired affect, current. The current source does not set the voltage across it: it will be whatever it has to be to produce the desired current. The ideal current source shown in Figure 4.11 shows the symbol used to represent a current source. Current direction is given as part of the current specification. Because a current source does not regulate the voltage (except indirectly), the symbol does not specify a voltage polarity. As with the voltage across a current source, the voltage polarity will be whatever it has to be to produce the desired current.

Again, these are idealizations, and real current and voltage sources usually fall short. Real voltage sources care about the current they have to produce and their voltages will decrease if the current requirement becomes too high. Similarly, real current sources care about the voltage across them, and their current output will decrease if the voltage required to produce the desired current gets too large. More realistic representations for voltage and current sources are given in Chapter 7 under the topic of Thévenin and Norton equivalent circuits.

Table 4.4 summarizes the various electrical elements giving the associated units, the defining equation, and the symbol used to represent that element in a circuit diagram.

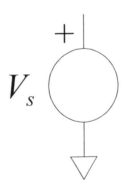

Figure   4.10
Schematic representation of a voltage source, *Vs*. This element specifies only the voltage, including the direction or polarity. The current value and direction are unspecified and depend on the rest of the circuit. Voltage sources are often used with one side connected to ground as shown.

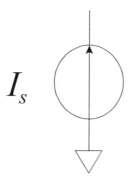

Figure 4.11 Schematic representation of a current source, *Is*. This element specifies only the current; the voltage value and polarity are unspecified and depend on the rest of the circuit.

TABLE 4.4 Electrical Elements: Basic Properties

Element	Units	Equation $v(t) = f[i(t)]$	Symbol
Resistor (R)	Ω (volt/ampere)	$v(t) = R\,i(t)$	
Inductor (L)	Henry (weber turns/ampere)	$v(t) = L\,di/dt$	
Capacitor (C)	Farad (coulombs/volt)	$v(t) = \dfrac{1}{C}\int i\,dt$	
Voltage source (V_s)	Volt (joules/coulomb)	$v(t) = V_s(t)$	
Current source (I_s)	Ampere (coulombs/second)	$i(t) = I_s(t)$	

4.2.4 The Fluid Analogy

One of the reasons analog modeling is popular is that it parallels human intuitive reasoning. To understand a complex notion, we often describe something that is similar but easier to comprehend. Some intuitive insight into the characteristics of electrical elements can be made using an analogy based on the flow of a fluid such as water. In this analogy, the flow rate of the water would be analogous to the flow of charge in an electric circuit (i.e., current), and the pressure behind that flow would be analogous to voltage. In this analogy, a resistor would be a constriction, or pipe, inserted into the path of water flow. As with a resistor, the flow through this pipe would be linearly dependent on the pressure (voltage) and inversely related to the resistance offered by the constructing pipe. The equivalent of Ohm's law would be: flow equals pressure/resistance. As with a resistor, the resistance to flow generated by the construction would increase linearly with its length and decrease with its cross-sectional area, as in Eq. 4.9 (pipe resistance is proportional to length/area).

The fluid analogy to a capacitor would be that of a container with a given cross-sectional area. The pressure at the bottom of the container would be analogous to the voltage across the capacitor, and the water flowing into or out of the container would be analogous to current. As with a capacitor, the height of the water (proportional to the amount of charge) would be proportional to the pressure at the bottom, and would be linearly related to the integral of water flow and inversely related to the area of the container (Figure 4.12). A container with a larger area (i.e., a larger capacity) would have the ability to accept larger amounts of water (charge) with little change in pressure (voltage), whereas a vessel with a small area would fill quickly. Just as in a capacitor, it would be impossible to change the height of the water, and therefore the pressure at the bottom, instantaneously (unless you had an infinite flow of water). With a high flow, you could change the height quickly, but not instantaneously. If the flow is outward, water will continue to flow until the vessel is empty. This is analogous to fully discharging a capacitor. In fact, even the time course of the outward flow would parallel that of a discharging capacitor. Also, for the pressure at the bottom of the vessel to remain constant, the flow into, or

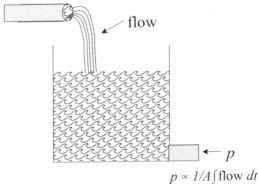

$$p \propto 1/A \int flow \ dt$$

Figure 4.12 Water analogy of a capacitor. Water pressure at the bottom is analogous to voltage across a capacitor and water flow is analogous to the flow of charge. The amount of water contained in the vessel is analogous to the charge on a capacitor.

out of, the vessel would have to be zero, just as the current must be zero for constant capacitor voltage.

A water container, like a capacitor, stores energy. In a container, the energy is stored as the potential energy of the contained water. Using a dam as an example, the amount of energy stored is proportional to the amount of water contained behind the dam and the pressure squared. A dam, or any other real container, would have a limited height, and if the flow of water into it continued for too long, it would overflow. This is analogous to exceeding the voltage rating of the capacitor where continuous charge going in would cause the voltage to rise until some type of failure occurred. It is possible to increase the overflow value of a container by increasing its height, but this would also lead to an increase in physical size just as in the capacitor. A real container might also leak; in which case water stored in the container would be lost rapidly or slowly, depending on the size of the leak. This is analogous to the leakage of current that exists in all real capacitors. Eventually all charge on the capacitor will be lost due to leakage, and a capacitor's voltage becomes zero.

In the fluid analogy, the element analogous to an inductor would be a large pipe with negligible resistance to flow, but where any change in flow would require some pressure just to overcome the inertia of the fluid. This parallel with inertial properties of a fluid again demonstrates why an inductor is sometimes referred to as an *inertial element*. The change in flow velocity (the time derivative of flow) would be proportional to the pressure applied to the water in the pipe, and the constant of proportionality would be related to the mass of the water. Hence, the relationship between pressure and flow in such an element would be:

$$p = k \text{ flow velocity} = kd(\text{flow})/dt \qquad \text{[Eq. 4.24]}$$

Energy would be stored in this pipe as kinetic energy of the moving water.

The greater the applied pressure, the faster the water would change velocity. However, just as with an inductor, it would not be possible to change the flow velocity of a mass of water instantaneously using finite pressures. Also, as with an inductor, it would be difficult to construct a large pipe holding a substantial mass of water without some associated or parasitic resistance, even if the size was quite large.

In the fluid analogy, a current source would be an ideal, constant-flow pump. It would generate whatever pressure was required to maintain its specified flow. A voltage source would be similar to a very large–capacity vessel, such as a dam. It would supply the same pressure stream, no matter how much water was flowing out of it, or even if water was flowing into it, or if there were no flow at all.

4.3 PHASOR ANALYSIS

If the system and its signals or variables are responding in a sinusoidal manner (or can be converted to sinusoids using the Fourier series or transform), then a technique known as *phasor analysis* can be used. As used here, the term phasor analysis is considerably more mundane than the name implies: the analysis of phasors, such as used on 'Star Trek' is, unfortunately, beyond the scope of this text. Phasor analysis is confined to systems whose variables or signals vary sinusoidally, have always been varying sinusoidally, and always will be varying sinusoidally (again, for all practical purposes). These *eternally sinusoidal* situations are referred to as *sinusoidal steady state* conditions. In this phrase, the term *steady-state* is used in a more general sense: it does not mean *static* because there is sinusoidal variation. Rather, sinusoidal steady state means that the sinusoid is unchanging: it has the same magnitude, phase, and frequency for all time. (Sometimes steady-state does mean static, so the phrase needs to be interpreted in context.)

Recall that one of the great virtues of the sinusoid was that its frequency is not modified by linear processes. Because all of the electrical elements discussed above are linear, circuits consisting of those elements are linear processes. Under sinusoidal steady state conditions, *all* system variables would be sinusoidal and could be described by the same general equation:

$$x(t) = A\cos(\omega t + \theta) = A\cos(2\pi f t + \theta) \qquad \text{[Eq. 4.25]}$$

where the values of A and θ can be modified by electric circuit elements, but the value of ω (or f) will be the same throughout the circuit. Thus, if the system is linear and the input to the system is a sinusoidal steady state signal, all variables/signals will be sinusoids at the same frequency. In fact, the input need not be sinusoidal, merely periodic, because periodic functions can be reduced to a series sinusoids using Fourier series analysis.

As mentioned earlier, the reason that linear elements and aggregations of linear elements do not alter the frequency of sinusoids relates to their mathematical definitions and the unique mathematical properties of a sinusoid. All linear elements can be defined in terms of integrals, differentials, or scalings, and such operations, when applied to sinusoids, do not change their frequency, only their amplitude and/or phase. The derivative or integral of a sinusoid is another sinusoid at the same frequency.

All sinusoids require three variables for complete description: amplitude, phase, and frequency (Eq. 4.25). However, if the frequency is always the same, as it would be for any variable of a linear system, then we need to keep track of only two variables: amplitude (or magnitude, an equivalent term[1]) and phase. This suggests that complex variables and complex arithmetic may be useful in simplifying the mathematics of sinusoids, because a single complex variable is actually two variables rolled into one $(a + jb)$. Perhaps a single complex number (or variable) could be used to describe the amplitude and phase of a sinusoid.

Returning again to the complex representation of sinusoids given up by Euler's equation:

$$e^{jx} = \cos x + j\sin x \qquad \text{[Eq. 4.26]}$$

---

[1] There is a tendency to use the word *amplitude* when referring to the peak-to-peak value of the sinusoid and the word *magnitude* when referring to the RMS value. Here the words amplitude and magnitude are used interchangeably.

or, in terms of the general equation for a sinusoid, Eq. 4.25:

$$Ae^{j(\omega t + \theta)} = A\cos(\omega t + \theta) + jA\sin(\omega t + \theta) \qquad \text{[Eq. 4.27]}$$

Comparing Eq. 4.25 with Eq. 4.27 shows that a general sinusoid can be represented by only the real part (i.e., Re) of e^{jx}:

$$A\cos(\omega t + \theta) = \text{Re } Ae^{j(\omega t + \theta)} = \text{Re } Ae^{j\theta}e^{j\omega t} \qquad \text{[Eq. 4.28]}$$

If all variables in an equation contain the real part of the complex sinusoid, the real terms can be eliminated, because if:

$$\text{Re } Ae^{j\theta_1}e^{j\omega t} = \text{Re } Be^{j\theta_2}e^{j\omega t} \text{ for all } t$$

then

$$Ae^{j\theta_1}e^{j\omega t} = Be^{j\theta_2}e^{j\omega t} \qquad \text{[Eq. 4.29]}$$

In general, if Re A = Re B, A does not necessarily equal B. However the only way the Re $Ae^{j\omega t}$ can equal the Re $Be^{j\omega t}$ at all values of t is if $A = B$. (Appendix E presents a review of complex arithmetic.) Because all variables in a sinusoidal steady-state system are the same except for amplitude and phase, they will all contain the 'Re' operator, and these terms can be removed from the equations as was done in Eq. 4.29 (they do not actually cancel, they are just unnecessary since the equality stands just as well without them). Similarly, because all variables will be at the same frequency, the $e^{j\omega t}$ term will appear in each variable and will cancel (once the Re's are eliminated). Therefore, the general sinusoid of Eq. 4.25 can be represented in a linear system by a single complex number:

$$A\cos(\omega t + \theta) \Leftrightarrow Ae^{j\theta} \qquad \text{[Eq. 4.30]}$$

where $Ae^{j\theta}$ is the *phasor* representation of a sinusoid. Equation 4.30 does not represent a mathematical equivalence, but a transformation from the standard sinusoidal representation to a complex exponential representation without loss of information. In the phasor representation, the frequency, ω, is not explicitly stated, but is understood to be conceptually a part of every term. Because the phasor, $Ae^{j\theta}$, is defined in terms of the cosine (Eq. 4.30), sinusoids defined in terms of sine waves must first be converted to cosine waves.

If phasors (i.e., $Ae^{j\theta}$) only offered a more succinct representation of a sinusoid, their usefulness would be limited. It is their *calculus-friendly* behavior that endears them to engineers. To determine the derivative of the phasor representation of a sinusoid, we return to the original complex definition of a sinusoid (i.e., Re $Ae^{j\theta}e^{j\omega t}$):

$$\frac{d(\text{Re } Ae^{j\theta}e^{j\omega t})}{dt} = \text{Re } j\omega Ae^{j\theta}e^{j\omega t} \qquad \text{[Eq. 4.31]}$$

The derivative of a phasor is the just the original phasor, but multiplied by $j\omega$. Hence, in phasor representation, taking the derivative is accomplished by multiplying the original term by $j\omega$, and a calculus operation has been reduced to a simple arithmetic operation:

$$d/dt \Leftrightarrow j\omega \qquad \text{[Eq. 4.32]}$$

Similarly, integration can be performed in the phasor domain simply by dividing by $j\omega$:

$$\int \mathrm{Re}\, A e^{j\theta} e^{j\omega t} dt = \mathrm{Re}\, \frac{A e^{j\theta} e^{j\omega t}}{j\omega} \qquad \text{[Eq. 4.33]}$$

and the operation of integration becomes, again, an arithmetic operation, in this case division:

$$\int dt \Leftrightarrow 1/j\omega \qquad \text{[Eq. 4.34]}$$

The basic rules of complex arithmetic are covered in Appendix E; however, a few properties of the complex operator j will be noted here. Note that $1/j$ is the same as $-j$, because:

$$\frac{1}{j} = \frac{1}{\sqrt{-1}} = \frac{-\sqrt{-1}}{-\sqrt{-1}\sqrt{-1}} = \frac{-\sqrt{-1}}{-(-1)} = -\sqrt{-1} = -j \qquad \text{[Eq. 4.35]}$$

So, Eq. 4.34 could also be written as:

$$\int dt \Leftrightarrow -j/\omega \qquad \text{[Eq. 4.36]}$$

Multiplying by j in complex arithmetic is the same as shifting the phase by 90 degrees, which follows directly from Euler's equation:

$$j e^x = j(\cos x + j \sin x) = j \cos x - \sin x = -\sin x + j \cos x$$

Substituting in $\cos(x + 90)$ for $-\sin x$, and $\sin(x + 90)$ for $\cos x$; $j e^x$ becomes:

$$j e^x = \cos(x + 90) + j \sin(x + 90)$$

This is the same as e^{x+90}, which equals $e^x e^{90}$:

$$j e^x = e^x e^{90} \qquad \text{[Eq. 4.37]}$$

Similarly, dividing by j is the equivalent of shifting the phase by -90 degrees:

$$\frac{e^x}{j} = \frac{(\cos x + j \sin x)}{j} = \frac{\cos x}{j} + \sin x = -j \cos x + \sin x = \sin x + j \cos x$$

Substituting in $\cos(x - 90)$ for $\sin x$, and $\sin(x - 90)$ for $\cos x$; e^x/j becomes:

$$\frac{e^x}{j} = \cos(x - 90) + \sin(x - 90) = e^x e^{-90} \qquad \text{[Eq. 4.38]}$$

Equations 4.32, 4.34, and 4.36 demonstrate the benefit of representing sinusoids by phasors: the calculus operations of differentiation and integration become the algebraic operations of multiplication and division. Moreover, the bilateral transformation that converts between the time domain and phasor domain (Eq. 4.30) is very easy to implement going in either direction.

Example 4.3:   Find the derivative of $x(t) = 10\cos(2t + 20)$ using phasor analysis.

Solution: Convert $x(t)$ to phasor representation [represented as $X(j\omega)$], multiply by $j\omega$, then take the inverse phasor transform:

$$10\cos(2t + 20) \Leftrightarrow 10e^{j20}$$

$$\frac{dx(j\omega)}{dt} \Leftrightarrow j\omega(10e^{j20}) = j2(10e^{j20}) = j20e^{j20}$$

$$j20e^{j20} = 20e^{j20}e^{j90} \Leftrightarrow 20\cos(2t + 20 + 90) = 20\cos(2t + 110)$$

Because $\cos(x) = \sin(x + 90) = -\sin(x - 90)$, this can also be written as $-20\sin(2t + 20)$, which would be obtained from straight differentiation.

A shorthand notation is common for the phasor description of a sinusoid. Rather than write $Ve^{j\theta}$, we simply write $V\angle\theta$. When a time variable such as $v(t)$ is converted to a phasor variable, it is common to write it as a function of ω: $V(\omega)$. This acknowledges that phasors represent sinusoids at a specific frequency even though the $e^{j\omega t}$ term is not explicitly shown in the phasor itself. (Although the frequency is part of the phasor equations for inductors and capacitors and will be embedded in the phasor values of these components.) When discussing phasors and implementing phasor analysis, it is common to represent frequency in radians per second, ω, rather than in hertz (Hz), f, even though hertz is more likely to be used in practical settings. It is also common to use capital letters for the phasor variable, a convention followed here. Hence, the time–phasor transformation for variables $v(t)$ and $i(t)$ can be stated as:

$$v(t) \Leftrightarrow V(\omega)$$

$$i(t) \Leftrightarrow I(\omega) \qquad\qquad \text{[Eq. 4.39]}$$

In this notation, the phasor representation of $20\cos(2t + 110)$ would be written as $20\angle110$ rather than $20\ e^{j110}$. Often, the phasor representation of a sinusoid expresses the amplitude of the sinusoid in RMS values rather than peak-to-peak values, in which case the phasor representation of $20\cos(2t + 110)$ would be written as: $(0.707)\,20\angle110 = 14.14\angle110$. In this text, peak-to-peak values are used simply because it saves converting back and forth between RMS and peak-to-peak.

The phasor approach is an excellent method for simplifying the mathematics of linear systems. It can be applied to all systems that are driven by sinusoids, or with the use of the Fourier series analysis or the Fourier transform, to systems driven by any periodic signal.

To formalize the analysis of circuits and analog models using the phase representation of sinusoids, electrical elements will be reintroduced, but now the mathematical description of their voltage–current relationships will be given in the phasor domain. After describing electrical elements, mechanical elements will be covered.

4.3.1   **Phasor Representation: Electrical Elements**

Using phasor analysis, it is possible to recast the differential/integral equations defining an inductor and a capacitor (Eq. 4.12 and Eq. 4.18) into algebraic equations. Differentiation becomes simply multiplication by $j\omega$ (or $j2\pi f$) and integration becomes division by $j\omega$.

Converting the voltage–current equation for a resistor from time to phasor domain is not difficult, nor is it particularly consequential because the time domain equation (Eq. 4.8) is already an algebraic relationship. Accordingly, the conversion is only a matter of restating the voltage and current variable in phasor notation:

$$V(\omega) = RI(\omega)\text{V} \qquad \text{[Eq. 4.40]}$$

Rearranged as a voltage–current ratio:

$$R = \frac{V(\omega)}{I(\omega)}\Omega \qquad \text{[Eq. 4.41]}$$

Because the voltage–current relationship of a resistor was algebraic, nothing is gained by converting it to the phasor domain. However, converting the voltage–current equation of an inductor or capacitor to phasor notation does make a considerable difference as the differential or integral relationships become algebraic. For an inductor, the voltage–current equation in the time domain is given in Eq. 4.12 and repeated here:

$$v_L = L\,di(t)\big/dt \text{ V} \qquad \text{[Eq. 4.42]}$$

But in phasor notation, the derivative operation becomes multiplication by $j\omega$:

$$di(t)\big/dt \Leftrightarrow j\omega I(\omega);$$

so the voltage–current operation for an inductor becomes:

$$V_L(\omega) = Lj\omega I(\omega) = j\omega LI(\omega)\text{V} \qquad \text{[Eq. 4.43]}$$

Now it is possible to solve Eq. 4.43 using algebra to obtain a voltage-to-current ratio similar to that for the resistor (Eq. 4.41):

$$\frac{V(\omega)}{I(\omega)} = j\omega L\Omega \qquad \text{[Eq. 4.44]}$$

The ability to express the voltage–current relationship as a ratio is part of the power of the phasor domain method. Thus, the term $j\omega L$ is something like the equivalent resistance of an inductor. It is termed the *impedance*, represented by the letter Z, and has the units of ohms (volts per ampere), the same as for a resistor:

$$Z_L(\omega) = \frac{V_L(\omega)}{I_L(\omega)} = j\omega L\Omega \qquad \text{[Eq. 4.45]}$$

Impedance, the ratio of voltage-to-current, is not defined for inductors or capacitors in the time domain because the voltage–current relationships for these elements contain integrals or differentials and it is not possible to determine a V/I ratio. In general, the impedance will be a function of frequency, ω (except for resistors), although often the impedance is written simply as Z with the frequency term understood. Because impedance is a generalization of the concept of resistance (it is the V/I ratio for any passive element), the term is often used in discussion of any V/I relationships, even when only resistances are involved.

For the moment, the concept of impedance will be limited to the phasor domain, and hence to circuits that involve only sinusoidal signals. In Chapter 8, we will see how to extend this concept to a broader class of signals using the Laplace transform. As with many other concepts presented in this text, impedance is a useful and broadly used concept found in electrical, mechanical, and thermal processes.

To extend the concept of impedance and algebraic operations to the capacitor, we start with the basic voltage–current equation for a capacitor (Eq. 4.17), repeated here:

$$v_C(t) = \frac{1}{C}\int i_C(t)dt \qquad \text{[Eq. 4.46]}$$

and noting that integration becomes the operation of dividing by $j\omega$ in the phasor domain:

$$\int i(t)dt \Leftrightarrow \frac{I(\omega)}{j\omega}$$

so the phasor voltage–current equation for a capacitor becomes:

$$V(\omega) = \frac{I(\omega)}{j\omega C}\,\text{V} \qquad \text{[Eq. 4.47]}$$

The capacitor impedance then becomes:

$$Z_C(\omega) = \frac{V_C(\omega)}{I_C(\omega)} = \frac{1}{j\omega C} = \frac{-j}{\omega C}\,\Omega \qquad \text{[Eq. 4.48]}$$

These voltage–current relationships and impedances are summarized in Table 4.5.

Remember that the phasor notation only applies if all voltage and currents are sinusoids at the same frequency, so the relationships given in Table 4.5 only hold for these conditions. Table 4.5 also gives the impedance, Z, the voltage current ratio. In the time domain, impedance can only be defined for a resistor (i.e., Ohm's law), but in the phasor domain, impedance is defined for inductors and capacitors as well.

TABLE 4.5 V/I Relationships and Impedance for Electrical Elements

Element	v/i Time Domain	V/I Phasor Domain	Impedance $Z(\omega)$ Phasor Domain
Resistor	$v = Ri$	$V(\omega) = R\,I(\omega)$	$R\,\Omega$
Inductor	$v = L\,di/dt$	$V(\omega) = j\omega L\,I(\omega)$	$j\omega L\,\Omega$
Capacitor	$v = 1/C\int idt$	$V(\omega) = I(\omega)/j\omega C$	$1/j\omega C\,\Omega$

(Impedance is also defined for these elements in the Laplace domain.) Using phasors, it is possible to treat the so-called *reactive* elements, inductors and capacitors, as if they were effectively resistors. This allows us to introduce a generalization of Ohm's law that applies to inductors and capacitors as well: voltage is equal to a constant times current, only in the case of inductors and capacitors the constant is imaginary ($j\omega L$, $1/j\omega C$). This allows the construction of relatively simple algebraic descriptions of circuits that involve these elements.

Active elements producing sinusoidal voltages or currents can also be represented in the phasor domain by returning to the original phasor description of sinusoid, Eq. 4.30, repeated here:

$$A\cos(\omega t + \theta) = Ae^{j\theta} = A \angle \theta \qquad\qquad \text{[Eq. 4.49]}$$

Using this equation, the phasor representation for a voltage source becomes:

$$V_S(t) = V_s \cos(\omega t + \theta) \Leftrightarrow V_s e^{j\theta} \equiv V_s \angle \theta \quad V \qquad\qquad \text{[Eq. 4.50]}$$

and for a current source:

$$I_S(t) = I_s \cos(\omega t + \theta) \Leftrightarrow I_s e^{j\theta} \equiv I_s \angle \theta \quad A \qquad\qquad \text{[Eq. 4.51]}$$

(Again, capital letters are used for phasor variables.) Thus converting active sources to phasor representation is straightforward and easy. These principles are demonstrated in the example below. If a source produces a current of:

$$i_S(t) = 0.1\sin(5t + 30) = 0.1\cos(5t + 30 - 90) = 0.1\cos(5t - 60)A$$

The phasor representation of the current source will be:

$$I_S(\omega) = 0.1 \angle -60A$$

Again note that the frequency, ω, is not explicitly shown in this representation, but it will be expressed in the circuit impedances.

Example 4.4:   Find the current through the capacitor in the circuit below.

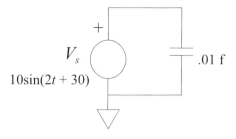

Solution: Because the voltage across the capacitor is known (it is the voltage of the voltage source, V_s), the current through the capacitor can be determined by a phasor extension of Ohm's law: $V = I\,Z$:

$$v_c = i_c Z_c; \text{ solving for } i_c\text{: } i_c = v_c/Z_c \text{ A}$$

The voltage across the capacitor is the same as the source voltage, V_S:

$$v_c = 10\sin(2t + 30)\text{V} \Rightarrow v_c(\omega) = 10 \angle -60\text{V}$$

Next, we find the phasor notation of the capacitor:

$$Z_c = \frac{1}{j\omega C} = \frac{1}{j2(0.01)} = -j50\Omega = 50 \angle -90\Omega$$

Then solving for I_C using Ohm's law:

$$I_C(\omega) = \frac{V_c(\omega)}{Z_c(\omega)} = \frac{10 \angle -60}{50 \angle -90}$$

To divide two complex numbers, convert then to polar notation (these already are in polar notation) and divide the magnitudes and subtract the denominator angle from the numerator angle. (See Appendix E for details.)

$$I_C(\omega) = \frac{10 \angle -60}{50 \angle -90} = 0.2 \angle 30\text{A}$$

Converting back to the time domain $i_c(t) = 0.2\cos(2t + 30)\text{A}$

The solution to the problem requires only algebra, but it does involve complex algebra. Again, the rules for arithmetic operations (addition, subtraction, multiplication, and division) involving complex numbers are given in Appendix E.

4.4   MECHANICAL ELEMENTS

The mechanical properties of a material often vary across and through the material so that a mechanical analysis involving this material must be made using continuous mathematical methods known as *continuum mechanics*. If only the overall behavior of an element, or collection of elements, is needed, the properties of each element can be lumped together and a *lumped-parameter* analysis can be performed. An intermediate approach facilitated by high-speed computers is to apply lump-parameter analysis to small segments of the material, then compute how each of these segments is to interact with its neighbors; a technique known as *finite element analysis*. Lumped-parameter mechanical analysis is similar to that used for electrical elements and most of the mathematical techniques described above and developed in the next several chapters can be applied to this type of mechanical analyses. In lumped-parameter mechanical analysis, the major variables are force and velocity, and the mechanical element produces a well-defined relationship between these variables that is similar to the voltage–current relationship defined by electrical elements. In mechanical systems, the flowlike variable analogous to current is velocity, whereas the potential energy variable analogous to voltage is force. Thus, mechanical elements define a relationship between force and velocity. Mechanical elements can be active or passive, and passive elements can dissipate or store energy.

4.4.1   Passive Elements

Dynamic friction is the only mechanical element that dissipates energy and, as with the resistor, that energy is converted to heat. The force–velocity relationship for a friction element is also similar to a resistor: the force generated by the friction element is proportional to its velocity:

$$F = k_f v \qquad\qquad \text{[Eq. 4.52]}$$

where k_f is the constant proportionality and is termed simply *friction* (F) is force, and v is velocity.

In the cgs metric system used in this text, the unit of force is dynes and the unit of velocity is centimeters per second (cm/sec), so the units for friction are dyne/cm/sec (force/velocity) or dyne-sec/cm. The other commonly used measurement system is the *mks* (meters, kilograms, seconds) system preferable for systems having larger forces and velocities. Conversion between the two is straightforward (see Appendix D).

The equation for the power lost as heat in a friction element is analogous to that of a resistor:

$$P = Fv \qquad\qquad \text{[Eq. 4.53]}$$

The symbol for such a friction element is termed a *dash-pot*, and is shown in Figure 4.13. Friction is often a parasitic element, but a device that is specifically constructed to produce friction can be made using a piston that moves through a fluid (or air); for example, the shock absorbers on a car or some door-closing mechanisms. This construction approach, a moving piston, forms the basis for the schematic representation of the friction element shown in Figure 4.13.

As with passive electrical elements, passive mechanical elements have a specified directional relationship between force and velocity: specifically, the direction of positive force is *opposite* that of positive velocity. Again, the direction of one of the variables can be chosen arbitrarily after which the direction of the other variable is determined.

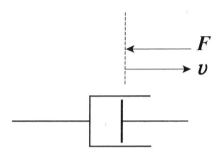

Figure 4.13 The schematic representation of a friction element showing the convention for the direction of force and velocity.

In addition to elements specifically designed to produce friction (such as shock absorbers), friction occurs in association with other elements, just as resistance is unavoidable in other electrical elements (particularly inductors). For example, a mass sliding on a surface would exhibit some friction no matter how smooth the surface. Regardless of whether friction arises from a dash-pot element specifically designed to create friction or is associated with another element, it is usually represented by the dash-pot schematic shown in Figure 4.13.

There are two mechanical elements that store energy just as there are two electrical elements. The *inertial type* element corresponding to inductance is, not surprisingly, inertia associated with mass. It is termed simply *mass*, and is represented by the letter m. The equation for the force–velocity relationship associated with mass is as follows:

$$F = ma = m\frac{dv}{dt}$$ [Eq. 4.54]

The mass element is schematically represented as a rectangle, again with force and velocity in opposite directions (Figure 4.14).

A mass element stores energy as kinetic energy following the well-known equation given in Eq. 4.2 and repeated here:

$$E = \frac{1}{2}mv^2$$ [Eq. 4.55]

The parallel between inertial electrical elements and the analogous mechanical element, mass, extends to variable continuity. Just as current moving through an inductor must be continuous and cannot be changed instantaneously, moving objects tend to continue moving (to paraphrase Newton), and the velocity of a mass cannot be changed instantaneously without applying infinite force. Hence, the velocity of a mass is continuous, so that $v_m(0-) = v_m(0+)$. It is possible to change the force on a mass instantaneously, just as it is possible to change the voltage applied to an inductor instantaneously, but not the velocity.

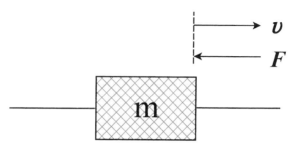

Figure 4.14 The schematic representation of a mass element that features the same direction conventions as the friction element.

The capacitor-like energy storage element in mechanical systems, the element analogous to a capacitor, is a spring, and it has a force–velocity equation that is in the same form as the equation of a capacitor:

$$F = k_e \int v dt = k_e x(t) \qquad \text{[Eq. 4.56]}$$

where k_e is the spring constant in dyne/cm. A related term frequently used is the *compliance*, C_k which is just the inverse of the spring constant $(1/k_e)$ and its use makes the equations of spring and capacitor even more similar in form:

$$F = \frac{1}{C_k} \int v dt = \frac{1}{C_k} x(t) \qquad \text{[Eq. 4.57]}$$

Although the spring is analogous to a capacitor, the symbol used for a spring is similar to that used for an inductor as shown in Figure 4.15. (Springs in schematics look like inductors, but they act like capacitors.)

As with a capacitor, a spring stores energy as potential energy. A spring that is stretched or compressed generates a force that can do work if allowed to move through distance. The work or energy stored in a spring is:

$$E = \int F dx = \int k_e x dx = \frac{1}{2} k_e x^2 \qquad \text{[Eq. 4.58]}$$

because displacement, x, is analogous to charge, q, in the electrical domain, the equation for energy stored in a spring is analogous to a form of the equation for energy stored in a capacitor (Eq. 4.22).

As with a capacitor, it is impossible to change the force on a spring instantaneously using finite velocities. This is because force is proportional to length ($F_s = k_e x$) and the length of a spring cannot change instantaneously. Using high velocities, it is possible to change spring force quickly, but not instantaneously; hence for a spring, $F_S(0-) = F_S(0+)$.

Because passive mechanical elements have defining equations similar to those of electrical elements, the same useful analysis techniques, such as phasor analysis, can be applied under the same constraints (i.e., sinusoidal steady state conditions). Moreover, the rules for analytically describing combinations of elements (i.e., mechanical systems) are similar to those for describing electrical circuits. Table 4.6

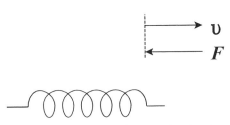

Figure 4.15 The symbol for a spring showing the direction conventions for this passive element.

TABLE 4.6 Energy Storage and Response to Discontinuous and Direct Current Variables in Mass and Elasticity

Element	Energy Stored	Continuity Property	DC Property
Mass	$E = \frac{1}{2}mv^2$	Velocity continuous $v_m(0-) = v_m(0+)$	If v_m = constant (DC velocity) $F \to 0$
Elastic element	$E = \frac{1}{2}k_e x^2$	Force continuous $F_e(0-) = F_e(0+)$	If F_e = constant (DC force) $v_m \to 0$

DC, direct current.

is analogous to Table 4.3 and shows the energy, continuity, and DC properties of mass and elasticity.

4.4.2 Elasticity

Elasticity is most often distributed through or within a material and is defined by the relationship between *stress* and *strain*. Stress is force normalized by the cross-sectional area:

$$Stress = \Delta F / A \qquad \text{[Eq. 4.59]}$$

Strain is elongation, or stretching, that is normalized by the rest length, the length the material would assume if no force were applied:

$$Strain = \Delta \ell / \ell \qquad \text{[Eq. 4.60]}$$

The ratio of stress to strain is a normalized measure of the ability of a material to stretch and is an elastic coefficient termed *Young's modulus*:

$$Y_M = Stress / Strain = \frac{\Delta F / A}{\Delta \ell / \ell} \qquad \text{[Eq. 4.61]}$$

If a material is stretched by a load or weight produced by a mass, m, then the equation for Young's modulus can be written as:

$$Y_M = \frac{mg / \pi r^2}{\Delta \ell / \ell} \qquad \text{[Eq. 4.62]}$$

where g is the gravitational constant, 980.665 cm/sec^2. Values for Young's modulus for a wide range of materials can be found in traditional references such as the *Handbook of Physics and Chemistry* (Lide, 2004). Some values for typical materials are shown in Table 4.7. The examples below illustrate applications of Young's modulus and related equations.

Example 4.5: A 10-lb. weight is suspended by a no. 12 (AWG) wire 10 inches long. How much does the wire stretch?

TABLE 4.7   Young's Modulus of Selected Materials

Material	Y_M (dyne/cm²)
Steel (drawn)	19.22×10^{10}
Copper (wire)	10.12×10^{10}
Aluminum (rolled)	$6.8–7.0 \times 10^{10}$
Nickel	$20.01–21.38 \times 10^{10}$
Constantan	$14.51–14.89 \times 10^{10}$
Silver (drawn)	7.75×10^{10}
Tungsten (drawn)	35.5×10^{10}

Solution: To find the new length of the wire, use Eq. 4.62 and solve for $\Delta\ell$. First convert all constants to cgs units:

$$m = 10\,\text{lb} = 10\,\text{lb}\left(\frac{1\,\text{kg}}{2.04\,\text{lb}}\right)\left(\frac{1,000\,\text{gm}}{1\,\text{kg}}\right) = 4,902\,\text{gm}$$

$$l = 10\,\text{in} = 10\,\text{in}\left(\frac{2.54\,\text{cm}}{1\,\text{in}}\right) = 25.4\,\text{cm}$$

To find the diameter of 12-gauge (AWG) wire used in the table for wire gauges in Appendix D:

$$d = 0.081\,\text{in from Table 4}$$

$$r = d/2 = 0.081/2\,\text{in}\left(\frac{2.54\,\text{cm}}{1\,\text{in}}\right) = 0.103\,\text{cm}$$

Then solve for Δl:

$$Y_M = \frac{mg/A}{\Delta l/l} = \frac{mg/\pi r^2}{\Delta l/l} \qquad \Delta l = \frac{mgl}{\pi r^2 Y_M} = \frac{(4,902)(980.6)(25.4)}{\pi(0.103)^2(10.12 \times 10^{10})} = 0.036\,\text{cm}$$

Example 4.6:   Find the elastic coefficient of a steel bar with a diameter of 0.5 mm and length of 0.5 m.

Solution: From Eq. 4.56: $F = K_e x$; $k_e = F/x$, where $x = \Delta\ell$. Find k_e in terms of Young's modulus, then solve the equations below:

$$Y_M = \frac{\Delta F/A}{\Delta\ell/\ell} = \frac{F\ell}{\Delta\ell A}; \qquad F = \frac{Y_M A \Delta\ell}{\ell}; \qquad k_e = \frac{F}{\Delta\ell} = \frac{Y_M A}{\ell} = \frac{Y_M \pi(d/2)^2}{\ell}$$

$$k_e = \frac{19.22 \times 10^{10}\,\pi(0.05/2)^2}{50} = \frac{3.77 \times 10^8}{50} = 7.55 \times 10^6\,\text{dynes/cm}^2$$

4.4.3 **Sources**

Sources supply mechanical energy and can be sources of force, velocity, or displacement. Displacement is another word for a change in position, and is just the integral of velocity: $x = \int v dt$. As mentioned earlier, displacement is analogous to charge in the electrical domain because $q = \int i dt$. While sources of constant force or constant velocity do occasionally occur in mechanical systems, most sources of mechanical energy are much less ideal than their electrical counterparts. Sometimes the same mechanical source can look like a velocity (or displacement) generator or a force generator depending on the characteristics of the load; that is, the mechanical properties of the elements connected to the source. For example, consider a muscle contracting under a light, constant load, a so-called *isotonic contraction* because the force (i.e., *tonus*) opposing the contraction is constant (i.e., *iso*). Under these load conditions, the muscle would appear to be a velocity generator, although the velocity would not be constant throughout the contraction. However, if the muscle's endpoints were not allowed to move, a so-called *isometric contraction* because the muscle's length (i.e., *metric*) is constant (again, *iso*), the muscle would look like a force generator.

In fact, a muscle is neither an ideal force generator nor an ideal velocity generator. An ideal force generator would put out the same force no matter what the conditions; however, the maximum force developed by a muscle depends strongly on its initial length. Figure 4.16 shows the classic length–tension curve for skeletal

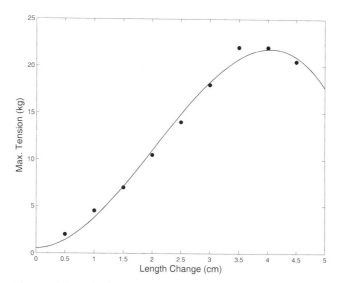

Figure 4.16 The length–tension relationship of skeletal muscle. The relationship between the maximum force a muscle can produce depends strongly on its length. An ideal force generator would produce the same force regardless of its length (or velocity for that matter). (Based on historical data from the human triceps muscle.)

muscle and the strong relationship between maximum force and the change in position from rest length. (The rest length is the position the muscle assumes where there is no force applied to the muscle.) When operating as a velocity generator under constant load, muscle is far from ideal as the velocity generated is highly dependent on the load. As shown in the also-classic force–velocity curve, as the force resisting the contraction of a muscle is increased, its velocity decreases and can even reverse if the opposing force becomes great enough (Figure 4.17). Of course, electrical sources are not ideal either, but they are generally more nearly ideal than mechanical sources. The characteristics of real sources, mechanical and electrical, are explored in Chapter 7.

With these practical considerations in mind, a force generator is usually represented by a circle or simply an F with a directional arrow (Figure 4.18).

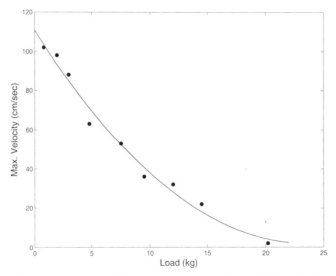

Figure 4.17   As a velocity generator, muscle is hardly ideal. As the load increases the maximum velocity does not stay constant as would be expected of an ideal source, but decreases with increasing force and can even reverse direction if the force becomes too high. This is known as the force–velocity characteristics of muscle. (Based on historical data from the human pectoralis major muscle.)

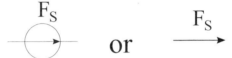

Figure 4.18   Two schematic representations of an ideal force generator showing direction of force.

In addition to the property of inertia described by Eq. 4.54, a mass generates a force when placed in a gravitational field. The inertial properties of a mass, or inertial mass, and its gravitational properties, gravitational mass, need not necessarily be coupled if they are the result of two separate physical mechanisms. However, experimentally they have been shown to be linked down to the most sensitive measurement resolution indicating that they are somehow related to the same underlying physics. The force is proportional to the value of the mass and the gravitational constant:

$$F = mg \qquad\qquad \text{[Eq. 4.63]}$$

where m is the mass in grams and g is the gravitational constant in cm/sec². Note that a gm-cm/sec² equals a dyne of force. The value of g at sea level is 980.665 cm/sec² Provided the mass does not change significantly in altitude (which would vary g), the force produced by a mass due to gravity is nearly ideal: the force produced would be independent of velocity (as long as it is in a vacuum so there is no wind resistance at higher velocities).

In some mechanical systems that include mass, the force due to gravity must be considered; in other systems, it is canceled by some sort of supporting structure. In Figure 4.19, the system on the left side has a mass supported by a surface (either a frictionless surface or with the friction incorporated in k_f) and exerts only the inertial force defined in Eq. 4.54. In the system on the right-hand side, the mass is under the influence of gravity and produces both an inertial force that is a function of velocity (Eq. 4.54) and a gravitational force that is constant and defined by Eq. 4.63. This additional force would be represented as a force generator acting in the downward direction with a force of mg.

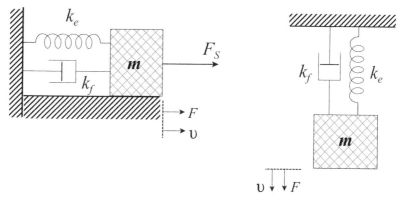

Figure 4.19 Two mechanical systems containing mass, m. In the left-hand system, the mass is supported by a surface so the only force involved with this element is the inertial force. In the right-hand system, gravity is acting on the mass so that it produces two forces: a constant force due to gravity (mg) and its inertial force.

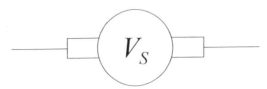

Figure 4.20 Symbol used to represent a motor. A motor can be either a force (torque) generator or a velocity [rotations per minute (rpm)].

TABLE 4.8 Mechanical Elements

Element (Units)	Equation $F(t) = f[v(t)]$	Phasor Equation	Impedance $Z(\omega)$	Symbol
Friction (k_f) (dyne-sec/cm)	$F(t) = k_f v(t)$	$F(\omega) = k_f v(\omega)$	k_f	
Mass (m) (gm)	$F(t) = m\, dv/dt$	$F(\omega) = j\omega m v(\omega)$	$j\omega m$	m
Elasticity (k_i) (spring) (dyne/cm)	$F(t) = k_e \int i\, dt$	$F(\omega) = k_e v(\omega)/j\omega$	$k_e/j\omega$	
Force generator (F_S)	$F(t) = F_S(t)$	$F(\omega) = F_S(\omega)$	—	F_S
Velocity or displacement generator (V_S or X_S)	$v(t) = V_S(t)$ $x(t) = X_S(t)$	$v(\omega) = V_S(\omega)$	—	V_S

Motors, which can be sources of torque (i.e., force generators) or velocities (in rotations per minute [rpm] or rad/sec), are represented as shown in Figure 4.20. A motor can be regarded as either a velocity or displacement generator, so a similar symbol could be used, but the letters used would be either V_S if it were a velocity generator or X_S for a displacement generator.

The mechanical elements, their differential and integral equations, and their phasor representations are summarized in Table 4.8 just as the electrical elements are summarized in Table 4.4. Note that impedance can also be defined for mechanical elements and has the units of dyne-cm/sec.

Example 4.7: Find the velocity of the mass in the mechanical system below. The force, F_S, is $5 \cos(10t)$ dyne and the mass is 5 gm. The mass is supported by a frictionless surface.

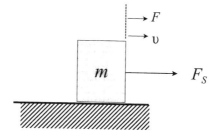

Solution: Convert the force to a phasor and apply the appropriate phasor equation from Table 4.5. Solve for $v(\omega)$.

Converting the force to phasor notation:

$$5\cos(10t) \Leftrightarrow 5\angle 0 \text{ dynes}$$

$$F(\omega) = j\omega_m v(\omega); \quad v(\omega) = \frac{F(\omega)}{j\omega_m}$$

$$v(\omega) = \frac{5\angle 0}{j(10)5} = \frac{5\angle 0}{50\angle 90} = 0.1\angle -90 \text{ cm/sec}$$

Converting back to the time domain (if desired):

$$v(t) = 0.1\cos(10t - 90) = 0.1\sin(10t)\text{cm/sec}$$

The analysis of more complicated systems is presented in the next chapter.

4.5 SUMMARY

The most complicated electrical and mechanical systems are constructed from a limited set of basic elements. These elements fall into two general categories: active elements, which usually supply energy to the system, and passive elements, which dissipate or store energy. In electrical systems, the passive elements are described by the relationship they enforce between voltage and current, whereas in mechanical systems the relationship is between force and velocity. These relationships are linear and involve scaling, differentiation, or integration. Active electrical elements supply either a specific well-defined voltage (i.e., voltage sources), or a well-defined current (i.e., current sources). Active mechanical elements are sources of force or sources of velocity. All elements are described as idealizations, and although many practical elements closely approach these idealizations, some of the major deviations have been noted.

These basic elements are combined to construct electrical and mechanical systems. Because some of the passive elements involve calculus operations, differential equations are required to describe most electrical and mechanical systems.

However, if we restrict our signals to sinusoids, it is possible to represent elements in such a manner that algebra can be used. This analysis is known as phasor analysis and uses complex variables to represent the basic electrical or mechanical variables. A complex variable can represent both the magnitude and phase of a sinusoid. Differentiation of a sinusoid just multiplies the sinusoid by the frequency and changes its phase by 90 degrees, an operation that can be achieved by multiplying a phasor by $j\omega$. Similarly, integration becomes division by $j\omega$ in the phasor domain. Hence, when circuits are represented in the phasor domain, only algebraic equations are required to analyze electrical and mechanical systems.

PROBLEMS

1. A resistor is constructed of thin copper wire wound into a coil (a wire-wound resistor). The wire has a diameter of 1 mm.
   a. How long is the wire required to be to make a resistor of 12 Ω?
   b. If this resistor is connected to a 5-V source, how much power will it dissipate as heat?
2. a. A length of no. 12 (AWG) copper wire has a resistance of 0.05 Ω. It is replaced by no. 16 (AWG) wire. What is the resistance of this new wire?
   b. Assuming both wires carry 2 A of current, what is the power lost in the two wires?
3. The figure below shows the current passing through a 2-h inductor.
   a. What is the voltage drop across the inductor?
   b. What is the energy stored in the inductor after two seconds?

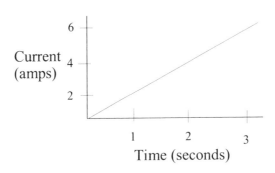

4. The voltage drop across a 10-h inductor is measured as 10 cos(20t) V. What is the current through the inductor?
5. A parallel-plate capacitor has a value of 1 μf(10^{-6} f). The separation between the two plates is 0.2 mm. What is the area of the plates?
6. The current waveform shown in Problem 3 passes through a 0.1-f capacitor.
   a. What is the equation for the voltage across the capacitor?
   b. What is the charge, q, contained in the capacitor after 2 seconds (assuming it was unchanged or $t = 0$)?

7. A current of 1 A has been flowing through a 1-f capacitor for 1 second.
   a. What is the voltage across the capacitor, and what is the total energy stored in the capacitor?
   b. Repeat for a 100-f capacitor.
8. Convert the following to phasor representation:
   a. $10 \cos(10t)$;
   b. $5 \sin(5t)$;
   c. $6 \sin(2t + 60)$;
   d. $2 \cos(5t) + 4 \sin(5t)$ (*Hint:* see Example 2.2):
   e. $\int 5 \cos(20t)dt$
   f. $\dfrac{d}{dt}(2\cos(20t + 30))$
9. Add the following real, imaginary, and complex numbers to form a single complex number: $6, j10, 5 + j12; 8 + j3, 10\angle 0, 5\angle -60, 1/(j.1)$
10. Evaluate the following expressions:
    a. $(10 + j6) + 10\angle -30 - 10\angle 30$
    b. $6\angle -120 + \dfrac{5 - j10}{j4}$
    c. $\dfrac{10 + j5}{15 - j6} - \dfrac{8 - j8}{12 + j4}$
    d. $\dfrac{(6 + j5)(3 - j4)j}{(8 + j3)(10 \angle 260)}$
11. Find the value of the current through the inductor using the phasor extension of Ohm's law.

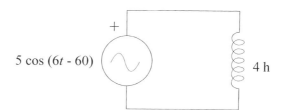

12. A constant force of 10 dyne is applied to a 5-gm mass. The force is initially applied at $t = 0$ when the mass was at rest.
    a. At what value of t does the speed of the mass equal 6 dyne/second?
    b. What is the energy stored in the mass after 2 seconds?
13. A force of $10 \cos(6t + 30)$ dynes is applied to a spring having a spring constant of 20 dyne/cm.
    a. What is the equation for the velocity of the spring?
    b. What is the instantaneous energy stored in the spring at $t = 2$ seconds?
14. A 100-foot length of silver wire having a diameter of 0.02 inches is stretched by 0.5 inches. What is the tension (stretching force) on the wire?

MATLAB Problem

15. Find the velocity of the mass in Example 4.7 for the 5-dyne cosine source where frequency varies from 1 to 40 rad/sec in increments of 1 rad/sec. Plot the magnitude of the velocity as a function of frequency. (*Hint:* Make frequency a vector between 1 and 40 using the MATLAB command `w = 1:40`, then solve for a velocity vector, `vel`, by dividing the source value, 5, by `j*5*w`. Since w is a vector you will need to do point-by-point division using the `./` command. Also you will need to plot the magnitude of the velocity vector [`plot(w,abs(vel))`] since it will be imaginary.)

5 ANALYSIS OF ANALOG MODELS AND PROCESSES

5.1 CONSERVATION LAWS: KIRCHHOFF'S VOLTAGE LAW

In this chapter, we continue to learn how to analyze systems composed of analog elements, be they electric circuits or physiological models. Eventually, these analysis techniques will not only describe behavior to specific inputs or stimuli but also enable us to estimate the response to any input. The analysis will also give us general information about the system. For now, we are still restricted to sinusoidal sources, but even with that restriction, much useful information can be extracted. Having defined the players (i.e., the elements) in the last chapter, we now need to set the rules of the game: the rules that describe the interactions between elements. For both mechanical and electrical elements, the rules are based on conservation laws: conservation of energy and conservation of mass (or charge). For electrical elements, related rules are termed *Kirchhoff's voltage law* (KVL) and *Kirchhoff's current law* (KCL). Either of these laws can be used in the analysis of electric circuits (or networks) and physiological models based on electrical analogs.

KVL is based on conservation of energy: the total energy in a closed system must be zero. Because voltage is related to potential energy, the law implies that voltage increases or decreases around a closed loop must sum to zero. Simply stated: what goes up must come down (in voltage):

$$\sum_{\text{Loop}} v = 0 \qquad \text{[Eq. 5.1]}$$

This law will allow us to write an equation for all electrical elements connected in a loop, and since current can only flow in a *closed circuit* or loop, all elements that do anything must be in some kind of loop. Figure 5.1 illustrates a loop containing three generalized elements to which KVL applies. The voltages across the three elements must sum (algebraically) to zero: $V_1 + V_2 + V_3 = 0$. Circuits that are more complicated may contain a number of loops and some elements may be involved in more than one loop, but KVL still applies to each of the loops and the analysis for any number of loops becomes a straightforward extension of the analysis for a single loop. The first example applies KVL to a single-loop problem.

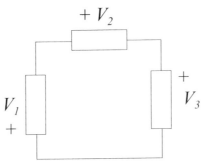

Figure 5.1 Illustration of Kirchhoff's voltage law. The three voltages must sum to zero: $V_1 + V_2 + V_3 = 0$.

While all circuits can be analyzed using only KVL, in some situations the analysis is simplified by using the other conservation law that is based on the conservation of charge. This law is known as KCL and states that the sum of currents into a connection point (otherwise known as a *node*) must be zero:

$$\sum_{\text{Node}} i = 0 \qquad \text{[Eq. 5.2]}$$

In other words, what goes in, must come out (with respect to charge at a connection point). For example, consider the three currents going into the connection point or node in Figure 5.2. According to KCL, the three currents must sum to zero: $i_1 + i_2 + i_3 = 0$. Of course we know that one, or maybe two, of the currents are actually flowing out of the node, but this just means that one (or two) of the current values will be negative.

Either of these laws can be employed in the service of network analysis. When KVL is applied the analysis is termed *mesh analysis* (a mesh is a specific type of loop) while when KCL is applied the analysis is termed *nodal analysis*. Mesh analysis is introduced here first and is better for circuits with many connection points, but not so many loops.

5.1.1   Mesh Analysis: Single Loops

Which law is most appropriate depends on which results in the fewest number of equations that need to be solved, and this depends on the circuit configuration. If there are many loops in a circuit but only a few nodes, using KCL will lead to fewer equations. A circuit consisting of many nodes but few loops is best approached by applying KVL in mesh analysis. The terminology makes more sense when you understand that a *mesh* is just a technical word for circuit loop. A detailed, step-by-step example of mesh analysis based on KVL is given in Example 5.1.

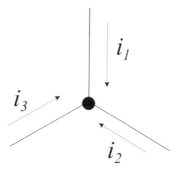

$$i_1 + i_2 + i_3 = 0$$

Figure 5.2 Illustration of Kirchhoff's current law. The sum of the three currents flowing into the connection point, or node, must be zero. In reality, one (or two) of these currents will actually be flowing out of the node and these currents will have negative values.

Example 5.1: Find the voltage across the capacitor in the network below.

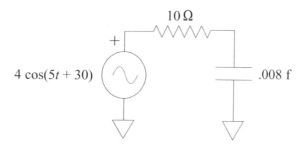

Solution: The circuit has one mesh (i.e., loop) and three nodes. Nodal analysis would require the (simultaneous) solution of two equations: the number of nodes −1 because one node can be assumed to be at 0.0 V. Mesh analysis would require the solution of only one equation, the number of meshes or loops, making it the obvious choice. In mesh analysis, you write an equation based on voltages in the mesh (applying KVL), but actually solve for the mesh *current*. Once you find the mesh current you can find the voltage across any element from its voltage-current relationship. (As you might guess, in nodal analysis it is the opposite: you write an equation or equations based on currents, but end up solving for node voltages.) The only trick in mesh analysis is to keep straight the direction, or polarity, of the voltage changes (up or down corresponding to voltage increases or decreases). Actually, this is not too difficult since mesh analysis can be approached in an algorithmic manner, proceeding in a series of simple steps.

Step 1: Apply a network transformation so that all elements are represented by their phasor domain notation. Use Table 4.4 to get the phasor representations of the various elements. In the *transformed circuit*, sources (which must be sinusoidal or phasor techniques do not apply) will be represented by phasor variables such as $V_S \angle\theta$ whereas passive elements will be given their respective phasor impedances: R Ω, $j\omega L$ Ω, or $1/j\omega C$ Ω. Converting elements to the phasor representation would be the first step in any analysis involving phasor techniques. Sometimes voltage sources use RMS values in the phasor domain, but peak-to-peak values will be used in this text as is more common. It really does not matter as long as you are consistent and know which units are being used.

Step 2: In mesh analysis, step 2 consists of defining the mesh current, or currents if more than one loop is involved. The loop in this example is *closed* by the two groundsthat are essentially connected. The mesh current will go completely around the loop in either a clockwise or counterclockwise direction, theoretically your choice. For consistency, in this text we assume current always travels clockwise around the mesh. (Of course, the current might actually be traveling in a direction opposite to that assigned; however, this means that the value obtained for this current will be negative.) Defining the current then defines the voltage polarities for the passive elements, since current must flow into the positive side of a passive element. Remember, the voltage *source* does not care about current and comes with its polarity already assigned. Application of Steps 1 and 2 lead to the modified circuit shown.

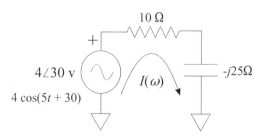

Example 5.1B   The circuit after steps 1 and 2. The elements and source have been converted to their phasor representation (*ohm symbol*) and the mesh current, $I(\omega)$, has been assigned.

Step 3: Now we simply go around the mesh summing the voltages, but it is an *algebraic* summation. Assign positive values if there is an increase in voltage and negative values if there is a decrease in voltage. Start at the lower left corner (below the source) and mentally proceed around the loop in a clockwise direction: traversing the source leads to a voltage rise so this entry is positive, the next two components encountered have a voltage drop (from + to −) so their entries will be negative:

$$V_S - V_R - V_C = 0$$

Substituting in: $V_S = 4 \angle 30; \ V_R = iR;$ and $V_C = \dfrac{I}{j\omega C}$

$$4 \angle 30 - I(\omega)R - \frac{I(\omega)}{j\omega C} = 4 \angle 30 - \left(R + \frac{1}{j\omega C} \right)I(\omega) = 4 \angle 30 - (10 - j25)I(\omega) = 0$$

Step 4: Solve for the current. Put the source(s) on one side and the terms for the passive elements on the other. Then solve for $I(\omega)$:

$$4 \angle 30 = (10 - j25)I(\omega); \quad I(\omega) = \frac{4 \angle 30}{10 - j25}$$

To divide (or multiply) complex numbers, put terms in polar notation:

$$I(\omega) = \frac{4 \angle 30}{27 \angle -68} = 0.148 \angle 98 \, \text{A}$$

Step 5: Solve for the voltage of interest. We want the voltage across the capacitor which, from the equation that defines a capacitor, is $V_C(\omega) = (1/j\omega C) \, I(\omega)$. Substituting our solution for $I(\omega)$ above:

$$V_C = \left(\frac{1}{j\omega C} \right)I(\omega) = (-j25)0.148 \angle 98 = (25 \angle -90)0.148 \angle 98 = 3.7 \angle 8 \, \text{V}$$

There are times when you do not know the specific values of R, L, or C or when several different values may be substituted into the circuit. In such situations, it is necessary to write the equations in terms of general R's, L's, or C's, and possibly a general V_S. However, when the values are known, it is advantageous to substitute these values into the equations as early as possible, as this simplifies the complex arithmetic and leads to fewer errors.

Example 5.2: Example of the general solution of an *RLC* circuit. Find the general solution for V_{out} in the circuit below. The arrows on either side of V_{out} indicate that this voltage is to be taken as the voltage across the capacitor. As is often the case, the connection between ground voltage points is not explicitly shown.

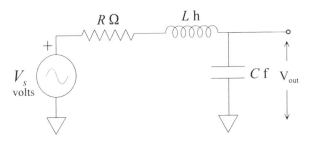

Steps 1 and 2   lead to the circuit below. Passive elements carry the units of ohm, which help remind us that we are now in the phasor domain.

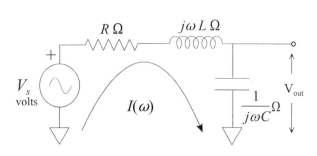

Step 3:   Now write the basic equation going around the loop:

$$V_S(\omega) - RI(\omega) - j\omega LI(\omega) - \frac{I(\omega)}{j\omega C} = V_S(\omega) - I(\omega)\left(R + j\omega L + \frac{1}{j\omega C}\right) = 0$$

Step 4:   Solve for $I(\omega)$:

$$I(\omega) = \frac{V_S(\omega)}{R + j\omega L + \dfrac{1}{j\omega C}}$$

To clean things up a bit, clear the fraction in the denominator and rearrange it into real and imaginary parts. Note $j^2 = -1$:

$$I(\omega) = \frac{V_S(\omega)j\omega C}{R(j\omega C) + j\omega L(j\omega C) + 1} = \frac{V_S(\omega)j\omega C}{1 - \omega^2 LC + j\omega RC}\ A$$

Step 5:   Now to find V_{out}, multiply $I(\omega)$ by the capacitance impedance, $1/j\omega C$:

$$V_{out}(\omega) = \frac{V_S(\omega)j\omega C}{1 - \omega^2 LC + j\omega RC}\left(\frac{1}{j\omega C}\right) = \frac{V_S(\omega)}{1 - \omega^2 LC + j\omega RC}\ V$$

To find a specific value for V_{out}, it is necessary to put in specific values for R, L, and C as well as for V_S (which would also specify ω). However, much can be learned from the general equation above as will be shown in Chapter 6.

The network in this example can also be viewed as a linear process or input–output function, where the voltage source is the input and V_{out} is the output (Figure 5.3).

From this viewpoint, it is possible to represent the network by a single equation that quantitatively describes the output $V_{out}(\omega)$ to any input $V_S(\omega)$. This equation relates the output to the input as a ratio and is known as the *transfer function* of the linear process. If we really want to be precise, the term *transfer function* should only be used for a function that is written in terms of the Laplace variables as described in Chapter 8. However, the concept is so powerful that it is used to describe almost any input–output relationship, even qualitative relationships. To find the transfer function for this network when viewed as a linear process (Figure 5.3), simply divide both sides of the equation above by $V_S(\omega)$.

$$\frac{V_{out}(\omega)}{V_S(\omega)} = \frac{1}{1 - \omega^2 LC + j\omega RC} \qquad \text{[Eq. 5.3]}$$

The transfer function completely defines the input–output relationship of this process. Although this transfer function is limited to sinusoid and periodic functions, the concept will be expanded to cover just about any input function. Implicit assumptions in the transfer function concept are that the input is an *ideal voltage source* and nothing is connected to the output: *nothing* meaning that *no current flows* out of the output terminals (also known as an *ideal load*). Of course, in practice, neither of these assumptions can be true, but in many electronics circuits the conditions are close enough to these idealizations that 'for all practical purposes' the assumptions are valid. The transfer function is fully explored in the next chapter.

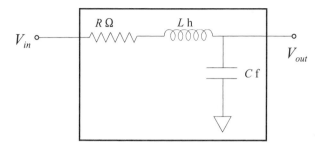

Figure 5.3 The network used in Example 5.2 viewed as a linear process or *input–output mechanism*. The input and output voltages are assumed to be referenced to ground.

5.1.2 Mesh Analysis: Multiple Loops

Any single-loop circuit can be solved using this five-step process and a surprising number of useful circuits consist of only a single loop. Nonetheless, it is not difficult to extend the approach to contend with two or more loops, although the complex arithmetic can become tedious for three loops or more. (This is not really a problem because MATLAB can handle the necessary math for a large number of loops without breaking a sweat.) The example below uses the five-step approach to solve a two-loop network and indicates how larger networks can be solved.

Example 5.3: Find the voltage across the capacitor in the network below. In this circuit, there are no grounds shown; the elements are simply shown as connected together. This convention is less common than the one that uses ground connections.

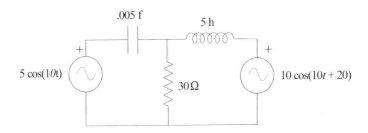

Step 1: Represent all elements by their equivalent phasor representation. This step is always the same in any analysis. Note that the capacitor impedance in phasor notation becomes:

$$\frac{1}{j\omega C} = \frac{1}{j(10)0.005} = -j20\Omega$$

Step 2: Define the *mesh* currents. This step is essentially the same as for single loop circuits. The only trick is that there is a mesh current for each loop and that current is limited, by definition, to its specific loop. So the left hand most loop has a mesh current labeled $I_1(\omega)$ and is strictly limited to go around that loop. Of course, real currents would not be limited to individual loops in such an organized fashion, but these currents are simply constructs that aid in solving the multi-loop problems. Nonetheless, in a two-mesh circuit, the two mesh currents can account for all possible currents in the circuit. For example, the current through the 30-Ω resistor would be the difference between the two mesh currents, $I_1(\omega) - I_2(\omega)$. This is no problem as long as this difference current is used when solving for the voltage across that resistor.

Steps 1 and 2 lead to the circuit below.

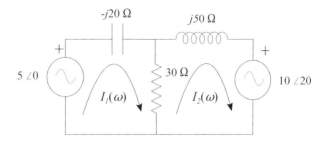

Step 3:   Apply KVL around each loop, keeping in mind that the voltage drop (or rise) across the resistor shared by both meshes will be due to two currents, and since the currents are flowing in opposite directions their voltage contributions will have opposite signs: $I_1(\omega)$ will produce the usual voltage drop, but $I_2(\omega)$ will give rise to a voltage increase (going clockwise around the loop) because it is flowing into the bottom of the resistor. The equations developed from each loop become:

Mesh 1 KVL following standard procedure, beginning in the lower left-hand corner.

$$5 \angle 0 - (-j20)I_1(\omega) - 30I_1(\omega) + 30I_2(\omega) = 0$$

Mesh 2 KVL using the same procedure:

$$30I_1(\omega) - 30I_2(\omega) - j50I_2(\omega) - 10 \angle 20 = 0$$

Step 4:   Solve for the current(s). Rearranging the two equations, placing current on the right side and sources on the left, and separating the coefficients of the two current variables, gives us two equations to be solved simultaneously. Pay particular attention to keeping the signs straight.

$$5 = (30 - j20)I_1(\omega) - 30I_2(\omega)$$

$$-10 \angle 20 = -30I_1(\omega) + (30 + j50)I_2(\omega)$$

With only two equations, it is possible to solve for the currents using substitution, avoiding matrix methods. However, with more than two meshes, matrix methods are easier and lend themselves to computer solutions. (The solution of a three-mesh circuit using MATLAB is given in Example 5.4.) The spacing facilitates transforming the equation into matrix notation for easier solution:

$$\begin{vmatrix} 5 \\ -10 \angle 20 \end{vmatrix} = \begin{vmatrix} 30 - j20 & -30 \\ -30 & 30 + j50 \end{vmatrix} \begin{vmatrix} I_1(\omega) \\ I_2(\omega) \end{vmatrix}$$

Solve for $I_1(\omega)$ using the method of determinants (see Appendix G):

$$I_1(\omega) = \frac{\begin{vmatrix} 5 & -30 \\ -10 \angle 20 & 30 + j50 \end{vmatrix}}{\begin{vmatrix} 30 - j20 & -30 \\ -30 & 30 + j50 \end{vmatrix}} = \frac{5(30 + j50) + 30(-10 \angle 20)}{(30 - j20)(30 + j50) - 30(30)}$$

$$I_1(\omega) = \frac{150 + j250 + 30(-9.4 - j3.4)}{900 + j1,500 - j600 - j^2 1,000 - 900}$$

$$I_1(\omega) = \frac{-132 + j148}{1,000 + j900} = \frac{198 \angle 132}{1,345 \angle 42} = 0.147 \angle 90 \, \text{amps}$$

Even this relatively simple two-mesh circuit involved considerable complex arithmetic with multiple conversions between polar and rectangular form and some tricky sign changes [e.g., $-j20(j50) = -j^2 1,000 = +1,000$].

Step 5: Solve for the voltage of interest, the capacitor voltage.

$$V_C(\omega) = \frac{1}{j\omega C} I_1(\omega) = -j20(0.147 \angle 90) = 2.94 \angle 0 \, \text{volts}$$

This solution could easily be converted to the time domain if required.

5.1.2.1   Shortcut Method for Multimesh Circuits

A shortcut method enables writing the matrix equation directly from inspection of the circuit. Regard the matrix equation and circuit of the example above:

$$\begin{vmatrix} 5 \\ -10 \angle 20 \end{vmatrix} = \begin{vmatrix} 30 - j20 & -30 \\ -30 & 30 + j50 \end{vmatrix} \begin{vmatrix} I_1(\omega) \\ I_2(\omega) \end{vmatrix}$$

Note that it has the general form of $V = ZI$, that is, the left-hand vector contains only the sources, and the right-hand side contains a matrix of impedances and a vector containing only the mesh currents. Moreover, the source vector has the source of mesh 1 in the upper position and the source of mesh 2 in the lower position (with appropriate sign). A similar arrangement holds for the current vector with the mesh 1 current above the mesh 2 current. The impedance matrix also relates topographically to the circuit: the upper left entry is the sum of impedances in mesh 1, the lower right is the sum of impedances in mesh 2, and the off-diagonals (upper right and lower left) contain the *negative* of the sum of impedances common to both loops. In this circuit there is only one element common to both elements, so the off diagonals contain the negative of that one element, but other circuit arrangements could have several elements common to the two meshes. Putting this verbal description into mathematical form:

$$\begin{vmatrix} \sum V_{s\text{mesh}1} \\ \sum V_{s\text{mesh}2} \end{vmatrix} = \begin{vmatrix} \sum Z'_{s\text{mesh}1} & -\sum Z'_{s\text{mesh}1} \\ -\sum Z'_{s\text{mesh}1\&2} & \sum Z'_{s\text{mesh}1} \end{vmatrix} \begin{vmatrix} i_1 \\ i_2 \end{vmatrix} \qquad \text{[Eq. 5.4]}$$

This shortcut method still requires some care, since the summations must take the signs into consideration, particularly the voltage sources. For example, the source in mesh 2 of the example above had a negative sign because it represented a voltage drop when going around the loop clockwise. This shortcut rule can easily be extended to circuits having any number of meshes, although the subsequent calculations become tedious for three or more meshes unless computer assistance is used. The extension to three meshes is given in the example below, but the solution is determined using MATLAB.

5.1.3 Mesh Analysis: MATLAB Implementation

Example 5.4: Solve for V_{out} in the three-mesh network below. This circuit uses more realistic values for R, L, and C.

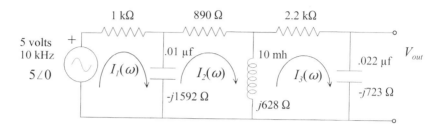

Solution: Follow the steps used in the previous examples, but use the shortcut method in Step 3, In Step 4, solve for the currents using MATLAB.

Steps 1 and 2: The figure shows both the original circuit and the circuit after steps 1 and 2 with the elements represented in phasor representation and the phasor currents defined.

Recall that the impedances for L and C are determined as:

$$Z_L = j\omega L = j2\pi f L = j2\pi 10^4 (10 \times 10^{-3}) = j628 \ \Omega$$

$$Z_{C2} = \frac{1}{j\omega C_2} = \frac{1}{j2\pi f C_2} = \frac{1}{j2\pi 10^4 (0.022 \times 10^{-6})} = \frac{1}{j1.38 \times 10^{-3}} = -j723 \ \Omega$$

$$\text{Similarly: } Z_{C2} = \frac{1}{j2\pi f C_2} = \frac{1}{j2\pi 10^4 (0.01 \times 10^{-6})} = -j1592 \ \Omega$$

Step 3: The matrix equation for step 3 is an extension of Eq. 5.4 where the voltage and current vector would have three elements each and the impedance matrix would be extended to a 3×3 matrix written as:

$$\begin{vmatrix} \sum Z_{mesh1} & -\sum Z_{mesh1,2} & -\sum Z_{mesh1,3} \\ -\sum Z_{mesh1,2} & \sum Z_{mesh2} & -\sum Z_{mesh2,3} \\ -\sum Z_{mesh1,3} & -\sum Z_{mesh2,3} & \sum Z_{mesh3} \end{vmatrix} \qquad \text{[Eq. 5.5]}$$

where ΣZ_{mesh1}, ΣZ_{mesh2}, and ΣZ_{mesh3} are the sum of impedances in the three meshes; $\Sigma Z_{mesh1,2}$ is the sum of impedances common to meshes 1 and 2; $\Sigma Z_{mesh1,3}$ is the sum of impedances common to meshes 1 and 3; and $\Sigma Z_{mesh2,3}$ is the sum of impedances common to meshes 2 and 3. The impedance matrix has symmetry about the diagonals. Such symmetry is often found in matrix algebra and a matrix with this symmetry is termed a *Toplitz matrix*. In this particular network, there are no impedances common to meshes 1 and 3, but the MATLAB section of the problems has an example of a three-mesh circuit in which all three meshes have elements in common.

Applying Eq. 5.5 and the extension of Eq. 5.4 that includes voltages and currents gives rise to the matrix equation for this network:

$$\begin{vmatrix} 5 \\ 0 \\ 0 \end{vmatrix} = \begin{vmatrix} 1{,}000 - j1{,}592 & j1{,}592 & 0 \\ j1{,}592 & 890 - j964 & -j628 \\ 0 & -j628 & 2{,}200 - j95 \end{vmatrix} \begin{vmatrix} I_1(\omega) \\ I_2(\omega) \\ I_3(\omega) \end{vmatrix}$$

Step 4:   Solve for the currents, in this case $I_3(\omega)$. The MATLAB program first defines the voltage vector and impedance matrix:

```
V = [5 0 0]';               % Note the use of the
                            % transpose symbol
Z= [1000 - j*1592, j*1592, 0; j*1592, 890-j*964;...
   -j*628; 0, -j*628, 2200-j*95];
I = Z\V                     % Solve for the currents
Vout = I(3)*(-j*723)        % Output the requested voltage
Vmag = abs(Vout)            % also as magnitude and phase
Vphase = angle(Vout)
```

Results: The output of this program is:

```
I =  0.0021 + 0.0004i
     0.0019 - 0.0015i
     0.0004 + 0.0005i
Vout = 0.3954 - 0.2829i Vmag = 0.4862 Vphase = -0.6211
```

Some of the extra lines and spaces generated by MATLAB have been eliminated. MATLAB accepts i or j to represent an imaginary number, but outputs using i. If you use either i or j as program variables, this will take priority over their representation as imaginary, so it is best to use capital I for current and avoid using i or j as program variables. Finally, most versions of MATLAB do not require a multiplier sign if the i or j follows the number (i.e., j*1234 ≡ 1234j).

The time domain output is directly determined from the phasor output given above. Note that the output of the MATLAB routine `angle` is in radians and remember that $\omega = 2\pi \times 10^4$ rad/sec:

$$V_{out}(t) = 0.486 \sin(2\pi 10^4 t - 0.62(360/2\pi)) = 0.486 \sin(62{,}832t - 36)\text{V}$$

Analysis: In the MATLAB program the voltage vector is written as a row vector, but it should be a column vector as in Eq. 5.4, so the MATLAB transpose operator (single quote) is used. Alternatively, the voltage vector could have been entered as a column vector directly: `V = [5; 0; 0]`. The second line defines the impedance matrix using standard MATLAB notation. The third line solves for the three currents by matrix inversion, implementing the equation $I = Z^{-1} V$ using the backslash (\) operator. The fourth line multiplies the third mesh current, $I_3(\omega)$, by the capacitor impedance to get V_{out}. The next two lines convert V_{out} into polar form.

5.2 CONSERVATION LAWS: KIRCHHOFF'S CURRENT LAW—NODAL ANALYSIS

Kirchhoff's current law can also be used to analyze circuits. This law, based on the conservation of charge, was given in Eq. 5.2 and is repeated here:

$$\sum_{\text{Node}} i = 0 \qquad \text{[Eq. 5.6]}$$

KCL is best suited to analyzing circuits with many loops, but only a few connections points. Figure 5.4 shows the Hodgkin–Huxley model for nerve membrane. The three voltage-resistor combinations represent the potassium membrane channel, the sodium membrane channel, and the chloride membrane channel while C is the membrane capacitance. Analyzing this circuit would require four mesh equations, but only one nodal equation. In this model, most of the components are nonlinear, at least during an action potential, so the model could not be solved analytically as is done with our linear processes. Nonetheless, the defining or *governing* equation(s) would be generated using nodal analysis and could be solved using computer simulation.

Another example of a circuit appropriate for nodal analysis is shown in Figure 5.5. This circuit has four meshes and mesh analysis would give rise to four

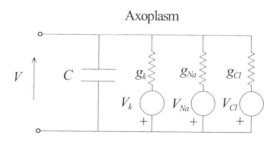

Figure 5.4 Model of nerve membrane developed by Hodgkin and Huxley. The three voltage-resistor combinations represent ion channels in the membrane that sustain the resting voltage and mediate an action potential. The equation describing this model is best developed using Kirchhoff's current law and nodal analysis.

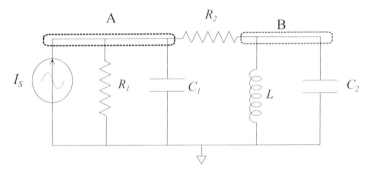

Figure 5.5 A circuit consisting of four meshes, but only two nodes. The nodes are the connection points (**A and B**). Nodes include all the connections that are at the same voltage as indicated by the dashed lines. The ground point (*line across the bottom*) is not considered an independent node since its voltage is, by definition, fixed at zero. Nodal analysis works with currents using Kirchhoff's current law and is easiest if the sources are current sources.

simultaneous equations. This same circuit only has two nodes (marked A and B, again ground points do not count), and would require solving only two nodal equations. If MATLAB is used, solving a four-equation problem is not much more difficult than solving a two-equation problem. However, when circuits are used as models representing physiological processes as in Figure 5.4, the more concise description given by nodal equations is of great value. This circuit contains a current source as opposed to the voltage sources that have become familiar. This is because nodal analysis is an application of a current law so it is easier to implement if the sources are current sources. A similar statement could be made about mesh analysis: mesh analysis involves voltage summation and it is easier to implement if all sources are voltage sources. The need to have only current sources may seem like a drawback to the application of nodal analysis, but we will see in Chapter 7 that it is easy to convert voltage sources to equivalent current sources and vice versa, so this requirement is not really a handicap. In this chapter, nodal analysis examples use current sources with the understanding that the technique can be applied equally well to voltage sources after a simple conversion.

Analyzing circuits using nodal analysis follows the same five-step procedure used in mesh analysis. In fact, steps 4 and 5 are the same. Step 1 could also be the same, but often elements are converted to $1/Z$, instead of simply Z. The inverse impedance, $Y = 1/Z$, is termed the *admittance*. In step 2, node voltages are assigned rather than the mesh currents, and in step 3 the equations are generated using KCL.

The equations developed from KCL have a sort of inverse symmetry with those of mesh analysis. In mesh analysis, we write matrix equations of the form:

$$v = Zi \qquad \text{[Eq. 5.7]}$$

where v is a voltage vector, i is a current vector, and Z is an impedance matrix (see Eq. 5.4). In nodal analysis, we are writing matrix equations in the form:

$$i = Yv \qquad \text{[Eq. 5.8]}$$

where Y is a matrix, termed the *admittance matrix* containing the inverse of the impedances. The terms v and I are vectors as in Eq. 5.7.

Example 5.5: Find the voltage, V_A, in the circuit below.

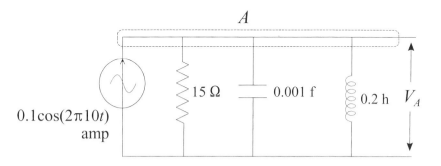

Solution: This circuit would require two mesh equations and the conversion of the current source to an equivalent voltage source (as explained in Chapter 7). Using nodal analysis, no conversion is required and only one nodal equation is required to find the required voltage. There are four currents flowing into or out of the single node at the top of the circuit labeled A. The current in the current source branch (left side) will be $0.1 \cos(2\pi 10t)$ and the current in the other three branches will be equal to the voltage, V_A, divided by the impedance of the branch [i.e., $I(\omega) = V_A(\omega)/Z_{Branch}$]. By KCL, these four currents will sum to zero.

After applying steps 1 and 2, the network becomes as shown below. If we define $V_A(\omega)$ as a positive voltage, the current though the passive elements will be flowing downward as shown due to the voltage-current polarity rule for passive elements. In addition, the frequency becomes: $\omega = 2\pi f = 62.8$ rad/sec.

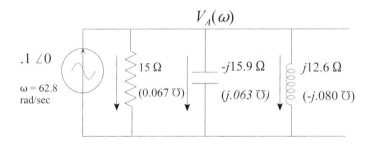

Step 3:   Apply the fact that the four currents sum to zero:

$$i_{S(\omega)} - i_{R(\omega)} - i_{C(\omega)} - i_{L(\omega)} = 0 \quad \text{(KCL)}$$

$$I_S - \frac{V_A(\omega)}{R} - \frac{V_A(\omega)}{1/j\omega C} - \frac{V_A(\omega)}{j\omega L} = 0$$

$$0.1 - \frac{V_A(\omega)}{15} - \frac{V_A(\omega)}{-j15.9} - \frac{V_A(\omega)}{j12.6} = 0$$

Now we can solve this single equation for $V_A(\omega)$. The equation would be easier if written in terms of admittances: $Y = 1/Z$. The values of the admittances are shown in parentheses in the circuit above. Using admittances, the nodal equation becomes:

$$I_S - Y_R V_A(\omega) - Y_C V_A(\omega) - Y_L V_A(\omega) = 0$$

$$I_S - V_A(\omega)(1/R + j\omega C + 1/j\omega L) = 0$$

$$0.1 - V_A(\omega)(0.067 + j0.063 - j0.080) = 0$$

$$V_A(\omega) = \frac{0.1}{0.067 + j0.063 - j0.080} = \frac{0.1}{0.067 - j0.017}$$

$$V_A(\omega) = \frac{0.1}{0.069 \angle -14} = 1.45 \angle 14 \,\text{volts}$$

Moving to multinodal systems, we go directly to the shortcut, matrix equation. If we were to apply KCL to circuits we would find that equations would fall into a pattern similar to that described by Eq. 5.4, except that it would have the specific form of Eq. 5.8, where: the admittance matrix would consist of the summed admittances common to each node along the diagonal and the negative summed admittances between nodes on the off-diagonals. This general format is shown here for a three-node circuit:

$$\begin{vmatrix} \sum I_1 \\ \sum I_2 \\ \sum I_3 \end{vmatrix} = \begin{vmatrix} \sum Y_{node1} & -\sum Y_{node1,2} & -\sum Y_{node1,3} \\ -\sum Y_{node1,2} & \sum Y_{node2} & -\sum Y_{node2,3} \\ -\sum Y_{node1,3} & -\sum Y_{node2,3} & \sum Y_{node3} \end{vmatrix} \begin{vmatrix} V_1 \\ V_2 \\ V_3 \end{vmatrix} \qquad \text{[Eq 5.9]}$$

The application of Eq. 5.9 is straightforward and follows the same pattern as in mesh analysis. An example of nodal analysis to the two-node circuit is given in Example 5.6.

Example 5.6:   Find the voltage, v_2 in the circuit shown here. This circuit is similar to the one shown in Figure 5.5 except an additional component has been added between the two nodes.

Solution: Apply nodal analysis to this two-node circuit. Follow the step-by-step procedure outlined above, but in Step 3 write the matrix equation directly as given in Eq. 5.9. Implement Step 4 to solve for v_2 using MATLAB.

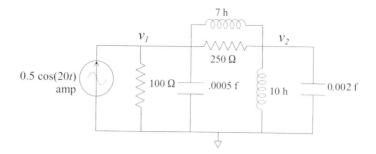

Step 1: Convert all the elements to phasor admittances. Note that $\omega = 20$ rad/sec.

Step 2: Assign nodal voltages. This has already been done in the circuit. Following these two steps, the circuit becomes:

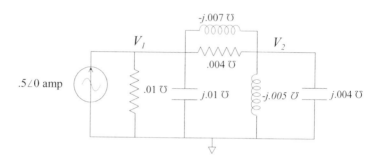

Step 3: Generate the matrix equations directly following Eq. 5.9. Since we are using admittances, inductors now have $-j$ values while conductors have a $+j$ values. In addition, because the two nodes share two components, the shared admittance will be the sum of the admittances from each component:

$$\sum Y_{node1,2} = 0.004 - j.007$$

Hence, the circuit equation becomes:

$$\begin{vmatrix} 0.5 \\ 0 \end{vmatrix} = \begin{vmatrix} 0.01 + 0.004 + j.01 - j.007 & -0.004 + j.007 \\ -0.004 + j.007 & 0.004 - j.005 - j.007 + j.004 \end{vmatrix} \begin{vmatrix} V_1 \\ V_2 \end{vmatrix}$$

$$\begin{vmatrix} 0.5 \\ 0 \end{vmatrix} = \begin{vmatrix} 0.014 + j.003 & -0.004 + j.007 \\ -0.004 + j.007 & 0.004 - j.008 \end{vmatrix} \begin{vmatrix} V_1 \\ V_2 \end{vmatrix}$$

Step 4: This matrix equation could be solved using determinants, but is even easier to solve using MATLAB as illustrated by the code below.

```
% Example 5.6 Solution of 2 node matrix equation
%
%Assign current vector and admittance matrix
i = [0.5; 0];
G = [.014+.003j -.004+.007j; -.004+.007j .004-.008j];
v = G\i;
disp([abs(v(2)) angle(v(2))*360/(2*pi)])
```

The output of the is program, the magnitude and phase of v_2 is:

```
32.2425 -38.9801
```

Hence, in the time domain:

$$v_2(t) = 32.2\cos(20t - 139)\text{V}$$

This approach could be extended to three-nodes or even higher node circuits without a great deal of additional difficulty. A three-node problem is given at the end of the chapter.

The basic five-step approach to analyzing electrical circuits can be used to analyze lumped-parameter mechanical systems as described in the next section.

5.3   CONSERVATION LAWS: NEWTON'S LAW–MECHANICAL SYSTEMS

The analysis of lumped-parameter mechanical systems also uses a conservation law, one based on conservation of energy; specifically, the sum of the forces around any one connection point must be zero. This is a form of the classic law associated with Newton:

$$\sum_{\text{Point}} F = 0 \qquad\qquad \text{[Eq. 5.10]}$$

In this application, a connection point includes all connections between mechanical elements that are at the same *velocity*. Figure 5.6 shows a linear model for skeletal muscle similar to the one presented in Chapter 1 (Figure 1.6). Skeletal muscle has two different elastic elements, a parallel elastic element, C_p and a series elastic element, C_s. (Note the symbol C indicates they are given as compliances, $1/k_e$.) The force generator, F_o, represents the active contractile element and k_f represents viscous properties inherent in the muscle tissue. The muscle model has two connection points that could have different velocities, labeled *Point 1* and *Point 2*. The positive force is defined inward reflecting the fact that muscles can only generate contractile force. This is the reason they are so often found in agonist–antagonist pairs.

Since this system has two different velocities, its analysis would require the simultaneous solution of equations similar to those of Eq. 5.6. The system is the mechanical equivalent of a two-mesh electrical circuit. The two equations would be written around points 1 and 2: the sum of forces around each point must be zero. The graphic on the left represents a zero-velocity point or a solid wall, the analog of a ground point in an electrical system.

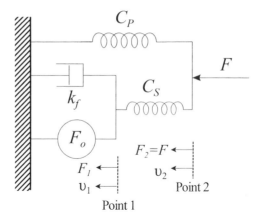

Figure 5.6 Linear mechanical model of skeletal muscle. F_o is the force produced by the active contractile element, C_p and C_s are known as the parallel and series elasticity, and k_f is viscous damping associated with the tissue.

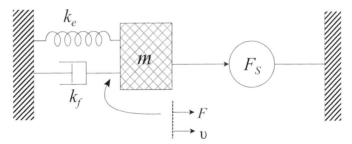

Figure 5.7 A mechanical system with one independent velocity v used in Example 5.7. The connections on either side of the mass are essentially the same point since they are at the same velocity.

The muscle model will be analyzed in a subsequent example, but for a first example of the application of Newton's law (Eq. 5.10), we will analyze a less complicated, one-equation system with single velocity point shown in Figure 5.7. Recall that in mechanical systems, friction elements are analogous to resistors, elastic elements are analogous to capacitors, and masses are analogous to inductors.

Example 5.7: Find the velocity and displacement of the mass in the mechanical system shown in Figure 5.7. The force, $F_S(t) = 5 \cos(2t + 30)$ dyne. The following parameters also apply:

$$k_f = 5\,\text{dyne-sec/cm}; \quad k_e = 8\,\text{dyne/cm}; \quad m = 3\,\text{gm}.$$

Solution: In this example, the units are comparable and are in the cgs (centimeters, grams, dynes) metric system, the standard for this text. (In Example 5.9, considerable conversion of units is required.) To analyze this system, we follow the same five-step plan developed for electric circuits.

Step 1:   Convert variables to phasor notation and represent the passive elements by their phasor impedances. Because $\omega = 2$:

$$Z_f = k_f = 5; \quad Z_e = \frac{k_e}{j\omega} = \frac{8}{j2} = -j4; \quad Z_m = j\omega m = j2(3) = j6$$

(The units for all these impedances would be dyne-sec/cm.)

Step 2:   Assign variable directions. In mechanical systems, we will use the convention of assigning the force and velocity in the *same* direction, but the direction (right or left) will be arbitrary (in this example it is to the right). This is analogous to assigning currents as counterclockwise and keeping track of voltage polarities by going in the same direction. By assigning force and velocity in the same direction, the force polarity of passive elements will *always* be negative just as in electric circuits, so the equations will look similar. Note that the velocity on both sides of the mass is the same.

Step 3:   Apply Newton's law about the point(s) around the mass:

$$\sum F = 0; \quad F_S(\omega) - k_f v(\omega) - \frac{k_e}{j\omega} v(\omega) - j\omega m v(\omega) = 0$$

$$5 \angle 30 - v(\omega)(5 - j4 + j6) = 0$$

The first three steps follow a path parallel to that followed in circuit analysis while the last two steps are essentially identical: solve for the velocity, then any other variable of interest such as a force or, in this case, a displacement.

Step 4:   Solve for the phasor velocity:

$$v(\omega) = \frac{5 \angle 30}{5 - j4 + j6} = \frac{5 \angle 30}{5 + j2} = \frac{5 \angle 30}{5.39 \angle 21.8} = 0.93 \angle 8.2 \, \text{cm/sec}$$

Step 5:   Solve for displacement. Since $x(t) = \int v \, dt$; and integration in the phasor domain is division by $j\omega$; then: $x(\omega) = v(\omega)/j\omega$.

$$x(\omega) = \frac{v(\omega)}{j\omega} = \frac{0.93 \angle 8.2}{j2} = \frac{0.93 \angle 8.2}{2 \angle 90} = 0.46 \angle -81.8 \, \text{cm}$$

Both $v(\omega)$ and $x(\omega)$ can be converted to the time domain if desired:

$$v(t) = 0.93 \cos(2t + 8.2) \, \text{cm/sec}$$

$$x(t) = 0.46 \cos(2t - 81.8) \, \text{cm}$$

The next example is more complicated because there are two summation points in the problem so there will be two equations that must be solved simultaneously.

Example 5.8: Find the force out of the skeletal muscle model in Figure 5.6.

Solution: After converting to the phasor domain, write Newton's law (Eq. 5.10) around points 1 and 2. In this solution, the algebra is a bit tedious because the various parameters remain as variables, but the procedure is otherwise straightforward.

Steps 1 and 2: The force and velocity assignments are given in Figure 5.7. In this model the letter C, which stands for compliance, is used to represent elastic components. To be consistent with other section of the text, k_{eP} and k_{eS} will be used. The various components transformed to the phasor domain become:

$$F_o \to F_o(\omega); \quad k_f \to k_f; \quad k_{ep} \to \frac{1}{j\omega k_{ep}}; \quad k_{eS} \to \frac{1}{j\omega k_{eS}}$$

Step 3a: Write the equations about the two points. The equation around point 1 is:

$$F_o - k_f v_1(\omega) - \frac{1}{j\omega k_{eS}}(v_1(\omega) - v_2(\omega)) = 0$$

The force generated by the elastic element k_{eS} depends on the difference in velocities between points 1 and 2. (Specifically, it depends on difference in the positions of the points, which is the integral of the difference in the two velocities.) The force that is generated by $v_1(\omega) - v_2(\omega)$ will be in the opposite direction of F_1, which accounts for the negative sign in front of this term.

Step 3b: The equation around point 2 is:

$$0 - \frac{1}{j\omega k_{eP}}v_2(\omega) - \frac{1}{j\omega k_{eS}}(v_2(\omega) - v_1(\omega)) = 0$$

The leading zero term just states that there are no active force generators connected to this node, only passive elements. Rearranging to separate coefficients of $v_1(\omega)$ and $v_2(\omega)$:

$$F_o = \left(k_f + \frac{1}{j\omega k_{eS}}\right)v_1(\omega) - \frac{1}{j\omega k_{eS}}v_2(\omega)$$

$$0 = \frac{1}{j\omega k_{eS}}v_1(\omega) + \left(\frac{1}{j\omega k_{eS}} + \frac{1}{j\omega k_{eP}}\right)v_2(\omega)$$

Step 4:   Solving $v_2(\omega)$:

$$v_2(\omega) = \frac{\begin{vmatrix} k_f + \dfrac{1}{j\omega k_{eS}} & F \\[2ex] -\dfrac{1}{j\omega k_{eS}} & 0 \end{vmatrix}}{\begin{vmatrix} k_f + \dfrac{1}{j\omega k_{eS}} & -\dfrac{1}{j\omega k_{eS}} \\[2ex] -\dfrac{1}{j\omega k_{eS}} & \dfrac{1}{j\omega k_{eS}} + \dfrac{1}{j\omega k_{eP}} \end{vmatrix}}$$

$$= \frac{F_o\left(\dfrac{1}{j\omega k_{eS}}\right)}{\dfrac{k_f}{j\omega k_{eS}} + \dfrac{k_f}{j\omega k_{eP}} + \left(\dfrac{1}{j\omega k_{eS}}\right)^2 + \dfrac{1}{(j\omega)^2 k_{eP} k_{eS}} - \left(\dfrac{1}{j\omega k_{eS}}\right)^2}$$

Canceling the two terms and multiplying through by j^2 and ω^2:

$$v_2(\omega) = \frac{j\omega F_o\left(\dfrac{1}{k_{eS}}\right)}{\dfrac{1}{k_{eP} k_{eS}} + j\omega\left(\dfrac{k_f}{k_{eS}} + \dfrac{k_f}{k_{eP}}\right)} = \frac{j\omega F_o(k_{eP})}{1 + j\omega(k_f k_{eS} + k_f k_{eP})}$$

Step 5:   Now to find the force at the output, multiply $v_2(\omega)$ by the impedance of the parallel elastic element.

$$F(\omega) = v_2(\omega)\left(\dfrac{1}{j\omega k_{eP}}\right) = \frac{j\omega F_o(k_{eP})}{1 + j\omega(k_f k_{eS} + k_f k_{eP})}\left(\dfrac{1}{j\omega k_{eP}}\right) = \frac{F_o}{1 + j\omega(k_f k_{eS} + k_f k_{eP})}$$

In the next chapter, we will find that this has the properties of a lowpass filter. While real muscle can be fairly well described by the elements in Figure 5.6, the equation does not take into account the component nonlinearities. Nonetheless, this linear analysis provides a starting point for models that are more complicated.

The next example solves a two-equation system that includes a mass. Unlike previous examples, the values of the elements are not all given in cgs units so it is first necessary to convert all parameter to the cgs system.

Example 5.9:   Find the velocity of the mass in the system shown in the figure below. Assume that $F_S(t) = 0.001 \cos(20t)$ newtons and that the following parameter assignments apply:

$$m = 1.0\,\text{oz}; k_{f1} = 200\,\text{dyne-sec/cm};$$

$$k_{e1} = 6,000\,\text{dyne/cm}; k_{e2} = 0.5\,\text{lb/in}.$$

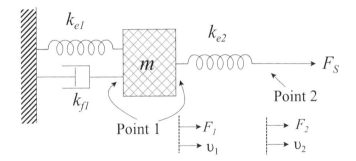

Solution: We first need to convert F_S, m, and k_{e2} to cgs units.[1] Converting F_S is relatively straightforward since it is already in MKS metric units. Using the conversion factors in Appendix D: 1 newton = 10^5 dyne. Hence, $F_S = 100 \cos(20t)$ dyne. To convert m from English units (oz) to cgs metric units, use conversion factors in Appendix D in conjunction with dimensional analysis:

$$m = 1.0 \, oz \frac{28.35 \, gm}{1 \, oz} = 28.35 \, gm$$

$$F_S = 0.001 N \frac{10^5 \, dynes}{1 N} = 100 \, dynes \quad (F(t) = 100 \cos(20t) dynes)$$

$$k_{e2} = 0.5 \, lb/in \frac{1 \, in}{2.54 \, cm} \frac{453.59 \, gm}{1 \, lb} = 89.3 \, gm/cm$$

Both pound and gram are measures of mass, but are often used as a measure of force. The assumption is that a pound weight, or force, is the mass of a pound accelerated by gravity (i.e., $F = mg$). Similarly, a 1 gram weight (gm wt) equals 1 gm × g where $g = 980.665$ cm/sec^2 so as a measure of force, 1 gm wt = 980.665 dyne. (Similarly, a kilogram weight is 9.807 newtons). Applying the conversion factors from pound (lb) weight to dyne and inches to centimeters:

$$k_{e2} = 89.3(980.665) = 87,573 \, dynes/cm$$

Step 1: Following the conversion to cgs units, the determination of the phasor impedances is straightforward. With $\omega = 20$ rad/sec:

$$Z_{f1} = k_{f1} = 200 \, dyne\text{-}sec/cm; \quad Z_m = j\omega m = j20(28.35) = j567 \, dyne\text{-}sec/cm$$

$$Z_{e1} = \frac{k_{e1}}{j\omega} = \frac{6,000}{j20} = -j300 \, dyne\text{-}sec/cm; \quad Z_{e2} = \frac{k_{e2}}{j\omega} = \frac{87,573}{j20} = -j4,378 \, dyne\text{-}sec/cm$$

---

[1] To add to the confusion of unit conversion, there are two units of mass in the English system termed *pounds* and abbreviated lb. The most commonly used pound is termed the *commercial* or *avoirdupois pound* whereas a less commonly used measure is the *troy* or *apothecary pound*. To convert: 1 troy lb = 0.822857 avoirdupois lb. In this text, only avoirdupois pounds are used, but conversions for both can be found in Appendix D.

Step 2:   The force and velocity directions have already been assigned with F_1 and F_2 as positive to the right. After the first two steps, the system becomes:

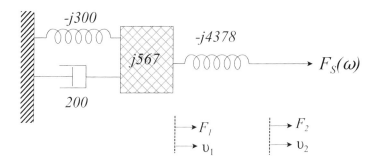

In writing the equations about the two points, we must take into account the fact that the spring on the right side of the mass has nonzero velocities on both sides. Therefore, the net velocity across these two elements is $v_2 - v_1$. Thus, the force across the right-hand spring is $k_{e2}/j\omega(v_2 - v_1)$. With this in mind, the equation around point 2 becomes:

$$F_S - \frac{k_{e2}}{j\omega}(v_2 - v_1) = F_S - (-j4,378)(v_2 - v_1) = 0$$

For point 2, the force exerted by the spring has the same magnitude, but is oppo-site in direction (i.e., positive with respect to F_1). Thus the equation for the force around point 1 is:

$$0 + \frac{k_{e2}}{j\omega}(v_2 - v_1) - j\omega m - k_f - \frac{k_{e1}}{j\omega} = 0$$

$$0 + j4,378(v_2 - v_1) - j567v_1 - 200v_1 - (-j300)v_1 = 0$$

Rearranging the two equations as coefficients of v_1 and v_2:

$$0.0 = (-j4,378 + j567 - j300 + 200)v_1(\omega) + j4,378v_2(\omega)$$

$$100 \angle 0 = j4,378v_1(\omega) - j4,378v_2(\omega)$$

and in matrix notation:

$$\begin{vmatrix} 0 \\ 100 \end{vmatrix} = \begin{vmatrix} 200 - j4,111 & j4,378 \\ j4,378 & -j4,378 \end{vmatrix} \begin{vmatrix} v_1(\omega) \\ v_2(\omega) \end{vmatrix}$$

Solving for v_1, the velocity of the mass:

$$v_1(\omega) = \frac{\begin{vmatrix} 0 & j4,378 \\ 100 & -j4,378 \end{vmatrix}}{\begin{vmatrix} 200 - j4,111 & j4,378 \\ j4,378 & -j4,378 \end{vmatrix}} = \frac{-j437,800}{-j875,600 - 17,997,958 + 19,166,884}$$

$$v_1(\omega) = \frac{j437,800}{1,168,926 - j875,600} = \frac{437,800 \angle 90}{1,146,501 \angle -37} = 0.30 \angle 127 \, \text{cm/sec}$$

Of course, this solution could also be obtained with greater ease using MATLAB.

Example 5.10: Find the output of the network below to a 1-V input at three different frequencies: $f = 6$, 60, and 600 Hz.

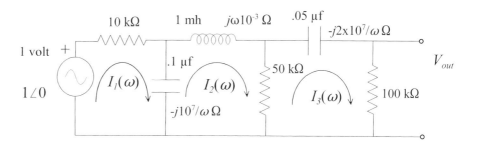

Solution: Since the circuit contains three meshes and must be solved three times at three different frequencies, using MATLAB is a clear choice. The first two steps are included in the figure, and the equations required in step 3 can be written directly into matrix form.

$$\begin{vmatrix} 1 \\ 0 \\ 0 \end{vmatrix} = \begin{vmatrix} 10^4 - \dfrac{j10^7}{\omega} & \dfrac{j10^7}{\omega} & 0 \\ \dfrac{j10^7}{\omega} & \dfrac{-j10^7}{\omega} + 50 \times 10^3 + j\omega 10^{-3} & -50 \times 10^3 \\ 0 & -50 \times 10^3 & 50 \times 10^3 + 10^5 - \dfrac{j2 \times 10^7}{\omega} \end{vmatrix} \begin{vmatrix} I_1(\omega) \\ I_2(\omega) \\ I_3(\omega) \end{vmatrix}$$

The appropriate MATLAB code is as follows:

```
%Example 5.10 Three mesh problem to be solved at three
frequencies
%
clear all; close all;
wn = 2*pi*[6 60 600];                % Desired frequencies
V = [1; 0; 0;];
for k = 1:length(wn)                 % Solve for each frequency
  w = wn(k);                         % Set w
  Z= [10^4 - j*10^7/w, j*10^7/w, 0; j*10^7/w,...
     -j*10^7/w + 50*10^3 + j*w*10^-3, -50*10^3;...
     0, -50*10^3, 50*10^3 + 10^5 - j*2*10^7/w];
  v = Z\V;
```

```
    Vout(k) = v(3)*10^5;              % Save desired output
  end
  disp([abs(Vout) angle(Vout)...    % Output results
    *360/(2*pi)])
```

Results: The output for Vout produced by the code is shown below in tabular form.

Frequency	Magnitude (V)	Phase (deg)
6 Hz	0.1538	76.6656
60 Hz	0.6569	9.8193
600 Hz	0.2501	−67.9720

As expected, the output voltage depends on the frequency and is higher for a 60-Hz signal than for frequencies above or below this frequency. For signals composed of a range of frequencies, this circuit could be used to enhance frequencies around 60 Hz. With only a minor modification, the MATLAB code in this example could be modified to determine Vout over a wide range of frequencies. This task is given for a single mesh circuit in Problem 5.16. The way in which circuits and other systems behave over a range of frequencies is often of considerable interest to bioengineers and is the subject of the next chapter.

5.4 SUMMARY

Conservation laws are invoked to generate an orderly set of descriptive or governing equations from any collection of mechanical or electrical elements. In electric circuits, the law of conservation of energy leads directly to KVL, which states that the voltages around the loop must sum to zero. Combining this rule with the phasor representation of network elements leads to an analysis technique known as mesh analysis. In mesh analysis, equations are constructed in the general from of $v = Zi$ that can be solved for the mesh or loop currents, i. These currents can then be used to determine any voltage in the system. The law of conservation of charge leads to KCL, which can also be used to find the voltages and currents in any network. Application of KCL leads to an equation of the form $i = v/Z$, which is solved for the node voltages, v, and from the node voltages any desired current. This analysis, termed *nodal analysis*, leads to fewer equations in networks that contain many loops, but only a few nodes.

The conservation law active in mechanical systems is Newton's law, which states that the forces on any element must sum to zero. Again using phasor representation of mechanical elements, this law can be applied to generate equations of the form $F = Zv$. These equations are solved for velocities and these velocities can be used to determine all of the forces in the system.

These conservation laws and the analysis procedures they lead to, allow us to develop equations for even very complex electrical or mechanical systems. These equations can then be used to provide very concise representations of even the most complex systems as described in the next chapter.

PROBLEMS

1. In the circuit shown, the voltage across element 1 is: $2 \cos(2t + 60)$
   What is the voltage, V_2, across element 2?

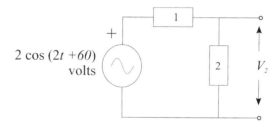

2. Find the voltage across the 50-Ω resistor.

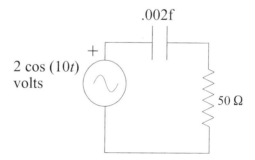

3. The loop current in the circuit below is $0.2 \cos(5t - 53)$. What is element 2 (i.e., R, L, or C) and what is its value?

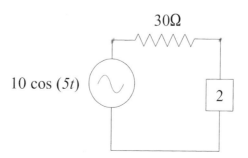

4. Find the voltage across the inductor, v_L, for $\omega = 5$ and 20 rad/sec.

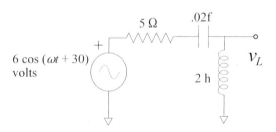

5. Find the voltage across the 5-h inductor in the network below.

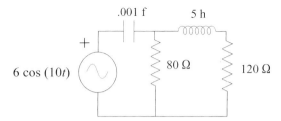

6. What is the voltage across the center 80-Ω resistor in the network given in Problem 5? (Note: The total current through the resistor is $i_R = i_1 - i_2$.)
7. Find the current through the 2-h inductor in the circuit below. Is the voltage source on the right, V_2, supplying energy or storing energy?

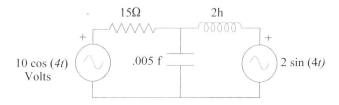

8. Find V_1.

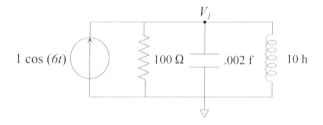

9. Find the voltage across the 10-Ω resistor.

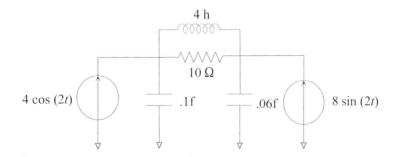

10. Given the mechanical system shown where $k_f = 8$ dyne-sec/cm; $k_e = 12$ dyne/cm; $m = 2$ gm; $F_s(t) = 10 \cos(2t)$ dyne.
    Find the length of the spring when $t = 0.5$ seconds. [*Hint:* Solve for $x(\omega)$ where $x(\omega) = \upsilon(\omega)/j\omega$. Then convert to time domain $x(t)$ and solve for $t = 0.5$ seconds.]

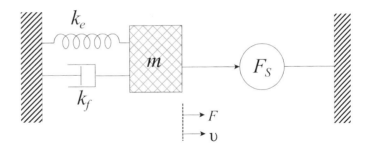

11. In the mechanical system of Problem 10, what must be the value of k_f to limit the maximum velocity to ± 1 cm/sec?

MATLAB Problems

12. Find V_{out} using MATLAB.
    Hint: modify the code in Ex. 5.4 appropriately.

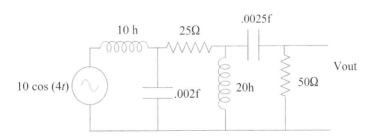

13. Find all voltages using MATLAB and nodal analysis.

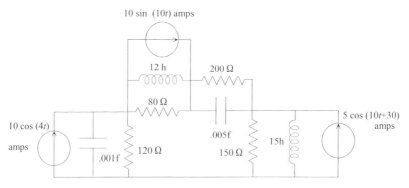

Remember: $10 \sin(10t) = 10 \cos(10t - 90)$.

14. Find v_{out} for the circuit below for frequencies ranging between 1 and 1,000 rad/sec. Plot the magnitude of v_{out} as a function of frequency. [*Hint:* write the MATLAB impedance matrix in terms of a variable frequency, set up ω as a vector, then solve for v_{out}. The vector ω should range between 1 and 1,000 rad/sec in increments of 1 rad/sec. Then plot abs(vout(ω)).] You are able to do this because of a principle known as *superposition*. Basically, you are solving the problem a number of times each at a different frequency. The fact that the overall answer can be obtained as just the sum of all the individual solutions is possible because superposition holds for linear system. This principle is used extensively in the next chapter.

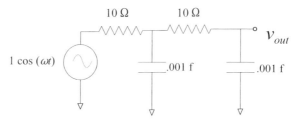

15. Repeat the analysis in Problem 5.14 for the mechanical system shown in Problem 5.10. Use the parameter values: $k_f = 10$ dyne-sec/cm; $k_e = 150$ dyne/cm; $m = 1.5$ gm; $F_s(t) = \cos(\omega t)$.

16. Find and plot $V_{out}(\omega)$ over a range of frequencies from 0.1 to 1000 rad/sec.

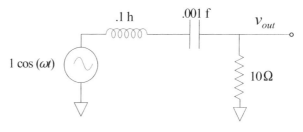

17. Modify Example 5.10 to solve for Vout over a range of frequencies between 0.1 and 500 Hz. Plot the magnitude voltage as a function of frequency. (*Hint:* Change the vector wn in the code of Example 5.10 to: wn = 2*pi*(.1:.1:500);).

6 FREQUENCY CHARACTERISTICS OF CIRCUITS AND ANALOG PROCESSES: THE TRANSFER FUNCTION

6.1 THE CIRCUIT OR MECHANICAL SYSTEM AS A PROCESS

The last chapter described circuits and mechanical systems as self-contained with their own sources and internal voltages. Yet, frequently they are used as modules in a larger system. In such cases, the circuit or mechanical system must have a well-defined input and output. The module can then be thought of in terms of information processing; that is, the module is a process that transforms the input signal to an altered output signal (Figure 6.1).

For electrical circuits, the input and output are usually voltages, whereas for mechanical systems they could be either forces or velocities. When the module represents a biological process, it is not uncommon for inputs and outputs to be different variable types, such as force in and velocity out, or even variables of different energy modalities. For example, the eye movement muscle model shown in Figure 1.7 uses mechanical elements to represent muscle properties, but the input is an electrical signal (agonistic and antagonistic neural signals) and the output is a mechanical variable (eye position).

Whether a circuit or mechanical system is thought of as an input–output process or as a self-contained system, it is analyzed the same way, using the methods

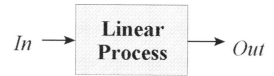

Figure 6.1 A circuit or mechanical system can be viewed as an information processing module, often part of a larger system. The module must have well-defined inputs and outputs, which could be voltages, forces, velocities, or even some mixture of these variables.

described in the previous chapter. When the system is analyzed as an input–output process, two critical assumptions are made: (a) The input is assumed to be an ideal, but otherwise unspecified, source such as a voltage source [the output is then determined by the rules based on conservation laws described previously (KVL, KCL, Newton's law)]; and (b) no energy is supplied by the output to any subsequent processes that may be connected to the output. For example, if the output were a voltage then there would be no current flowing in the output path. When a system is connected between other processes, these two assumptions will not be met exactly, but often they are very close to true. In Chapter 7, we will explore the consequences of interconnecting systems. Operating under these basic assumptions, the method for determining the output for a general input is shown in Example 6.1.

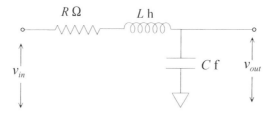

Figure 6.2 An RLC (resistor, inductor, capacitor) circuit arranged as an input–output system. The input is assumed to be an ideal source and the output is assumed to be unconnected.

Example 6.1: Find the output $V_{out}(\omega)$ for the RLC (resistor, inductor, capacitor) circuit shown in Figure 6.2. The input is given only as $V_{in}(\omega)$, but is assumed to be an ideal source. It is also assumed that there is no current flowing out of the circuit; in other words, $V_{out}(\omega)$ is connected to an open circuit (which is the same as saying it is connected to nothing). The values for R, L, and C are: $R = 10 \ \Omega$; $L = 3$ h; $C = 0.01$ f.

Solution: Under these conditions, it is easy to see that $V_{out}(\omega)$ is just the voltage across the capacitor. Under the assumption that $V_{in}(\omega)$ is an ideal source, we can apply KVL to the single loop produced by the source and R, L, and C.

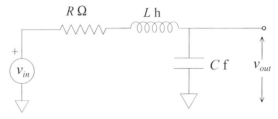

Example 6.1 The RLC circuit of Figure 6.2 with the ideal input explicitly shown. It is clear that this single-loop circuit can be solved with the application of Kirchhoff's voltage law.

Transforming the network into the phasor domain and writing the single-loop KVL equation:

$$V_{in}(\omega) - I(\omega)\left(R + j\omega L + \frac{1}{j\omega C}\right) = 0$$

Substituting in the values of R, L, and C and solving for $I(\omega)$:

$$I(\omega) = \frac{V_{in}(\omega)}{10 + j3\omega + \dfrac{1}{j.01\omega}} = \frac{V_{In}(\omega)j0.01\omega}{1 - 0.03\omega^2 + j.1\omega} \quad \text{(Recall: } j^2 = -1)$$

Substituting in: $V_{out}(\omega) = I(\omega)Z_C(\omega) = I(\omega)\dfrac{1}{j.01\omega}$

$$V_{out}(\omega) = \frac{V_{In}(\omega)j0.01\omega(1/j0.01\omega)}{1 - 0.03\omega^2 + j0.1\omega} = \frac{V_{In}(\omega)}{1 - 0.03\omega^2 + j0.1\omega} V$$

When working with phasors, it is common to normalize this type of equation so that the constant in the denominator is 1.0 as is done here. In addition, when working with analog equations in complex arithmetic, it is best to substitute in the component values, if available, as early as possible into the equations. This generally simplifies the arithmetic and results in fewer errors.

To find a specific $V_{out}(\omega)$ we need to assign a specific $V_{in}(\omega)$ with an associated value of ω. An example of this is shown below.

Example 6.2: Assume the $V_{in}(\omega)$ in the RLC circuit of Figure 6.2 is 4 cos(10t + 30) V. Find $V_{out}(\omega)$ and $v_{out}(\omega)$.

Solution: Substitute in the phasor representation of $v_{in}(t)$ into the solution of Example 6.1 and solve.

$$4\cos(10t + 30) \Leftrightarrow 4\angle 30$$

$$V_{out}(\omega) = \frac{4\angle 30}{1 - 0.03(10)^2 + j.1(10)} = \frac{4\angle 30}{-2 + j} = \frac{4\angle 30}{2.24\angle 180 - 27}$$

$$V_{out}(\omega) = 1.79\angle - 153V$$

Often the phasor representation is sufficient; however, it is easy to convert $V_{out}(\omega)$ to the time representation:

$$v_{out}(t) = 1.79\cos(10t - 153)V$$

6.1.1 Superposition

Up until now, we have dealt with problems in which there was only one input. In Example 6.1, the input, $V_{in}(\omega)$, was not specified, but it was assumed to be a single input. However, what if the input was connected to a number of different sources, or equivalently, what if $V_{in}(\omega)$ was not a sinusoid, but a periodic function containing sinusoids at many different frequencies? Intuitively, you might think we could solve for the output to each input separately, or each sinusoidal component separately and just add up the individual solutions to get the total response. In fact, this divide-and-conquer approach is valid if the system is linear.

For linear systems, the principle of *superposition* holds, which states that if two or more sources are active in a circuit, a valid solution can be obtained by solving for each source as if it were acting alone, then algebraically summing these partial solutions. In fact, the sources could be anywhere in the network, even at different locations. Multiple sources at different locations will be covered at the end of the next chapter. In this chapter, the multiple sources will be at only one location, the location defined as the input.

The superposition principle is a consequence of linearity and only applies to linear systems. It makes the tools that we have studied much more powerful since now they can be applied to any periodic function by treating each of the sinusoidal components as a separate source. Examples using the principle of superposition will be found throughout this text.

6.1.2   **The Transfer Function**

Representing circuits or analog models as processes with well-defined inputs and outputs opens up a new way of thinking about and analyzing these processes. Rather than solve the output of the processes to a specific input as in Example 6.2, we can just as easily develop a more general solution where the input is not specified. Armed with the principle of superposition, we know that such a solution applies not only to sinusoids at any given frequency, but also to any waveform that can be decomposed into sinusoids. Returning to the general solution obtained in Example 6.1:

$$V_{out}(\omega) = \frac{V_{In}(\omega)}{1 - 0.03\omega^2 + j.1\omega} \text{ V} \qquad \text{[Eq. 6.1]}$$

Dividing both sides by $V_{in}(\omega)$, we obtain a function of ω that is *independent* of the input:

$$\frac{V_{out}(\omega)}{V_{In}(\omega)} = \frac{1}{1 - 0.03\omega^2 + j.1\omega} \qquad \text{[Eq. 6.2]}$$

This function is known as the *transfer function* because it describes how the input, $V_{in}(\omega)$, is *transferred* to the output, $V_{out}(\omega)$. The concept of the transfer function is so compelling that it has been generalized to include many different types of processes or systems with different types of inputs and outputs. A general definition for a transfer function becomes:

$$Transfer\ Function(\omega) = \frac{Output(\omega)}{Input(\omega)} \qquad \text{[Eq. 6.3]}$$

While the transfer function itself is independent of the input (because it was divided out), it is, in general, a function of the frequency, ω. Later, we will see that this concept can be expanded beyond sinusoidal and periodic signals using the Laplace transform. In the Laplace transform, the sinusoidal frequency, ω, is replaced by a complex frequency, s. By strict definition, the transfer function should always be a function of either s or ω, but the idea of expressing a process by its transfer function is so powerful that it is sometimes used in a nonmathematical, conceptual sense.

6.1.3 Transfer Function Characteristics

Given the power of the transfer function it is worth exploring some of its general properties. One of the more useful features of the transfer function is its frequency or spectral characteristics. This is similar to the spectral analysis concept developed in Chapter 3 except that now we are describing spectral characteristics of a process, not a signal. The transfer function spectral characteristics will tell us how the process influences input signals at different frequencies. With the application of superposition, transfer function spectral characteristics also tell us how a process influences signals that contain components over a range of frequencies.

If the entire spectral range of a process is explored, the transfer function frequency characteristics provide a complete description of the way in which the process alters the input signal, including any and all of its frequency components, to produce an output signal. In Chapter 3, we learned that a signal could be equally well represented in either the time or the frequency domain. The same is true of a process: the frequency characteristics of the transfer function provide a complete description of the process's input–output characteristics. In fact, the input–output or transfer properties of a process are most often represented in the frequency domain. There is also a way to represent these properties in the time domain as shown in Chapter 9.

Because the transfer function is already a function of frequency, ω, its frequency characteristics can be displayed simply by plotting the function, $Out(\omega)/In(\omega)$, against frequency. In general, the transfer function is a complex function of frequency such as in Eq. 6.2, so a complete graphic description will require two plots just as in signal spectral analysis. As with signal spectra, it is more informative to plot the magnitude and phase of the transfer function than its real and imaginary parts. With the aid of a computer program such as MATLAB this is easy to do for transfer functions of almost any complexity. Figure 6.3 shows the magnitude and phase characteristics of the of the RLC circuit given in Figure 6.2. These plots were obtained simply by plotting the system transfer function (Eq. 6.2). The MATLAB code that generated these plots can be found in Example 6.7.

It is easy to generate the frequency characteristics of a transfer function using MATLAB software; yet, as is often the case, relegating the problem to a computer does not provide as much insight as working it out manually. To really understand the frequency characteristics of transfer functions, it is necessary to examine the typical components of a general transfer function. In so doing, we will learn how to plot transfer functions without the aid of a computer. More important, by examining the component structure of a typical transfer function, we also gain insight into what the transfer function represents. This knowledge will often be sufficient

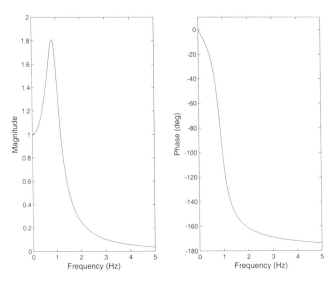

Figure 6.3 Magnitude (*left*) and phase (*right*) frequency characteristics of the transfer function of the RLC circuit shown in Figure 6.2.

to allow us to examine processes strictly within the frequency domain without the need to look at time domain responses.

6.2   TRANSFER FUNCTION FREQUENCY PLOTS: THE BODE PLOT

At first sight, it may appear that separating out and plotting the magnitude and phase of a transfer function equation such as Eq. 6.2 is quite difficult. However, a number of tricks make this task easy. These tricks also provide insight into the structure of the transfer functions.

We begin with a description of the most general form of a transfer function; specific transfer functions will be less complicated. In this framework, a transfer function consists of numerator and denominator terms and these terms, in the most general case, will be polynomials of ω. In the relatively simple transfer function of Eq. 6.2, the numerator was only a constant and the denominator was a quadratic polynomial of ω (note the ω^2 in the denominator). The numerator and denominator terms can always be factored, at least in theory, into combinations of quadratics or lower-order polynomials of ω. Under this assumption, the general expression for a transfer function becomes:

$$TF(\omega) = \frac{Bj\omega(1+jk_1\omega)(1-k_2\omega^2+jk_3\omega)\dots}{j\omega(1+jk_4\omega)(1-k_5\omega^2+jk_6\omega)\dots} \qquad \text{[Eq. 6.4]}$$

where B is a constant and k_1, k_2, k_3, and so forth, are coefficients of the various polynomials of ω. Equation 6.4 is a general representation: If the numerator and denominator both contained a $j\omega$, or any other identical terms, they would, of course, cancel. It is more common to use constants that are a little different from those in Eq. 6.4. These new coefficients are more directly linked to the frequency characteristics and to the behavior of the system as shown later.

$$TF(\omega) = \frac{B\angle j\omega(1+j\,\omega/\omega_1)\left(1-(\omega/\omega_{n1})^2+j2\delta_1\omega/\omega_{n1}\right)\ldots}{j\omega(1+j\,\omega/\omega_2)\left(1-(\omega/\omega_{n2})^2+j2\delta_2\omega/\omega_{n2}\right)\ldots}$$ [Eq. 6.5]

where ω_1, ω_2, ω_{n1}, ω_{n2}, δ_1 and δ_2 are the new constants. The parameter δ is called the *damping factor* and ω_n the *undamped natural frequency*. These terms relate to specific time domain behaviors and the names will take on more meaning when we look at this behavior in Chapter 8. Repeating the transfer function for the RLC circuit of Figure 6.2 (Eq. 6.2):

$$\frac{V_{out}(\omega)}{V_{In}(\omega)} = \frac{1}{1-0.03\omega^2+j.1\omega}$$

The appropriate values of these new coefficients can be found by equating coefficients:

$$B = 1$$

$$(1/\omega_n)^2 = 0.03; \quad \omega_n = \frac{1}{\sqrt{0.03}} = 5.77$$

$$2\delta/\omega_n = 0.1; \quad \delta = \frac{0.1\omega_n}{2} = \frac{(5.77)0.1}{2} = 0.29$$

The other constants in Eq. 6.5 do not exist for this particular transfer function, nor are there $j\omega$'s in the numerator or denominator.

The magnitude of a transfer function is usually plotted in decibels (dB); hence, the magnitude of the transfer function will be 20 times the log of the magnitude of the function:

$$|TF(\omega)|_{dB} = 20\log|TF(\omega)|$$ [Eq. 6.6]

Putting the magnitude of the transfer function in decibels may seem like an added complication, but it actually makes the analysis easier. Applying Eq. 6.6 to the general transfer function in Eq. 6.4 gives a formidable looking equation for the magnitude transfer function, but one that is surprisingly easy to dissect:

$$
\begin{aligned}
|TF(\omega)|_{dB} = {} & 20\log(B) + 20\log|j\omega| + 20\log|1+j\,\omega/\omega_1| \\
& + 20\log\left|1-(\omega/\omega_{n1})^2+j2\delta_1\omega/\omega_{n1}\right|\ldots \\
& - 20\log|j\omega| - 20\log|1+j\,\omega/\omega_2| \\
& - 20\log\left|1-(\omega/\omega_{n2})^2+j2\delta_2\omega/\omega_{n2}\right|\ldots
\end{aligned}
$$ [Eq. 6.7]

In this equation, the equivalent numerator and denominator terms are of similar form except for a sign: Numerator terms are positive and denominator terms are

TABLE 6.1   Bode Plot Primitives

Function	Function Name(s)	
	Numerator	**Denominator**
B	Constant	—
$j\omega$	Isolated zero	Isolated pole
$1 + j\omega/\omega_1$	Real pole or just pole	Real zero or just zero
	Lead element	Lag element
	First-order element*	First-order element*
$1 - (\omega/\omega_n)^2 + j2\delta\omega/\omega_n$	Complex zeroes[†]	Complex poles[†]
	Second-order element*	Second-order element*

* Name most commonly used in this text.
[†] Depends on the values of δ.

negative. In addition to the constant term, there are only three different types of polynomials: (a) a '20 log$|j\omega|$'; (b) a '20 log$|1 + j\omega/\omega_1|$'; and (c) the quadratic term, '20 log$|1 - (\omega/\omega_n)^2 + j2\delta\omega/\omega_n|$'. Moreover, each of the terms is either added to, or subtracted from, the other terms to make up the overall magnitude of transfer function. If we were to plot each of these terms separately, then we could construct a plot for the entire magnitude transfer function by graphically combining the individual plots. This approach is termed the *Bode plot* technique, and the individual terms are call *Bode plot primitives* because they cannot be reduced into subcomponents. (Technically, the second-order term is not a primitive since it can be factored into two first-order terms, as is sometime done, but often it is easier to treat as a primitive.) As we will see below, the Bode plot approach produces a plot that is only approximate and often somewhat crude, but usually sufficient to represent the general characteristics of the transfer function.

The different primitives of the magnitude transfer function (and their associated phase characteristics described below) occur so often, and are of sufficient importance, that they are given names, sometimes more than one name. A primitive's name sometimes depends on whether it is found in the numerator or denominator. The various primitives and their names are given in Table 6.1. The rationale behind some of the stranger terms, such as *poles* and *zeros*, will be given later.

The phase portion of the transfer function can also be dissected into individual components:

$$\angle TF(\omega) = \frac{\angle\left[Bj\omega(1+j\omega/\omega_1)\left(1-(\omega/\omega_{n1})^2 + j2\delta_1\omega/\omega_{n1}\right)\right]}{\angle\left[j\omega(1+j\omega/\omega_2)\left(1-(\omega/\omega_{n2})^2 + j2\delta_2\omega/\omega_{n2}\right)\right]} \qquad \text{[Eq. 6.8]}$$

However, by the rules of complex arithmetic, the angles simply add if they are in the numerator or subtract if they are in the denominator:

$$\angle TF(\omega) = \angle B + \angle j\omega + \angle(1 + j\omega/\omega_1) + \angle\left(1 - (\omega/\omega_{n1})^2 + j2\delta_1\omega/\omega_{n1}\right)$$
$$- \angle j\omega - \angle(1 + j\omega/\omega_2) - \angle\left(1 - (\omega/\omega_{n2})^2 + j2\delta_2\omega/\omega_{n2}\right) \quad \text{[Eq. 6.9]}$$

As with the magnitude transfer function, the phase transfer function consists of individual components that add or subtract from one another. In the generalized-phase transfer function, there are only three different terms corresponding to the last three entries in Table 6.1. This is because the phase angle of the constant term, B, is zero. Again, if we are able to construct a frequency plot for these three terms, we could simply add them together graphically to get the overall phase plot. As with the magnitude plot, this approach will lead to a sometimes crude, but usually sufficient, approximation. In fact, often only the magnitude plot is of interest and it is not necessary to construct the phase plot. The magnitude and phase frequency characteristics of the various terms are developed in the following section.

6.2.1 Frequency Characteristics of Bode Plot Primitives

The magnitude and phase frequency characteristics of the four components in Table 6.1 are presented here.

6.2.1.1 The Constant Primitive: $TF(\omega) = B$

The constant primitive does not contribute to the phase plot because the angle of a real variable is zero. In the magnitude plot, it simply scales the vertical axis by:

$$|TF|_{Constant} = 20\log(B)\text{dB} \quad \text{[Eq. 6.10]}$$

When constructing the overall magnitude plot, it is easiest to leave the influence of this term until the end. After all the other terms have been plotted and summed, the vertical axis is rescaled so that the former zero line now equals $20\log(B)$. This rescaling will be shown in the examples below.

Again, the angle of a real constant is zero, so the constant term makes no contribution the phase plot.

$$\angle B = 0 \quad \text{[Eq. 6.11]}$$

6.2.1.2 Isolated Zero or Pole Primitive: $TF(\omega) = j\omega$ or $1/j\omega$

The magnitude frequency plot of $20\log|j\omega|$ is a logarithmic function of ω if plotted on a linear axis. However, it is common to plot transfer function spectra using a log function of ω, because the frequency range of interest often spans several orders of magnitude. Since the magnitude of $j\omega$ is just ω, the function $\log|j\omega|$ becomes $\log \omega$ which plots as a straight line against $\log \omega$ on the frequency axis. This line intersects the 0-dB line at 1.0 rad/sec because $20\log(\omega = 1) = 0$ dB. The magnitude plot of this term is shown in Figure 6.4.

To determine the slope of this line, note that when $\omega = 10$ rad/sec, $20\log|j\omega| = 20\log(10) = 20$ dB, and when $\omega = 100$, $20\log|j\omega| = 20\log(100) = 40$ dB. Therefore,

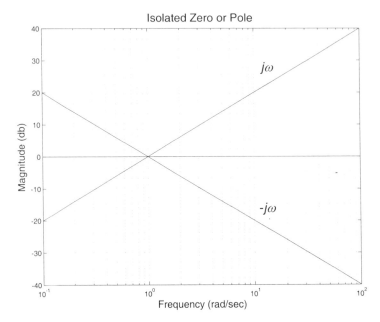

Figure 6.4 Frequency plot of an isolated zero ($j\omega$) or pole ($-j\omega$). These functions plot as a straight line when magnitude is plotted in decibels and frequency is plotted logarithmically.

for every order-of-magnitude increase in frequency, there is a 20-dB change in the value of the function. This leads to unusual dimensions for the slope; specifically 20 dB/decade, but this is the result of the logarithmic scaling of the horizontal and vertical axes. The slope will be positive 20 dB/decade if the primitive is in the numerator and −20 dB/decade if the primitive is in the denominator (Figure 6.4).

The angle of $j\omega$ is 90 degrees, so in the phase plot this just adds or subtracts 90 degrees. If the term is in the denominator, the angle is −90 degrees ($1/j\omega = -j\omega$, which has an angle of −90 degrees) otherwise it is +90 degrees.

$$\angle j\omega = 90 \text{ degrees;} \quad \angle 1/j\omega = -90 \text{ degrees} \qquad \text{[Eq. 6.12]}$$

As with the constant term, this simply adds plus or minus 90 degrees to the vertical axis of the phase plot, which can be accomplished by rescaling the vertical axis, when the rest of the plot is complete.

6.2.1.3 First-Order Primitive $TF(\omega) = (1 + j\omega/\omega_1)$ or $1/(1 + j\omega/\omega_1)$

The magnitude of the frequency plot of the first-order primitive is established by two asymptotes: a high-frequency asymptote and a low-frequency asymptote. The high-frequency asymptote is defined for the frequencies $\omega \gg \omega_1$, and the low-frequency asymptote is defined for frequencies $\omega \ll \omega_1$ (ω_{low}). To determine these two asymptotes:

$$|TF(\omega_{low})| = \lim_{\omega << \omega_1} [20\log|1 + j\omega/\omega_1|] = 20\log(1) = 0\,\text{dB} \qquad \text{[Eq. 6.13]}$$

$$|TF(\omega_{high})| = \lim_{\omega >> \omega_1} [20\log|1 + j\omega/\omega_1|] = 20\log|j\omega/\omega_1| = 20\log(\omega/\omega_1)\text{dB} \qquad \text{[Eq. 6.14]}$$

The low-frequency asymptote given by Eq. 6.13 plots as a straight line at 0 dB. The high-frequency asymptote given in Eq. 6.14 has the plot the same as $20\log|j\omega|$ term described above, except that the intercept with the 0-dB line will now be at $\omega = \omega_1$ because $20\log(\omega_1/\omega_1) = 0$ dB. Thus, the high-frequency asymptote plots as a straight line, intersecting 0 dB at frequency ω_1 and has a slope of 20 dB/decade. Errors between the actual curve and the asymptotes will occur when the assumptions are least true: when the frequency, ω, is neither much greater than, nor much less than, ω_1. In fact, the greatest error will occur when ω exactly equals ω_1. At that frequency the value of the magnitude of the first-order term will be:

$$|TF(\omega = \omega_1)| = 20\log|1 + j\omega_1/\omega_1| = 20\log|1 + j| = 20\log(\sqrt{2}) = 3\,\text{dB} \qquad \text{[Eq. 6.15]}$$

The high- and low-frequency asymptotes are plotted for a first-order numerator and denominator primitives in Figure 6.5 along with actual curves. Note that the actual

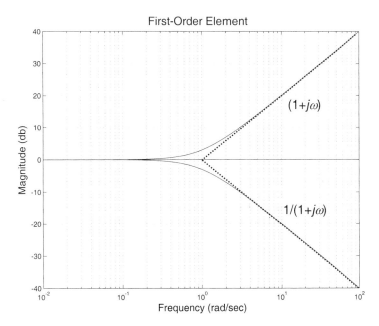

Figure 6.5 Magnitude plot of a first-order primitive in the numerator or denominator. In this plot, ω_1 has been normalized to 1.0 rad/sec. The low-frequency asymptote is valid for frequencies $\omega \ll \omega_1$ and is just a straight line at 0 dB. The high-frequency asymptote is valid for frequencies $\omega \gg \omega_1$ and plots as a line with a slope of 20 dB/decade. The high-frequency asymptote intersects the 0 dB line at $\omega = \omega_1$ Thus the two curves intersect at $\omega = \omega_1$ (1.0 rad/sec). The greatest deviation between the actual curves and their asymptotes occurs at ω_1 and this error is small (3 dB).

curves are very close to the asymptotes and deviate only slightly even at $\omega = \omega_1$. For frequencies much above ω_1, the curve closely follows the asymptote, increasing or decreasing in magnitude at a slope of 20 dB/decade. Often, only asymptotes are used to plot this primitive. For greater accuracy, a curve can be drawn freehand through the 3-dB point.

The phase plot of the first-order term can be estimated using the same approach by taking the asymptotes and the worst-case point, $\omega = \omega_1$:

$$|TF(\omega_{low})| = \lim_{\omega \ll \omega_1} [\angle(1 + j\omega/\omega_1)] = \angle 1 = 0 \text{ degrees} \qquad [\text{Eq. 6.16}]$$

$$|TF(\omega_{high})| = \lim_{\omega \gg \omega_1} [\angle(1 + j\omega/\omega_1)] = \angle(j\omega/\omega_1) = 90 \text{ degrees} \qquad [\text{Eq. 6.17}]$$

These are both straight lines at 0 and 90 degrees. In this case, the assumption is that *much much less than* means one order of magnitude, so that the low-frequency asymptote is assumed to be valid from 0.1 ω_1 on down. By the same reasoning the high-frequency asymptote is assumed to be valid from 10 ω_1 and up.

Again the greatest difference between the asymptotes and the actual curve will be when ω equals ω_1:

$$\angle(TF(\omega = \omega_1)) = \angle(1 + j\omega_1/\omega_1) = \angle(1 + j) = 45 \text{ degrees} \qquad [\text{Eq. 6.18}]$$

This value, 45 degrees, will be exactly between the high-frequency and low-frequency asymptote. What is usually done is to draw a straight line between the high end of the low-frequency asymptote at 0.1 ω_1 and the low end of the high-frequency asymptote at 10 ω_1, passing through 45 degrees. Although the phase curve is nonlinear in this range, the error induced by a straight-line approximation is surprisingly small as shown in Figure 6.6, which shows the actual phase plots and asymptotes of a first-order term in both the numerator and denominator.

6.2.1.4   Second-Order Term: $TF(\omega) = (1 - (\omega/\omega_n)^2 + j2\delta\omega/\omega_n)$ or the Inverse

As mentioned, the second-order term is not really a primitive since it can be factored into two first order terms. In cases where the roots are real, factoring can provide a more accurate frequency plot, particularly for phase. However, if the roots are complex, this term is best treated as a primitive. Treating the second-order term as a primitive will be used here for Bode plots regardless of the roots.

Not surprisingly, the second-order term is the hardest to plot. The same approach is used as for the first-order term except special care must be taken when $\omega = \omega_n$. Again we begin with the high- and low-frequency asymptotes. These occur when ω is either much greater or much less than ω_n.

$$|TF(\omega_{low})| = \lim_{\omega \ll \omega_n} \left[20\log\left|1 - (\omega/\omega_n)^2 + j2\delta\omega/\omega_n\right| \right] = 20\log(1) = 0\,\text{dB} \quad [\text{Eq. 6.19}]$$

$$|TF(\omega_{high})| = \lim_{\omega \gg \omega_n} \left[20\log\left|1 - (\omega/\omega_n)^2 + j2\delta\omega/\omega_n\right| \right]$$

$$= 20\log(\omega/\omega_n)^2 = 40\log(\omega/\omega_n)\text{dB} \qquad [\text{Eq. 6.20}]$$

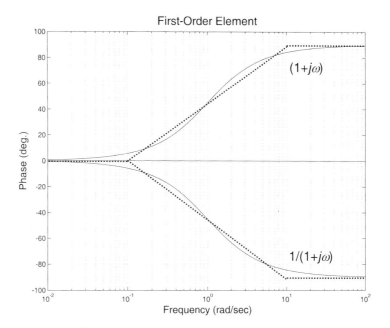

Figure 6.6 Phase plot of a first-order term in the numerator and denominator. For this plot, ω_1 has been normalized to 1.0 rad/sec. The low-frequency asymptotes are 0.0 degrees for both numerator and denominator curves up to 0.1 ω_1. The high-frequency asymptotes begin at 10 ω_1 and are at $\pm$ 90 degrees. The intermediate asymptotes are drawn as straight lines that pass through $\pm$ 45 degrees at $\omega = \omega_1$ and these asymptotes approximate the actual curves shown as solid lines.

The low-frequency asymptote is the same as for the first-order term, while the high-frequency asymptote is similar, but with double the slope: 40 dB/decade instead of 20 dB/decade. However, a major difference occurs when $\omega = \omega_n$:

$$|TF(\omega = \omega_n)| = 20\log|1 - (\omega/\omega_n)^2 + j2\,\delta\omega/\omega_n| = 20\log|j2\delta| = 20\log(2\delta)\text{dB} \quad \text{[Eq. 6.21]}$$

The magnitude function at $\omega = \omega_n$ is not a constant, but depends on δ: specifically, it is 20 log (2δ). If δ is less than 1.0 then the log operation will produce a negative value and the curve will dip below the 0.0 dB axis. If the second-order term is in the denominator as is usually the case, a value of δ is less than 1.0 will result in a *positive* value for the log operation (the inverse of the log of a number less than 1.0 is positive). In this case the magnitude curve will have a peak above the 0.0 dB axis and the smaller the δ the greater the peak. Hence, the value of δ can radically alter the shape of the magnitude curve so it must be taken into account when plotting. Usually the magnitude plot is determined by calculating the value of $TF(\omega)$ at $\omega = \omega_n$ using Eq. 6.21, plotting that point, then drawing a curve freehand through that point converging smoothly with the high and low-frequency asymptotes above

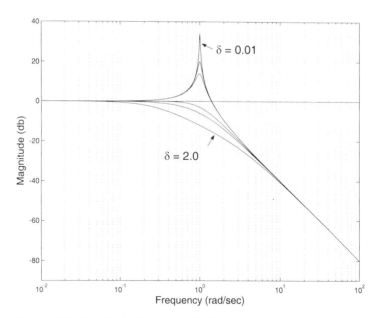

Figure 6.7 Magnitude frequency plot for a second-order term in the denominator for several values of δ. The value of ω_n has been normalized to 1.0 rad/sec. The values of δ range between 0.01 and 2.0 (0.01, 0.5, 1, 0.707, 1.0, and 2.0). Only the denominator curves are shown for clarity.

and below ω_n. The second-order magnitude plot is shown in Figure 6.7 for various values of δ. The plot is drawn only for a second-order *denominator* term; if the term were in the numerator it would plot as the inverse of the curves in Figure 6.7.

The phase plot of a second-order system is also approached using the asymptote method. For phase angle the high- and low-frequency asymptotes are given as:

$$\angle|TF(\omega_{low})| = \lim_{\omega<<\omega_n} \left[1-(\omega/\omega_n)^2 + j2\delta\omega/\omega_n\right] = \angle 1 = 0 \text{ degrees} \qquad \text{[Eq. 6.22]}$$

$$\angle|TF(\omega_{high})| = \lim_{\omega>>\omega_n} \left[\angle 1-(\omega/\omega_n)^2 + j2\delta\omega/\omega_n\right] = \angle -\omega^2 = 180 \text{ degrees} \qquad \text{[Eq. 6.23]}$$

This is similar to the asymptotes of the first-order process except the high-frequency asymptote is at ± 180 degrees. The phase angle when $\omega = \omega_n$ can be easily determined:

$$\angle TF(\omega_n) = \angle\left(1-(\omega_n/\omega_n)^2 + j2\delta\omega_n/\omega_n\right) = \angle j2\delta = 90 \text{ degrees} \qquad \text{[Eq. 6.24]}$$

So the phase at $\omega = \omega_n$ is 90 degrees, halfway between the two asymptotes. Unfortunately, the shape of the phase curve between $0.1\ \omega_n$ and $10\ \omega_n$ is a function of δ and can no longer be approximated as a straight line except at larger values of δ (>2.0). Phase curves are shown in Figure 6.8 for the same range of values of δ used in Figure 6.7. Again, the plot is only for a denominator term. The curves for low values of δ have steep transitions between 0 and 180 degrees while the curves for high values of δ have gradual slopes approximating the phase characteristics of a

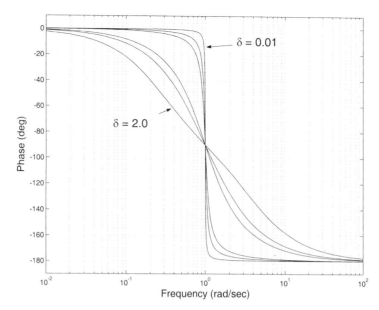

Figure 6.8 Phase curves for a second-order term in the denominator. The value of ω_n has been normalized to 1.0 rad/sec. The values of δ are the same as in Figure 6.7.

first-order primitive except for the larger phase change (0 to 180 degrees for the second-order term as opposed to 0 to 90 degrees for the first-order term). Hence, if δ is 2.0 or more, the phase curve can be approximated by a straight line between the low-frequency asymptote at $0.1\omega_n$ and the high-frequency asymptote at $10\omega_n$. If δ is much less that 2.0, the best that can be done is to approximate, freehand, the appropriate curve in Figure 6.8.

The frequency characteristics of the four Bode plot primitives are summarized in Table 6.2. If a transfer function contains only one primitive the plotting is straightforward, but usually a number of these primitives are found in a typical transfer function. In such cases, the magnitude plots of each primitive are plotted on the same graph and the same is done for the phase plots. Usually only the asymptotes and a few other important points are plotted (such as the value of a second-order term at $\omega = \omega_n$), then the individual plots are summed graphically. This graphic summation requires some care, but is usually not too difficult. This approach is illustrated in the next example.

Example 6.3: Find the magnitude and phase curves (Bode plots) for the transfer function.

$$TF(\omega) = \frac{100j\omega}{(1+j1\omega)(1+j.1\omega)}$$

TABLE 6.2   Bode Plot Primitives

Denominator Term	Magnitude Plot	Phase Plot
Constant	20 log B	—
$j\omega$		−90 degrees
$1 + j\dfrac{\omega}{\omega_1}$		
$1 - \left(\dfrac{\omega}{\omega_n}\right)^2 + j\dfrac{2\delta\omega}{\omega_n}$		

Solution: The transfer function contains four elements: a constant, an isolated zero (i.e., a $j\omega$ in the numerator), and two first-order terms in the denominator. For the magnitude curve, plot the asymptotes for all but the constant term. Add these asymptotes together graphically to get an overall asymptote. At the end, use the constant term to scale the value of the vertical axis. For the phase plot, construct the asymptotes for the two first-order denominator elements, then rescale the axis by +90 degrees to account for the $j\omega$ in the numerator. Recall that the constant term does not contribute to the phase plot.

Regarding the two first-order terms in the denominator, the general form for these primitives is $(1 + j\omega/\omega_1)$ where ω_1 is the point at which the high and low-frequency asymptotes intersect. In this transfer function, these two frequencies, ω_1 and ω_2, are $1/1 = 1.0$ rad/sec and $1/0.1 = 10$ rad/sec. Figure 6.9 shows the asymptotes obtained for the two first-order primitives and the $j\omega$ primitive.

Graphically adding the three asymptotes shown in Figure 6.9 gives the curve consisting of three straight lines shown in Figure 6.10. Note that the numerator and denominator asymptotes cancel out at $\omega = 1.0$ rad/sec, so the overall asymptote is flat until the additional downward asymptote comes in at $\omega = 10$ rad/sec. The actual

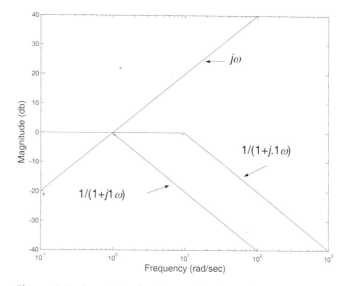

Figure 6.9   Asymptotes for the magnitude transfer function given in Example 6.3. These asymptotes are graphically added to get the overall asymptote in Figure 6.10.

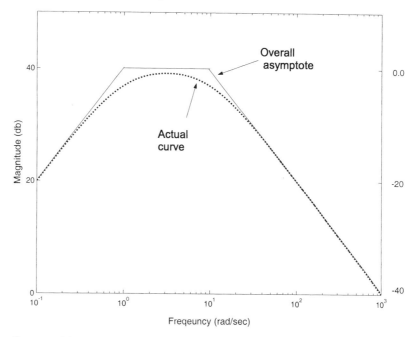

Figure 6.10   Magnitude plot for the transfer function given in Example 6.3. The *solid line* shows the graphical addition of the individual asymptotes shown in Figure 6.9. The *dashed line* shows the actual magnitude transfer function. The vertical axis has been rescaled by 40 dB on the right side to account for the 100 in the numerator [20 log (100) = 40 dB].

magnitude transfer function is also shown in Figure 6.10 and closely follows the overall asymptote. A small error (3 dB) is seen at the two breakpoints: $\omega = 1.0$ and $\omega = 10$ as expected. A final step in constructing the magnitude curve is to rescale the vertical axis so that 0 dB corresponds to $20 \log(100) = 40$ dB.

The asymptotes of the phase curve are shown in Figure 6.11 along with the overall asymptote that is obtained by graphical addition. Also shown is the actual phase curve, which, as with the magnitude curve, closely follows the overall asymptote. As a final step the vertical axis of this plot has been rescaled by +90 degrees on the right side to account for the $j\omega$ term in the numerator.

In both the magnitude and phase plots, the actual curves follow the overall asymptote closely and this will be true for all transfer functions that have terms no higher than first-order. Tracing freehand through the 3 dB points would further improve the match, but often the asymptotes are sufficient. As we will see in the next example this is not true for transfer functions that contain second-order terms.

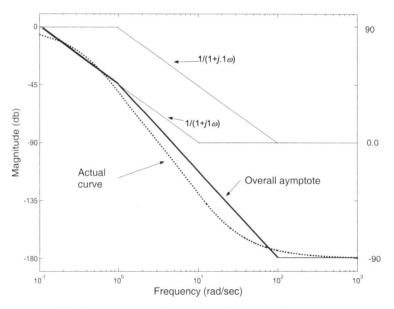

Figure 6.11 Phase frequency characteristics for the transfer function given in Example 6.3. The two first-order terms in the denominator produce downward curves that cross the 45-degree mark at $\omega = 1.0$ and $\omega = 10$. The actual curve closely follows the overall asymptote that is created by graphically adding the two first-order curves. To complete the plot, the vertical axis has been rescaled to account for the $j\omega$ term in the numerator by +90 degrees as shown on the right axis.

Example 6.4: Find the magnitude and phase curves (Bode plots) for the transfer function:

$$TF(\omega) = \frac{10(1 + j2\omega)}{j\omega(1 - 0.04\omega^2 + j.04\omega)}$$

Solution: This transfer function contains four primitives: a constant, a numerator first-order term, an isolated pole in the denominator, and a second-order term in the denominator. For the magnitude curve, plot the asymptotes for all primitives except the constant, then add these up graphically. Lastly, use the constant term to scale the value of the vertical axis. For the phase curve, plot the asymptotes of the first-order and second-order terms, then rescale the vertical axis by –90 degrees to account for the $j\omega$ term in the denominator. To plot the magnitude asymptotes, it is first necessary to determine ω_1, ω_n, and δ from the associated coefficients:

$$\text{For } \omega_1: \quad 1/\omega_1 = 2; \quad \omega_1 = 0.5 \text{ rad/sec}$$

$$\text{For } \omega_n: \quad 1/\omega_n^2 = 0.04; \quad \omega_n = 1/\sqrt{0.04} = 5 \text{ rad/sec}$$

$$\text{For } \delta: \quad 2\delta/\omega_n = 0.04; \quad \delta = 0.04\omega_n/2 = 0.04(5)/2 = 0.1$$

$$20\log(2\delta) = -14 \text{ dB}$$

Note that the second-order term is positive 14 dB when $\omega = \omega_n$. This is because the second-order term is in the denominator. Otherwise this term would have a value of –14 dB at $\omega = \omega_n$. Using these values, the asymptote magnitude plot becomes as shown in Figure 6.12.

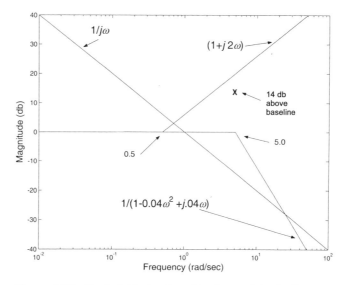

Figure 6.12   The individual and combined asymptotes for the magnitude plot of the transfer function given in Example 6.4. The value of the second-order term at $\omega = \omega_n$ is also shown. The vertical axis will be rescaled in the final plot. The position of 0.5 rad/sec and 5.0 rad/sec are indicated on the log frequency horizontal axis.

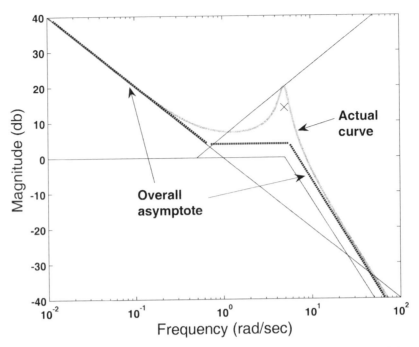

Figure 6.13 Overall asymptote of the magnitude frequency characteristic of the transfer function given in Example 6.4. The actual curve (dotted line) follows the overall asymptote fairly closely except for the region on either side of ω_n. To find the actual value at ω_n where the second-order term has a 14-dB peak, the value of the overall asymptote at that point (about 5 dB) must be added to the 14-dB peak. The vertical axis is rescaled on the right side to account for the constant term.

The log frequency scale can be a little confusing so the positions of 0.5 rad/sec and 5.0 rad/sec are indicated on the frequency axis. The overall asymptote and the actual curve for the magnitude transfer functions are shown in Figure 6.13. The second-order denominator system will reach a value that is 14 dB above 0-dB value. This is a positive value because the second-order term is in the denominator. This 14 dB should be added to the net asymptote to get the peak point for the summed curve. Note that the net asymptote at $\omega = \omega_n$ is approximately +5 dB. Finally, the vertical axis is rescaled by $20\log(10) = 20$ to account for the constant term.

The individual phase asymptotes are shown in Figure 6.14, whereas the overall asymptote and the actual phase curve are shown in Figure 6.15. Note that the actual phase curve is quite different from the overall asymptote because the second-order term has a small value for δ. Recall that this low value of δ means the phase curve has a sharp transition between 0 and −180 degrees and will deviate substantially from the asymptote.

The Bode plot approach may seem like a lot of effort to achieve a graph that could be better done with a couple of lines of MATLAB code. However, these techniques provide us with a mental map between the transfer function and the fre-

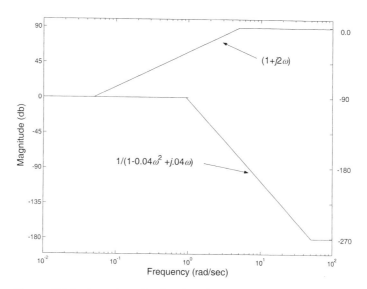

Figure 6.14 Asymptotes for the phase frequency characteristics on the transfer function in Example 6.4. Only two terms, the first-order term in the numerator and the second-order term in the denominator, contribute to the phase curve. The overall asymptote and actual phase curve are shown in Figure 6.15. The vertical axis has been rescaled on the right side to account for the $j\omega$ primitive in the denominator.

quency characteristics via the Bode plot primitives. In addition, Bode plot techniques can help go the other way, from frequency curve to transfer function. Given a desired frequency response curve, we can use Bode plot methods to construct a transfer function that will match that frequency curve. It is then only one more step to design a process that has this transfer function. Therefore, if we can get the frequency characteristics of a physiological process, we ought to be able to quantify its behavior with a transfer function and ultimately design a quantitative model for this system. These approaches are explored in subsequent sections.

6.3 FILTERS

The transfer function provides a mathematical description of how input signals are modified as they pass through a process and the Bode Plot describes this alteration in the frequency domain. Some processes are specifically designed to alter the frequency characteristics of an input signal in a well-defined manner. Their purpose is to produce an output signal that has reduced or enhanced frequencies within specific frequency ranges. Such processes are called *filters*. Filters can be constructed using electronic components or implemented in software on a digital computer. The former are usually referred to as *analog filters*, whereas the latter are termed *digital*

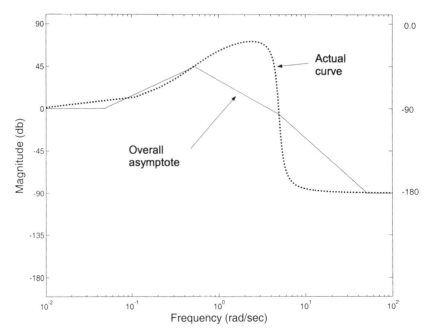

Figure 6.15 Overall asymptote and actual phase curve (*dotted line*) for the transfer function given in Example 6.4. The actual curve deviates from the asymptote due to the phase characteristics of the second-order term. The *jω* term in the denominator scales the axis as shown on the right side.

filters. This text covers only the analog variety. (In the spirit of full disclosure, some of the figures in this text were generated with the aid of MATLAB digital filters.)

6.3.1 Filter Types

Filters are usually named according to the range of frequencies they do *not* suppress. Thus, *lowpass* filters allow low frequencies to pass with minimum reduction or *attenuation*, while higher frequencies are attenuated. Conversely, *highpass* filters pass high frequencies, but attenuate low frequencies. *Bandpass* filters reject frequencies above and below a *passband* region. An exception to this terminology is *bandstop* filters that pass frequencies on either side of a range of attenuated frequencies. Figure 6.16 shows typical frequency characteristics for these four different types of filters.

Within each class, filters are defined by the frequency *range* that they pass, termed the filter *bandwidth*, and the *sharpness* with which they increase (or decrease) attenuation as frequency varies. Spectral sharpness is further specified in two ways: as an initial sharpness in the region where attenuation first begins and as a *slope* further along the attenuation curve. These various filter properties are best described graphically using Bode plots. The term *filter gain* is another name for the filter's

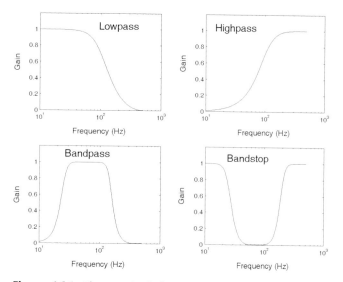

Figure 6.16 The magnitude frequency characteristics of the four basic filter types. **A:** A lowpass filter allows low frequencies to pass through, but attenuates (i.e., suppresses) higher frequencies. **B:** A highpass filter is the opposite of a lowpass filter, attenuating low frequencies while letting high frequencies pass through unattenuated. **C:** A bandpass filter combines features of the lowpass and highpass filters to allow a range of frequencies through, attenuating frequencies outside that range. **D:** A bandstop filter is the inverse of a bandpass filter: it attenuates frequencies within a certain range, but lets others pass through unattenuated.

transfer function, $V_{out}(\omega)/V_{in}(\omega)$ and is usually in decibels. Technically the frequency characteristic should be defined for all frequencies for which it is nonzero, but practically it is usually given only for the frequencies of interest.

6.3.2 Filter Bandwidth

All real filters make the transition between the unattenuated or passband frequency region and the attenuated or stopband region in a smooth manner such as seen in Figure 6.16. For purposes of discussion, a boundary is established between passband and stopband frequencies known as the *cutoff* frequency, and this boundary is the same as used for signal frequency characteristics. (See Chapter 3, Section 3.8.) Frequencies on one side of this cutoff are taken as unattenuated while frequencies on the other side are taken as attenuated. As for signals, the cutoff boundary is defined as the frequency when the attenuation is 3 dB; that is, the gain of the filter is 3 dB below the unattenuated gain. As stated in Section 3.8, this point is also known as *the half-power point* because the power of the signal out is half that of the input signal at 3-dB attenuation.

6.3.3 Filter Order

The slope of a filter's attenuation curve is related to the complexity of the filter: Filters that are more complex have a steeper slope, better approaching the ideal. In analog filters, complexity is proportional to the number of energy storage elements in the circuit (either inductors or capacitors, but usually capacitors for practical reasons). As you may have observed in some of the previous examples, each independent energy storage device (i.e., capacitor or inductor) leads to an additional order of ω in the denominator polynomial of the filter's transfer function. From Bode plot analysis, we find that with each additional order of ω in the denominator polynomial, there is an increase in the asymptote slope by 20 dB/decade.

In electrical engineering, it has long been common to call the roots of the denominator equation *poles*. Thus the terms *filter order* and *filter poles* are used synonymously and they relate directly to the order of ω in the denominator of the transfer function. The larger the filter order, the greater the number of poles, the steeper the attenuation slope, and the greater the filter's complexity in terms of number of elements. Figure 6.17 shows two circuits that are one-pole (left circuit) and two-pole (right circuit) lowpass filters. The single-pole filter will have a slope of 20 dB/decade (see Example 6.5), whereas the two-pole filter has a slope of 40 dB/decade (see Problem 6.7). As mentioned above, the number of poles correlates to the number of energy storage devices, which in these circuits are capacitors. Again, each additional filter pole (or order) increases the downward slope (sometimes referred to as the *rolloff*) by 20 dB/decade.

Figure 6.18 shows the frequency plot of a second-order or two-pole filter, which will have a slope of 40 dB/decade, the same slope produced by the right-hand circuit in Figure 6.17. Also plotted is the frequency curve of a 12th-order lowpass filter that has the same cutoff frequency as the two-pole filter, and, hence, the same bandwidth. The steeper slope, or rolloff, of the 12-pole filter is apparent. In principle, a 12-pole lowpass filter would have a slope of 240 dB/decade (12×20 dB/decade). In fact, this frequency characteristic is theoretical because in real analog filters, parasitic components and inaccuracies in the circuit elements limit the actual attenuation that can be obtained. All of the above arguments also apply to highpass filters

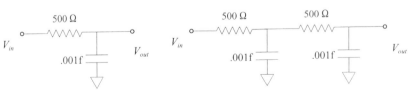

Figure 6.17 Two circuits that act as lowpass filters. The left circuit is a one-pole filter and has a high-frequency attenuation of 20 dB/decade (as will be shown in Example 6.5). The left circuit is a two-pole lowpass filter with a high-frequency attenuation of 40 dB/decade.

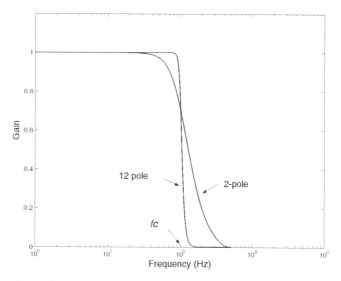

Figure 6.18 Frequency plot of a second-order (two-pole) and 12th-order lowpass filter with the same cutoff frequency. The higher order filter more closely approaches the sharpness of an ideal filter.

except that the frequency plot decreases with *decreasing* frequency at a rate of 20 dB/decade for each highpass filter pole.

6.3.4 Filter Initial Sharpness

Figure 6.18 shows that the both the slope and the initial sharpness increase with filter order (number of poles), but increasing filter order also increases the complexity, and hence the cost, of the filter. It is possible to increase the *initial* sharpness of the filter's attenuation characteristics without increasing the order of the filter if you are willing to make other compromises; specifically, if you are willing to accept some unevenness, or *ripple*, in passband. Figure 6.19 shows two lowpass, fourth-order (or four-pole) filters, differing in the initial sharpness of the attenuation. The one marked Butterworth has a smooth passband, but the initial attenuation is not as sharp as the one marked Chebyshev, which has a passband that contains ripples. However, most Biomedical filter applications require a smooth passband characteristic, so a Butterworth-type filter is required with its more gradual initial attenuation.

6.3.5 Evaluating Filter Frequency Characteritics

Calculating the transfer function and frequency characteristic of a filter is no more different than for any other circuit. Designing a filter (as opposed to analyzing an existing filter circuit) is a bit more complicated. Design approaches for some of the

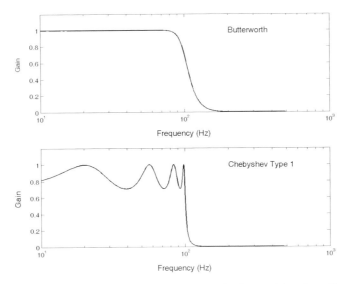

Figure 6.19   Two filters having the same order (four-pole) and cutoff frequency, but differing in the sharpness of the initial slope. The filter marked Chebyshev has a steeper initial slope or rolloff, but contains ripples in the passband.

more common filters are given in the Chapter 10. The analysis of a simple filter is given in the next example.

Example 6.5:   Determine the transfer function and frequency characteristics of the circuit shown in the left hand side of Figure 6.17. This circuit, consisting of only a resistor and a capacitor, is known as an RC circuit.

Solution: Because the circuit has one energy storage device (a capacitor), you already know it is a first-order filter and will have an attenuation slope of 20 dB/decade. You do not know what filter type this is (highpass and lowpass, etc.), but it is not difficult to find both the transfer function and the frequency characteristics using basic circuit analysis tools. Under the standard assumptions (an ideal source at V_{in} and no current flowing out at V_{out}), the circuit can be analyzed as a one-mesh problem. In this solution, R and C will be left as variables to show their relationship to the bandwidth of the filter. Applying KVL:

$$V_{in}(\omega) - I(\omega)\left(R + \frac{1}{j\omega C}\right) = 0$$

$$I(\omega) = \frac{V_{in}(\omega)}{\left(R + \dfrac{1}{j\omega C}\right)} = \frac{V_{in}(\omega)j\omega C}{(1 + j\omega RC)}$$

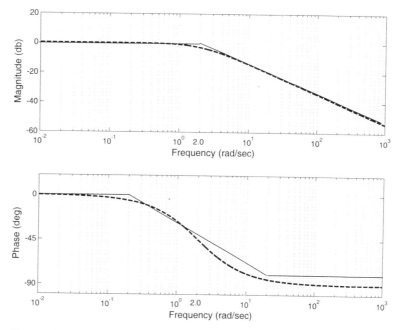

Figure 6.20 Magnitude frequency curve of the transfer function obtained from the RC circuit in Example 6.5. This curve shows that the circuit is a first-order, or one-pole, lowpass filter with a bandwidth of 2.0 rad/sec.

$$V_{out}(\omega) = Z_c I(\omega) = \frac{V_{in}(\omega)j\omega C(1/j\omega C)}{(1+j\omega RC)}$$

$$TF(\omega) = \frac{V_{out}(\omega)}{V_{in}(\omega)} = \frac{1}{(1+j\omega RC)} = \frac{1}{(1+j\omega/\omega)} \quad \text{where } \omega_1 = 1/RC$$

Substituting in the specific values for R and C: $\omega_1 = 1/RC = 1/500(0.001) = 1/0.5 = 2$ rad/sec.

To plot the frequency curves, note that the transfer function has only a constant in the numerator and one first-order primitive in the denominator. Thus the magnitude asymptote will be a straight line up to ω_1 (2 rad/sec) and then a straight line that falls off at 20 dB/decade (Figure 6.20). The bandwidth of this filter is simply ω_1 or $1/RC$.

The phase plot has a low-frequency asymptote of 0 degrees up to 0.2 rad/sec, a high-frequency asymptote of -90 degrees above 20 rad/sec, and passes through -45 degrees at ω_1 (i.e., 2.0 rad/sec).

6.3.6 Filter Design

Bode plot techniques provide a method for quickly plotting the frequency characteristics of a transfer function, provided it is not too complicated (in which case you would probably use a computer as described in the next section). These same techniques also allow you to go the other way: to estimate a transfer function that has frequency characteristics matching a given frequency curve. This transfer function can then be used to guide the development of an analog or digital filter having the desired frequency characteristics, at least for relatively simple filters. This approach can also be used to construct models of biological processes as shown later. The design of complex filters is a well-developed area with entire texts devoted to the subject. The design of a filter or other network is a process known as *synthesis*, whereas the analysis of an existing circuit is termed, logically, *analysis*. The following example illustrates the design of a simple filter using this approach.

Example 6.6:   Find the equivalent transfer function that produces the magnitude frequency curve shown below.

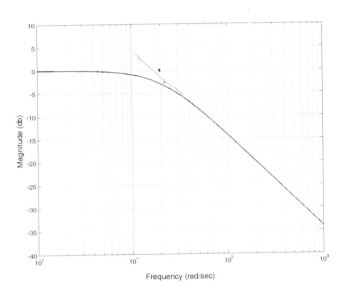

Solution: We begin by noting the obvious: the frequency curve of a lowpass filter. Next, we measure the high-frequency slope and find that it decreases at 20 dB/decade; hence, this must be the product of a first-order transfer function. Drawing the high-frequency asymptote, we find that it intersects the low-frequency asymptote (the 0-dB line) at 20 rad/sec. This gives us the cutoff frequency of the filter and the bandwidth (20 rad/sec). This information is all we need to reconstruct the equivalent transfer function:

$$TF(\omega) = \frac{1}{1 + j\omega/\omega_n} \quad \text{where } \omega_n = 20 \text{ rad/sec}$$

$$TF(\omega) = \frac{1}{1 + j\omega/20} = \frac{1}{1 + j0.05\omega}$$

To design a circuit that matches this transfer function, we rely on experience: knowledge of what circuits produce this type of transfer function. From the previous example, we know that an RC circuit produces this type of first-order transfer function. This may be the only transfer function for which you have this prior knowledge, but the ability to design circuits with more complicated transfer functions will be gained with experience.

To complete the design, we only need to give values to R and C. From the analysis of Example 6.5, we know that the cutoff frequency, ω_n, is given by:

$$\omega_n = 1/RC \qquad\qquad \text{[Eq. 6.25]}$$

In this case, we have two unknowns constrained by only one equation, so we can pick one of the element values arbitrarily. Usually the capacitor value is selected first because the range of commercial capacitor values is more limited than that of resistor values. We will choose a value for C of 1.0 μf (1×10^{-6} f) because this is a readily obtained value. (Again, experience helps.) In this case, the value of R becomes:

$$\omega_n = 1/RC; \quad R = 1/\omega_n C = \frac{1}{20(1 \times 10^{-6})} = 50 \times 10^3 \ \Omega = 50 \ \text{k}\Omega$$

So the desired frequency characteristic can be attained from an RC circuit as shown on the left hand side of Figure 6.17, with $C = 1 \ \mu F$ and $R = 50 \ \text{k}\Omega$.

The characteristics of other more complicated filters will be explored in the next section with the aid of MATLAB.

6.4 MATLAB IMPLEMENTATION

6.4.1 Transfer Function

Because transfer functions are already functions of frequency, it is easy to plot these functions using MATLAB. Simply define the function with the aid of a frequency vector, then plot. The plot can be either linear or logarithmic in the frequency axis. To plot logarithmically, use the MATLAB plotting function `semilogx`. To plot in decibels, simply take 20 times the log (the `log10` command in MATLAB) of the magnitude before plotting. (Recall the magnitude is obtained using the `abs` function and the phase using the `angle` function.) These points are illustrated in the example below, which plots the graphs in Figure 6.3.

Example 6.7: Plot the magnitude and phase frequency characteristics of the transfer function given as Eq. 6.2 (repeated below) and plotted in Figure 6.3.

$$\frac{V_{out}(\omega)}{V_{In}(\omega)} = \frac{1}{1 - 0.03\omega^2 + j.1\omega}$$

Solution: Because the transfer function was obtained from an RLC circuit that has two energy-storage devices, we know it should have a second-order denominator as, indeed, it does. We also expect a frequency characteristic with a slope of −40 dB/decade. The graph used in Figure 6.3 is plotted using linear horizontal and vertical axes, but the transfer function will be plotted in this solution using decibels and log frequency for variety.

To plot the transfer function using MATLAB we first generate a frequency vector over the range of frequencies desired (0.1 to 1,000 rad/sec in this example), then code the transfer function directly using this vector for ω.

```
% Example 6.7 Plots the transfer function given in Eq.
% 6.2 in both linear and dB/log coordinates
%
clear all; close all;
% Generate frequency vector between .1 and 100 rad/sec
w = (.1:.1:100);                       % Generate freq. Vector
%
% Generate the Transfer Function directly from Eq. 6.2
TF = 1./(1-.03*w.^2 + j*.1*w);         % Transfer Function
Mag = 20*log10(abs(TF));               % Take magnitude,
                                       % convert to dB
Phase = angle(TF)*360/(2*pi);          % Put phase in degrees.
%
% The remainder of the program is just plotting. First
% plot in linear coordinates (not shown), then dB/log
%
figure;                                % Plot in dB/log plots
subplot(2,1,1);
  semilogx(w/(2*pi),Mag,'k');          % Plot magnitude in Hz
  xlabel('Frequency (rad/sec)'); ylabel('Magnitude (db)');
subplot(2,1,2)
  semilogx(w/(2*pi),Phase,'k');        % Plot phase in Hz
  xlabel('Frequency (rad/sec)'); ylabel('Phase (deg)');
```

Analysis: Generating the transfer function, TF, requires only two lines of MATLAB code (in bold): One to produce the frequency vector w, and a second, which uses w to generate the transfer function. Because w is a vector, the transfer function division must be done using point-by-point division implemented with the MATLAB '. /' command. The same is true when squaring w. We are relying on the principle of superposition here, since we are actually solving the equation many times over (1,001 times to be exact), each using a different value of frequency. The magnitude and phase curves plotted using a linear horizontal and vertical axis are shown in Figure 6.3 while the dB/log plots are shown below.

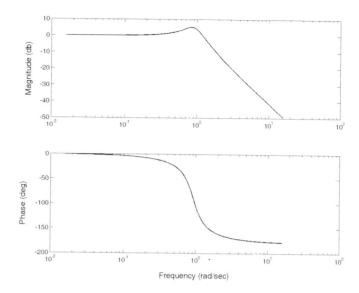

In this example, we were given the transfer function, but in most situations, we would have to determine the transfer function from the configuration of the circuit or mechanical system. If we are only interested in the system's frequency characteristics as is often the case, it is not necessary to determine the transfer function. The frequency characteristics can be determined directly from the system equations if we use a phasor domain input of 1.0 over all frequencies. With such an input, the system's output will be identical to the transfer function in the frequency domain:

$$TF(\omega) = \frac{Out(\omega)}{In(\omega)} = \frac{Out(\omega)}{1.0 \angle 0} = Out(\omega)\big|_{In(\omega)=1} \qquad \text{[Eq. 6.26]}$$

Therefore, in the frequency (or phasor) domain, the output of any system given a phasor input of $1.0\angle 0$ at all frequencies will be the transfer function of that system and the frequency characteristics of the system's transfer function. This is illustrated in the example below.

Example 6.8:   Find the frequency characteristics of the transfer function of the mechanical system below given $F(\omega)$ as the input and $v(\omega)$ as the output. Plot the magnitude and phase characteristics in log frequency with frequency in hertz. The element parameters are:

$$k_f = 0.5 \text{ dyne-sec/cm}; \quad k_e = 2 \text{ dyne/cm}; \quad m = 5 \text{ gm}$$

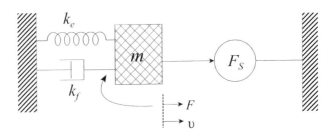

Solution: It may seem a little peculiar to have a system where force is the input and velocity is the output, but there is really nothing wrong with it: Biological systems have such mixed units all the time (neural signals in, movement out, and so forth). Applying Newton's law to this system:

$$F(\omega) - v(\omega)\left(k_f + \frac{k_e}{j\omega} + j\omega m\right) = 0$$

Assuming $F(\omega) = 1.0$, and solving for $v(\omega)$:

$$v(\omega) = \frac{1}{\left(k_f + \dfrac{k_e}{j\omega} + j\omega m\right)} = TF(\omega)$$

It would be good practice to put this equation into the standard transfer function format for a second-order denominator (such as in Eq. 6.2) by multiplying through by $j\omega$ and dividing by k_e. However, because we will be using MATLAB, and MATLAB does not care, the equation can be used as is. Programming the equation directly into MATLAB:

```
% Ex 6.8 Plot frequency characteristics of a second-order
% mechanical system directly
%
close all; clear all;
kf = 0.5;                        % Set m, kf, and ke values
m = 5;
ke = 2;
w = (.01:.01:100);               % Set frequency vector, w
%
% Now construct Impedance vector for this frequency
Z = kf + ke./(j*w) + j*w*m;
TF = 1./Z;                       % Solve for v = TF, (F =
                                 % 1.0)
%
```

```
Mag = 20*log10(abs(TF));        % Calculate magnitude in dB
Phase = angle(TF)*360/(2*pi);   % and phase in degrees
%
% Plot magnitude and phase using code similar to that in
   Example 6.7
```

Analysis: After assigning the parameters, the program constructs a vector, w, that ranges between 0.01 rad/sec and 100 rad/sec in steps of 0.01. Assigning the variables at the beginning rather than embedding them in the code as constants allows easy modification of these variables if need be. The impedance variable is constructed based on the equation for $\upsilon(\omega)$ above. (Again, note the use of the '. /' command.) Next, the transfer function is determined by solving for υ, with the force input set to 1.0 for all values of w. After converting the magnitude to decibels and the phase to degrees, the two are plotted using code similar to that in the last example.

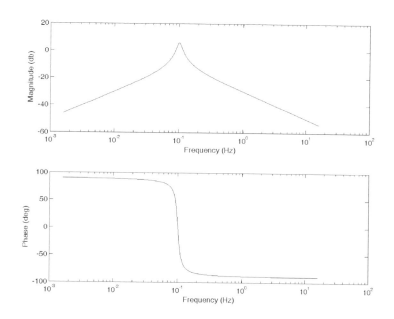

Again, this solution relies on superposition because we are essentially solving for $\upsilon(\omega)$ (i.e., the transfer function) at each frequency using MATLAB's vector arithmetic. When displaying all of these solutions on a single plot we are combining the different solutions to produce a seemingly continuous curve. Direct plotting of the frequency characteristics in MATLAB can be particularly advantageous in more complicated systems such as the two-mesh circuit given in the next example.

Example 6.9: Find the frequency characteristics of the transfer function, V_{out}/V_{in}, of the two-mesh circuit shown below.

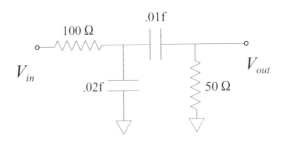

Solution: Circuit equations can be programmed directly to give the frequency characteristics; however, since this is a two-mesh circuit, matrix algebra will be required. The basic matrix equation applied using KVL was given in Chapter 5 and is repeated here:

$$V = ZI; \text{ specifically for a two-mesh circuit:}$$

$$\begin{bmatrix} \sum V_1 \\ \sum V_2 \end{bmatrix} = \begin{bmatrix} \sum Z_1 & -\sum Z_{12} \\ -\sum Z_{21} & \sum Z_2 \end{bmatrix} \begin{bmatrix} I_1 \\ I_2 \end{bmatrix} \qquad \text{[Eq. 6.27]}$$

If $V_{in} = 1$, V_{out} will equal the transfer function. To find V_{out}, use Eq. 6.27 to find I_2, then multiply by the output impedance (the 50 Ω resistor). As with the previous example, do this for each specific value of ω, using a loop to iterate ω over the desired range of values. In this solution, we must use a 'for loop' instead of MATLAB's vector math because of the matrix involved. The two previous examples could also be programmed this way, but they would execute slower. Some trial and error may be required to find an appropriate range of frequencies.

By inspection of the circuit, we find that:

$$Z_1 = 100 + \frac{50}{j\omega}; \quad Z_{12} = Z_{21} = \frac{50}{j\omega}; \quad Z_2 = 50 + \frac{50}{j\omega} + \frac{100}{j\omega}$$

```
% Ex 6.9 To plot frequency characteristics of a second-
order, two-mesh circuit directly
%
close all; clear all;
V = [1; 0];                        % Set Vin = 1.0
for k=1:10000
  w = .005 + .01*k;                % Set frequency, w
  % Now construct Impedance matrix for this frequency
  Z = [100 + (50/(j*w)), -50/(j*w); ...
       -50/(j*w), 50 + 50/(j*w) + 100/(j*w)];
  I = Z\V;                         % Solve matrix equation
  TF = I(2) * 50;                  % Find Vout from current I2
  Mag(k) = 20*log10(abs(TF));      % Convert to dB
```

```
    Phase(k)  = angle(TF)*360/(2*pi);
     f(k)= w/(2*pi);                    %freq. vector
 end
 %
 ...Plot as in Example 6.7.
```

The plot produced by this program is shown below.

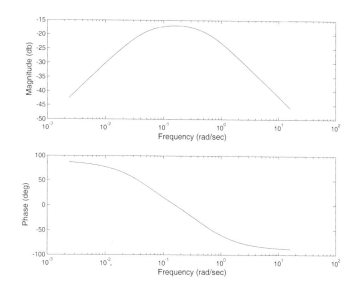

6.4.2 System Identification

Bioengineers are often faced with complex systems whose internal components are unknown. In such cases, it is clearly impossible to develop equations for system behavior to construct the frequency characteristics. However, if you can control the stimulus to the system, and measure its response, you should be able to determine the frequency characteristics experimentally, if the system is linear or can be taken as linear. To determine the frequency response experimentally, we take advantage of the fact that a sinusoidal stimulus will produce a sinusoidal response at the same frequency. By stimulating the biological system with sinusoids over a range of frequencies and measuring the change in amplitude and phase of the response, we can construct a plot of the frequency characteristics by simply combining all the individual measurements (Figure 6.21).

Figure 6.22 shows the magnitude frequency plots obtained in this way for the pupillary light reflex. In these classic studies, performed by L. Stark and colleagues in the 1960s, the eye was stimulated with light of sinusoidally varying intensity and the change in area of the pupil was measured using a technique based on infrared reflection. The response of the pupil to light is seen to decrease rapidly with

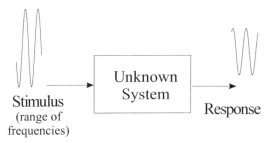

Figure 6.21 A If an unknown process can be stimulated with sinusoids and the resulting response measured, it is possible to determine its frequency characteristics experimentally. For each stimulus frequency, calculate the change in amplitude and phase (if desired) induced by the process. Combine the measurements over all frequencies to generate a frrequency plot.

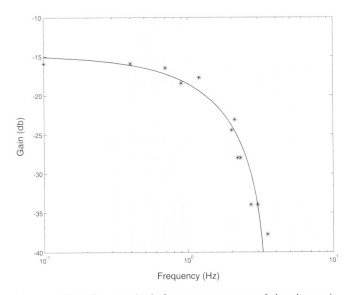

Figure 6.22 The magnitude frequency response of the change in pupil area when stimulated by sinusoidal variations of light intensity at different frequencies. (From L. Stark, Berkeley, CA, with permission.)

increasing frequency. The transfer function of the pupillary light reflex, Δarea/Δlight intensity, can be determined directly from these plots and is left as an exercise.

A taste of this experimentally based approach is given in the next example. To carry out what would normally be laboratory experiment, a MATLAB function termed X_perimental is used to *simulate* the unknown system. This function can be found on the CD. It takes in an input variable and produces an output variable, but what the function does is a mystery.

Example 6.10: Find the magnitude frequency characteristics of the process X_perimental.

Solution: Generate a sinusoid with a root-mean-squared (RMS) value of 1.0 (i.e., a peak-to-peak amplitude of 1.414). Input this sinusoid to the unknown process and measure the RMS value of the output. The RMS value is usually a more accurate measurement of a signal's value than the peak-to-peak amplitude as it is less susceptible to noise induced error. Repeat this protocol for a large number of frequencies (in this case up to 100 Hz). Plot the results as a single magnitude curve. If the phase was also desired, the phase of the input and output could be compared using cross-correlation.

```
% Example 6.10 Tests an unknown process termed
% 'X_perimental' by inputting
% sinusoidal at different frequencies and measuring the
% response amplitude.
% Combines the measurements into a single magnitude plot
%
clear all; close all;
N = 1000;                        % Input signal length
t = (0:N)/N;                     % Time vector ( 1 sec)
for k = 1:100
  f(k) = k;                      % Freq = k = 1 to 100Hz
  x = 1.414*cos(2*pi*f(k)*t);    % Sinusoid with Vrms = 1.0
  y = X_perimental(x);           % Input stimulus to process
  Y(k) = sqrt(sum(y.^2));        % Take RMS value as
                                 % magnitude
end
  Y = 20*log(Y);                 % Convert to dB
  semilogx(f,Y);                 % Plot magnitude as dB
                                 % versus log
  xlabel('Frequency (Hz)'); ylabel('Magnitude (dB)');
```

The results from this "experiment" are shown in the figure below.

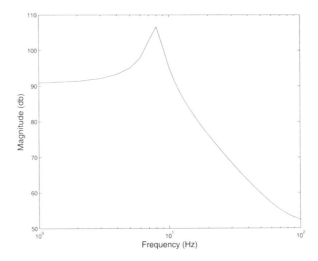

The process appears to be second-order with a damping factor less than 1. From the frequency characteristic, ω_n is approximately 8 Hz and the peak is approximately 15 dB above the low-frequency value. Hence, $20 \log(2\delta) = -15$ so δ is approximately 0.18. The gain term is around 91 dB, or 35,481 in relative units. From these numbers, a good estimate for the transfer function of this process would be:

$$TF(s) = \frac{35,481}{s^2 + 2\delta\omega_n s + \omega_n^2} = \frac{35,481}{s^2 + 1.44 + 64}$$

Sometimes it is not possible to stimulate a physiological system sinusoidally. In Chapter 9, we will extend the approach used in Example 6.10 to functions that are often easier to generate as physiological stimuli.

6.4.3   The Transfer Function and Fourier Series Decomposition

As mentioned previously, the transfer function can be used to find the output to any input, although in the phasor domain we are limited to sinusoidal or periodic signals. The next, and last, example shows how to combine MATLAB implementation of the transfer function with Fourier decomposition to find the output of any network to very complex signals.

Example 6.11: Find the output of the network below when the input is the electroencephalogram (EEG) signal first shown in Figure 2.4. Plot both input and output signals in both the time and frequency domain. For the circuit, R = 3 Ω, L = 2 h, and C = 0.001 f.

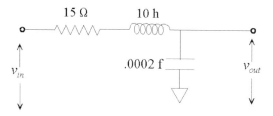

Solution: Using standard KVL analysis the transfer function of the circuit shown can be determined as:

$$\frac{V_{out}(\omega)}{V_{in}(\omega)} = \frac{1}{1 - 0.002\omega^2 + j0.003\omega}$$

In the MATLAB program, this transfer function will be applied to the EEG data after it is converted to the frequency domain using the MATLAB `fft` command. The output, $V_{out}(\omega)$ will be determined by multiplying the converted signal by the transfer function and will then be converted back into the time domain using the inverse Fourier transform command, `ifft`. The program plots the input EEG signal, the time domain reconstruction of the output, $v_{out}(t)$ and the frequency domain representation of both the input and output signals. Again, this approach takes advantage of the principle of superposition. This approach is summarized in the figure below followed by the MATLAB code. The conversions are highlighted in boldface type.

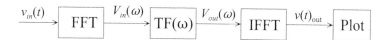

```
% Example 6.11 Applies an RLC Transfer Function to
% the EEG data stored in eeg_data.
%
clear all, close all;
%
% Section 1 Get and plot original EEG data
load eeg_data;
n = length(eeg);
fs = 50;                        % Sample frequency is 50 Hz
t = (1:n)/fs;                   % Construct time vector
subplot(2,1,1);
  plot(t,eeg,'k');              % Plot time data
  xlabel('Time (sec)'); ylabel ('EEG');
%
% Section 2 Decompose data
```

```
Vin = (fft(eeg));          % Vin to freq. domain
f = (1:n)*fs/n;            % Construct freq. vector(Hz)
%
% Section 3 Now solve for output using Vin components as
% input
Vout = Vin./(1 - .002*(2*pi*f).^2 + j *.003*2*pi*f);
%
% Section 4 Now reconstruct time data of output and plot
vout = ifft(Vout);
t = (1:length(vout))/fs;   % Construct new time vector
subplot(2,1,2);            % Plot time domain output data
  plot(t,real(vout),'k');
  xlabel('Time (sec)'); ylabel ('EEG');
%
% Section 5 Plot frequency domain data
figure;
nf = fix(length(Vout)/2);  % Plot only nonredundant
                           % points
subplot(2,1,1)             % Plot frequency domain data
                           % for both input and output
  plot(f(1:nf),...
  abs(Vin(1:nf)),'k');
  xlabel('Frequency (Hz)'); ylabel('Vin');
subplot(2,1,2);
  plot(f(1:nf),abs(Vout(1:nf)),'k');
  xlabel('Frequency (Hz)'); ylabel ('Vout');
```

The program is divided into five sections. The first section loads the data, constructs a time vector based on the sampling frequency, and plots these data. The second section uses the fft command to convert the EEG data to the frequency domain and constructs a frequency vector in Hz based on the sampling frequency (recall the highest frequency in the frequency data is f_s). The third section applies the transfer function to the Fourier Transform of the EEG data (Vin) and the fourth section converts the output of the transfer function back into the time domain using the inverse Fourier Transform program, ifft. The last section plots the frequency data. As with many MATLAB programs, much of the code is devoted to plotting. The Fourier series conversion, multiplication by the transfer function and the inverse Fourier series conversion require only one line of code each (shown in bold).

The plots generated by this program are shown in Figures 6.23 and 6.24. It is clear from the frequency plots that this circuit acts as a narrowband bandpass filter with a peak frequency around 4 Hz. This filter emphasizes the EEG activity in the region of 4 Hz as suggested by the time domain plots. Note how this filtering significantly alters the time domain appearance of the EEG waveform. Although an effective filter, this circuit would not make a practical filter due to the problems associated with real inductors, notably parasitic elements.

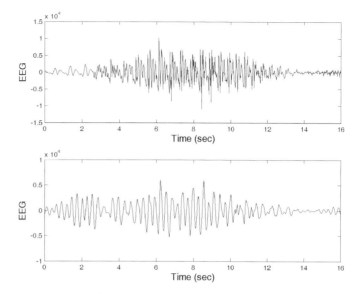

Figure 6.23 Electroencephalogram data before (*upper plot*) and after (*lower plot*) processing by the circuit of Example 6.11. Note the strong oscillatory behavior seen in the filtered data.

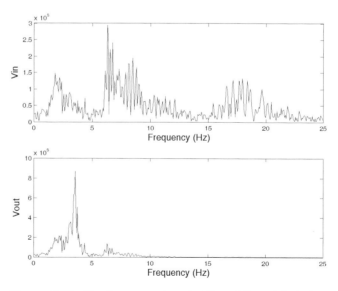

Figure 6.24 The frequency characteristics of the original electroencephalogram signal (*upper plot*) and the signal coming out of the RLC circuit (*lower plot*). A comparison of the two spectra shows the circuit acts as a filter removing frequencies outside a small passband range between approximately 3 and 6 Hz.

6.5   SUMMARY

The transfer function can represent complex networks or systems by a single equation. As presented here, the transfer function can only deal with signals that are sinusoidal or can be decomposed into sinusoids, but the transfer function concept will be extended to a wider class of signals in Chapter 8. The transfer function not only offers a succinct representation of the system, but also gives a direct link to the frequency characteristics of the system. Frequency plots can be constructed directly from the transfer function using MATLAB, or they can be drawn by hand using a graphical technique based on Bode plot primitives. With the aid of Bode plot primitives, we can also go the other way: from a given frequency plot to the corresponding transfer function. This would be the first step in the design of a network, system, or model with specific frequency characteristics.

Filters are circuits designed to modify the frequency characteristics of an input signal in a well-defined manner. Filters are categorized by their general type, lowpass, highpass, bandpass, and band stop, as well as by their bandwidth. Other important specifications include the filter order, which determines how steeply filter attenuation changes with frequency, and the initial sharpness. Filters are an important part of many analog circuit designs, and can also be implemented on a computer to process digitized signals.

The transfer function can provide the output of the system to any input signal as long as that signal can be decomposed into sinusoids. We first transform the signal into its frequency domain representation (i.e., its sinusoidal components) using Fourier series decomposition. We then multiply the frequency characteristics of the signal by those of the transfer function to obtain the frequency domain representation of the output signal. We then take the inverse Fourier transform to get the time domain output signal. In MATLAB, this three-step process takes only three lines of code.

When an unknown system is available to us for experimentation, we can input sinusoids at different frequencies, measure the magnitude and phase of the response, and plot the frequency characteristics of this unknown system. This strategy has been applied to many physiological systems. Using Bode plot primitives, we should be able to estimate a transfer function to represent the system. This transfer function can then be used to predict the system behavior to other input signals, including signals that might be difficult to generate experimentally. Thus, the transfer function concept coupled with Bode plot primitives can be a powerful tool for determining the transfer function representation of a wide range of physiological systems, and for predicting their behavior to an equally broad range of stimuli.

PROBLEMS

1. Find the transfer function, $V_{out}(\omega)/V_{in}(\omega)$, for the circuit below.

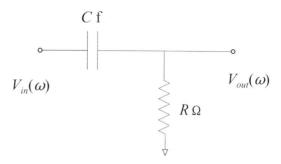

2. Find the transfer function of the circuit below.

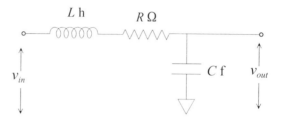

3. Plot the Bode plot (magnitude and phase) for the circuit in Problem 1 when $R = 100\ \Omega$ and $C = 0.0001$ f. Plot over three orders of magnitude from 1 to 1,000 rad/sec.

4. Plot the Bode plot (magnitude and phase) for the transfer function:

$$TF(\omega) = \frac{100(1+j.05\omega)}{(1+j.01\omega)}$$

5. Plot the Bode plot (magnitude and phase) for the transfer function:

$$TF(\omega) = \frac{100(1+j.001\omega)}{(1+j.005\omega)(1+j0.002\omega)}$$

6. Plot the Bode plot (magnitude and phase) for the circuit in Problem 6.2 where $R = 10\ \Omega$, $L = 2$ h, $C = 0.0002$ f. Plot between 1 and 1,000 rad/sec.

7. Determine the Bode plot of the circuit on the right side of Figure 6.17.

8. Plot the Bode plot (magnitude and phase) of the mechanical system shown below, where the input is $F_s(t)$ and the output is $x(t)$. $k_f = 10$ dyne-sec/cm, $k_e = 500$ dyne/cm, $m = 5$ gm. Plot between 1 and 1,000 rad/sec.

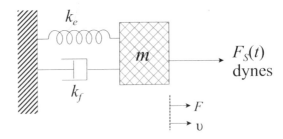

9. Find the transfer function that produces the magnitude frequency curve shown below.

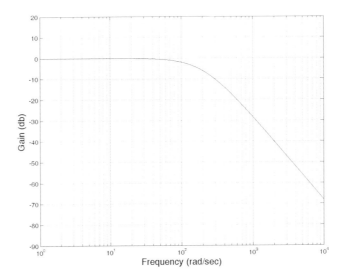

10. Find the transfer function that produces the magnitude frequency curve shown below. (*Hint:* This curve was produced by a second-order transfer function and could be represented by a single second-order term with a larger δ, but would be more accurately represented by two first-order terms.)

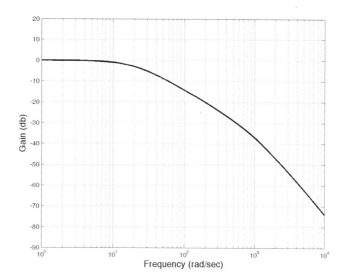

11. Estimate the magnitude transfer function of the pupil light reflex data in Figure 6.22. (*Hint:* Several answers are possible, but the original researchers used a series of first-order terms.)

MATLAB Problems

12. Plot the Bode plot (magnitude and phase) for the transfer function of the circuit in Problem 6.2 for values of $R = 1\ \Omega$, $10\ \Omega$, $70\ \Omega$, and $500\ \Omega$. Assume that $L = 1$ h and $C = 0.0001$ f. How does the resistance affect the shape of the frequency plot?

13. Plot the Bode plot (magnitude and phase) for the transfer function for the mechanical system in Problem 6.8 for values of $k_e = 2$ dyne/cm, $m = 3$ gm, and $k_f = 1$, 10, 70, and 500 dyne-sec/cm. How does the friction element affect the shape of the frequency plot? (*Hint:* Modify the code in Example 6.7.)

14. Plot the Bode plot (magnitude and phase) of the transfer function of the circuit below. Note the units for the various elements. What type of filter is it? Plot between 100 and 10^6 rad/sec, but plot in Hz. (*Hint:* Extend the approach used in Example 6.9.)

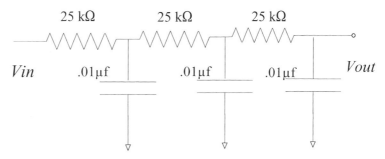

15. Plot the Bode plot (magnitude and phase) for the transfer function of the circuit in Problem 6.14, but reverse the position of the resistors and capacitors. That is, put the capacitors across the top and connect the resistors to ground. How does this change the filter?

16. Repeat Example 6.11 except change the values of the components to: $R = 2\ \Omega$, $L = 0.25$ h, and $C = 0.001$ f. You will have to derive the transfer function for the circuit in Figure 6.1 for the new values of R, L, and C. What change has this made in the properties of the filter?

17. The function unknown found on the disk represents a biological system. Determine the magnitude transfer function for this unknown process using the approach outlined in Example 6.10.

7 RELATIONSHIPS BETWEEN ANALOG ELEMENTS

7.1 SYSTEM SIMPLIFICATIONS: PASSIVE NETWORK REDUCTION

Sometimes it is desirable to simplify a circuit or system by combining elements. Many quite complex circuits can be reduced to just a few elements using an approach known as *network reduction*. Such simplifications can be very valuable: They can be used to simplify analysis and provide a summary-like representation of a complicated system. New insights can be had on the properties of a network after it has been simplified. Network reduction can be particularly useful when two networks or systems are to be connected together as the reduced forms show how their interconnection will affect each network and the passage of information between them. Finally, the principles of network reduction are useful in understanding the behavior of real sources.

Reduction principles are sometimes necessary to understand and analyze the biomedical measurements. Making a measurement on a biological system can be viewed as connecting two systems together: the biological system and the measurement system. All measurements require drawing some energy from the system being measured. How much energy is taken depends on a match between the biological and measurement system and this match may be defined in terms of impedance. This match between the biological and measurement system may be quantified in terms of a generalized concept of impedance; specifically, the difference between the output impedance of the biological system and the input impedance of the measurement system.

Before we can reduce complex networks or systems, we must first learn to reduce simple configurations of elements such as series and parallel combinations. Network reduction is based on a few simple rules for combining series and/or parallel elements. The approach is straightforward, although implementation can become quite tedious for large networks. After introducing the reduction rules for networks consisting only of passive elements, the guidelines will be expanded to include networks with sources. These reduction tools will then be applied to problems were two systems are interconnected.

7.1.1   Series Electrical Elements

Electrical elements are said to be in *series* when they are connected end-to-end and no other elements share that common connection as shown in Figure 7.1A. Although series elements are often drawn as in line with one another, they can be drawn in other configurations and still be in series. The three elements in Figure 7.1B are also in series as long as no other elements are connected between the elements. By simple application of Kirchhoff's voltage law (KVL) it can be demonstrated that when elements are in series, their impedances add: The voltage across three series elements is:

$$v_{total} = v_1 + v_2 + v_3 = (Z_1 + Z_2 + Z_3)i.$$

The total voltage can also be written as:

$$v_{total} = Z_{eq}i; \quad \text{where } Z_{eq} = Z_1 + Z_2 + Z_3.$$

Thus, series elements can be represented by a single element that is the sum of the individual elements (Eq. 7.1 and Figure 7.1A):

$$Z_{eq} = Z_1 + Z_2 + Z_3 + \cdots \qquad \text{[Eq. 7.1]}$$

If the series elements are all resistors or all inductors, they can be represented by a single resistor or inductor that is the sum of the individual elements:

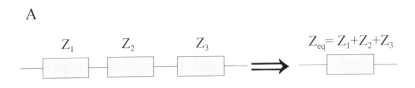

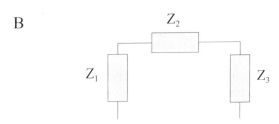

Figure 7.1   A: Elements in series, Z_1, Z_2, and Z_3, can be converted into a single equivalent element, Z_{eq} that is the sum of the three individual elements. Elements are in series when they share one node and no other elements share this node. B: Series elements are often drawn in line, but these elements are also in series. (As long as nothing else is connected between the elements.)

$$R_{eq} = R_1 + R_2 + R_3 + \cdots \qquad \text{[Eq. 7.2]}$$

$$L_{eq} = L_1 + L_2 + L_3 + \cdots \qquad \text{[Eq. 7.3]}$$

If the elements are all capacitors, then their reciprocals add, since the impedance of a capacitor is a function of $1/C$:

$$1/C_{eq} = 1/C_1 + 1/C_2 + 1/C_3 + \cdots \qquad \text{[Eq. 7.4]}$$

If the string of elements includes different element types, the individual impedances can be added using complex arithmetic to determine a single equivalent impedance. In general, this element will be complex, as shown in Example 7.1

Example 7.1: Series element combination. Find the equivalent single impedance, Z_{eq}, of the series combination below.

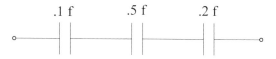

Solution: It is possible first to combine the two resistors into a single 25-Ω resistor, then combine the two inductors into a single 11-H inductor, and then add the three impedances (R_{eq}, $j\omega L_{eq}$, and $1/j\omega C$). Alternatively, simply convert each element to its equivalent impedances, then add these impedances.

$$Z_{eq} = 10 + j\omega5 + \frac{1}{j\omega.01} + 15 + j\omega6 = 25 + j\omega11 + \frac{100}{j\omega}$$

If a specific frequency is given, for example, $\omega = 2.0$ rad/sec, then Z_{eq} can be evaluated as a single complex number.

$$Z(\omega = 2) = 25 + j(11(2) - 100/2) = 25 - j28 \ \Omega = 37.5 \ \angle -48 \ \Omega$$

Example 7.2: Find the equivalent capacitance for the three capacitors in series below.

Solution: Because all the elements are capacitors, they add as reciprocals:

$$1/C_{eq} = 1/C_1 + 1/C_2 + 1/C_3 = 1/0.1 + 1/0.5 + 1/0.2 = 10 + 2 + 5 = 17$$

$$C_{eq} = 1/17 = 0.059 \text{ f}$$

7.1.2 Parallel Elements

Elements are in parallel when they share both connection points as shown in Figure 7.2. For parallel electrical elements, it does not matter if other elements share these mutual connection points, as long as *both* ends of the elements are connected to one another.

When looking at electrical schematics, it is important to keep in mind the definition of parallel and series elements because series elements may not be drawn in line (Figure 7.1B) and parallel elements may not be drawn as geometrically parallel. For example, the two elements, Z_1 and Z_2 on the left side of Figure 7.3 are in parallel because they connect at both ends even though they are not drawn in parallel geometrically. Conversely, elements Z_1 and Z_2 in the right hand figure are drawn parallel, but they are not in parallel electrically because they do not connect at each end (in this case, they are not connected at all).

As can be shown by Kirchhoff's Current Law (KCL), parallel elements combine as the reciprocal of the sum of the reciprocals of each impedance. Applying KCL to the upper node of the three parallel elements in Figure 7.2, the total current flowing through the three impedances would be: $i_{total} = i_1 + i_2 + i_3$. Substituting in v/Z for the currents through the impedances, the total current becomes:

$$i_{total} = v/Z_1 + v/Z_2 + v/Z_3 = v(1/Z_1 + 1/Z_2 + 1/Z_3).$$

This equation, restated in terms of an equivalent impedance, becomes:

$$i_{total} = v/Z_{eq}, \quad \text{where } 1/Z_{eq} = 1/Z_1 + 1/Z_2 + 1/Z_3.$$

Figure 7.2 Parallel elements share both end connection points. Three elements in parallel, Z_1, Z_2, and Z_3, can be converted into a single equivalent element, Z_{eq}, that is the reciprocal of the sum of the reciprocals of the three individual elements.

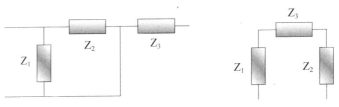

Figure 7.3 **Left circuit:** The elements Z_1 and Z_2 are connected at both ends and are therefore electrically in parallel even though they are not drawn parallel. **Right circuit:** Although they are drawn parallel, elements Z_1 and Z_2 are not electrically parallel because they are not connected at both ends.

Hence,

$$1/Z_{eq} = 1/Z_1 + 1/Z_2 + 1/Z_3 + \cdots \qquad \text{[Eq. 7.5]}$$

Eq. 7.5 also holds for the value of parallel resistors and inductors:

$$1/R_{eq} = 1/R_1 + 1/R_2 + 1/R_3 + \cdots \qquad \text{[Eq. 7.6]}$$

$$1/L_{eq} = 1/L_1 + 1/L_2 + 1/L_3 + \cdots \qquad \text{[Eq. 7.7]}$$

Parallel capacitors, however, simply add because C is in the denominator of the impedance term:

$$C_{eq} = C_1 + C_2 + C_3 + \cdots \qquad \text{[Eq. 7.8]}$$

Hence, if the three capacitors in Example 7.2 were in parallel, their equivalent capacitance would be simply the addition of the three values:

$$C_{eq} = 0.1 + 0.5 + 0.2 = 0.8 \, \text{f}$$

Example 7.3: Find the equivalent single impedance for the parallel RLC combination below.

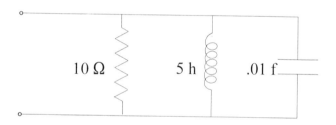

Solution: First take the reciprocals of the impedances:

$$1/R = 1/10 = 0.1\text{ʊ}; \quad 1/j\omega L = 1/j5\omega\text{ʊ}; \quad j\omega C = j0.01\omega\text{ʊ}$$

Then add, and invert:

$$\frac{1}{Z_{eq}} = Y_{eq} = 0.1 + \frac{1}{j\omega 5} + j\omega 0.01 = 0.1 + j\left(0.01\omega - \frac{0.2}{\omega}\right)$$

$$Z_{eq} = \frac{1}{Y_{eq}} = \frac{1}{0.1 + j\left(0.01\omega - \dfrac{0.2}{\omega}\right)}$$

Once a value of frequency, ω, is given, this equation could be solved for a specific impedance value. Alternatively, we could solve for Z_{eq} over a range of frequencies using MATLAB This is given as Problem 15 at the end of the chapter.

Example 7.4: Find the equivalent resistance of the parallel combination of three resistors: 10 Ω, 15 Ω, and 20 Ω.

Solution:　Calculate reciprocals, add them, then invert:

$$1/R_{eq} = 1/R_1 + 1/R_2 + 1/R_3 = 1/10 + 1/15 + 1/20 = 0.1 + 0.0667 + 0.05 = 0.217\mho$$

$$R_{eq} = 1/0.217 = 4.61\ \Omega$$

Note that the equivalent resistance of a parallel combination of resistors will always be less than the smallest resistor in the group. The same will be true for inductors while the opposite is true for parallel capacitors.

7.1.2.1　Combining Two Parallel Impedances

Combining parallel elements via Eq. 7.5 is mildly irritating with its double inversions. Most parallel element combinations involve only two elements, so it is useful to have an equation that directly states the equivalent impedance of two parallel elements without the inversion. Starting with Eq. 7.5 for two elements:

$$\frac{1}{Z_{eq}} = \frac{1}{Z_1} + \frac{1}{Z_2} = \frac{Z_2}{Z_2 Z_1} + \frac{Z_1}{Z_1 Z_2} = \frac{Z_1 + Z_2}{Z_1 Z_2}$$

$$\text{Inverting: } Z_{eq} = \frac{Z_1 Z_2}{Z_1 + Z_2} \qquad \text{[Eq. 7.9]}$$

Hence the equivalent impedance, Z_{eq}, of two parallel elements equals the product of the two impedances divided by their sum.

7.1.3　Network Reduction: Passive Networks

The rules for combining series and parallel elements can be applied to networks that include a number of elements. Even very involved configurations of passive elements can usually be reduced to single element. Obviously, it is easier to grasp the significance of a single element than a confusing combination of many elements.

7.1.3.1　Network Reduction: Successive Series—Parallel Combination

In the last section, we saw that it is possible to combine a number of series or parallel combinations. Even when most of the elements are not in either series or parallel configurations, it is possible to combine them into a single impedance using the techniques of network reduction. In this section, the networks being reduced consist only of passive elements, but in the next section, we will learn how to expand the concepts of network reduction to include networks with sources.

In the network in Figure 7.4, most of the elements are neither in series nor in parallel. It is important to realize that the combination of elements across the top of this network—inductor, resistor, inductor—are *not* in series because their connection points are shared by other elements, capacitors in this case. To be in series, the elements must not only share one connection point, they must be the only elements to share that point. If we could somehow eliminate the two capacitors (we

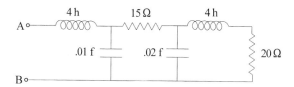

Figure 7.4 A network containing R, L, and C's where most of the elements are neither in series nor in parallel. Nonetheless, this network can be reduced to a single equivalent impedance (with respect to nodes A and B) as is shown in Example 7.5.

cannot), then these three elements would be in series. Nor are any of the elements in parallel since no elements share both connection points. If they did, they would be in parallel even if other elements shared these connection points. However, there are two elements in series: the 4-h inductor and the 20-Ω resistor on the right-hand side of the network. We could combine these two elements using Eq. 7.1. After combining these two elements, we now find that the new combined element is in parallel with the 0.02-f capacitor. We can then combine these two parallel elements using the parallel rule given in Eq. 7.9. Although the argument may become difficult to follow at this point, the newly combined parallel element is now in series with the 15-Ω resistor. Most reductions of passive networks proceed in this fashion: Find a series or parallel combination to start with, combine them, and look to see if the new combination produces a new series or parallel combination. Then just continue down the line.

   The example below uses the approach based on sequential series–parallel combinations to reduce the network in Figure 7.4 to a single equivalent impedance.

Example 7.5: Network reduction using sequential series-parallel combinations. Find the equivalent impedance between the nodes A and B in Figure 7.4. Find the impedance at only one frequency, $\omega = 5.0$ rad/sec. {Using network reduction, we could find the equivalent impedance leaving frequency as a variable [i.e., $Z(\omega)$], but this would make the algebra more difficult. With a specific frequency, we are able to use complex arithmetic instead of complex algebra.}

Solution: Convert all elements to their equivalent impedances at $\omega = 5.0$ rad/sec. (to simplify subsequent calculations). Then begin the reduction by combining the two series elements on the right-hand side together. As a first step, the elements are first converted to their phasor impedances (at $\omega = 5$ rad/sec). Then the two series elements are combined using Eq. 7.1 leading to the network shown below on the right-hand side.

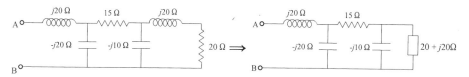

This combination puts two elements in parallel: the newly formed impedance and the $-j10\ \Omega$ capacitor. These two parallel elements can be combined using Eq. 7.9:

$$Z_{eq} = \frac{Z_1 Z_2}{Z_1 + Z_2} = \frac{(20 + j20)(-j10)}{20 + j20 - j10} = \frac{200 - j200}{20 + j10} = \frac{282.8\ \angle - 45}{22.36\ \angle\ 27}$$

$$Z_{eq} = 12.65\ \angle - 72\ \Omega = 4 - j12\ \Omega$$

This leaves a new series combination that can be combined as shown below. In the third step of network reduction, the newly formed element from the parallel combination ($4 - j12\ \Omega$) is now in series with the 15-Ω resistor, and this equivalent series element will be in parallel with the 0.01-f capacitor.

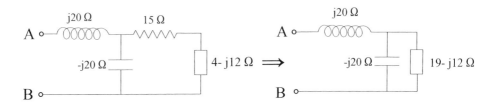

Combining these two parallel elements:

$$Z_{eq} = \frac{-j20(19 - j12)}{-j20 + 19 - j12} = \frac{-240 - j380}{19 - j32} = \frac{449.4\ \angle\ 238}{37.2\ \angle - 59}$$

$$Z_{eq} = 12.07\ \angle\ 297\ \Omega = 5.48 - j10.75\ \Omega$$

This leads to the final series combination and a single equivalent impedance as shown. In the final step, the two series elements are combined to a single equivalent impedance.

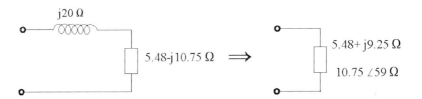

Network reduction is always done from the point of view of two nodes, such as nodes A and B in this example. In principle, any two nodes can be selected for analysis and the equivalent impedance can be determined between these nodes. Generally the nodes selected have some special significance; for example, the nodes that make up the input or output of the circuit. Network reduction usually follows the format of this example: sequential combinations of series elements, then parallel elements, then series elements, and so on. In a few networks, there are no elements that are either in series or in parallel to start with, and an alternative method described in the next section must be used. (In these cases, a transformation in the configuration of three elements called a pi to T transformation can be used, but the application of this transformation is rare in electronics and will not be covered here.) The voltage-current method described in the next section works for all networks and any combination of two nodes, but it is usually more computationally intensive. On the other hand, it does lend itself to computer solution using MATLAB.

7.1.3.2 Network Reduction: Voltage–Current Method

The other way to find the equivalent impedance of a network follows the approach that would be used given an actual network in a laboratory setting. Suppose you were asked to determine the impedance between two nodes of a network (a classic so-called *two-terminal* problem). Perhaps the actual network was inaccessible to you and all you had available were the two nodes as shown in Figure 7.5.

The actual network could contain any number of nodes, but we assume that only two are accessible for measurement, or are of interest. Even without *opening the box*, it is possible to determine the equivalent impedance between these two terminals. What you would do in this situation (or, at least, what I would do) is to apply a known voltage to the two terminals, measure the resulting current, and calculate the equivalent impedance Z_{eq}, using Ohm's law:

$$Z_{eq} = \frac{V_{known}}{I_{measured}}$$

[Eq. 7.10]

Of course, V_{known} would have to be a sinusoidal source of known amplitude, phase, and frequency unless you knew, a priori, that the network was purely resistive, in which case a DC (direct current) source would suffice. Moreover, you would be limited to determining Z_{eq} at only one frequency, the frequency of the voltage source, but most laboratory sinusoidal sources offer a range of selectable frequencies so the impedance could be determined over a range of frequencies. If the unknown network contained sources, a slightly different strategy developed in the next section could be used instead.

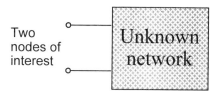

Figure 7.5 An unknown network that has only two terminals available for use or for measurements. This is a classic two-terminal device.

This same approach can be applied to a network that exists only on paper, such as the network in the last example. Using the tools that we have acquired thus far, we simply connect a voltage source of our choosing to the network and solve for the current into the network. The source could be of any voltage, frequency, or phase. Moreover, this approach can be applied to simplify any network, even one that does not have any series or parallel elements. This method is applied to the network in Example 7.6 and, in the subsequent example, to an even more challenging network.

Example 7.6:   Passive network reduction using the source-current method. Find the equivalent impedance between nodes A and B in the network of Figure 7.4 for a frequency of $\omega = 5$ rad/sec.

Solution: Apply a known source across nodes A and B and solve for the resulting current using mesh analysis. For a source, we could chose any sinusoidal source at 5 rad/sec, but why not choose something simple like 1 volt at 0.0 degree phase. The desired impedance Z_{ab} will then be $Z_{ab} = 1/I_1(\omega)$ and mesh analysis can be used to solve for $I_1(\omega)$. Going directly to phasor notation, the network as a standard 3 mesh is shown below.

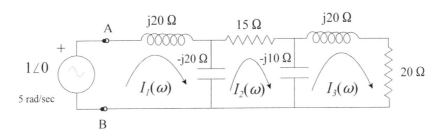

To find the input current, we could analyze this as a straightforward three-mesh problem solving for I_1. Several alternatives are also possible: using network reduction to convert the last three elements to a single element and solving as a two-mesh problem; converting the voltage source to an equivalent current source (as shown later in this chapter) and solving as a two-node problem; or combining these approaches and solving as a one-node problem. In this example, we will chose the first suggestion and solve the three-mesh problem directly with the aid of MATLAB. Applying standard mesh analysis to the network above, the basic matrix equation can be written as:

$$\begin{vmatrix} 1.0 \\ 0 \\ 0 \end{vmatrix} = \begin{vmatrix} j20 - j20 & +j20 & 0 \\ +j20 & 15 - j20 - j10 & +j10 \\ 0 & +j10 & 20 + j20 - j10 \end{vmatrix} \begin{vmatrix} i_1 \\ i_2 \\ i_3 \end{vmatrix}$$

Simplifying by complex addition:

$$\begin{vmatrix} 1.0 \\ 0 \\ 0 \end{vmatrix} = \begin{vmatrix} 0 & j20 & 0 \\ j20 & 15-j30 & j10 \\ 0 & j10 & 20+j10 \end{vmatrix} \begin{vmatrix} i_1 \\ i_2 \\ i_3 \end{vmatrix}$$

Solving for I_1, then Z_{eq} using MATLAB:

```
% Example 7.6 Find the equivalent impedance of a network
% by applying a source and solving for the resultant
% current.
%
% First assign values for v and Z
v = [1 0 0];
Z = [0 20j 0; 20j 15-30j 10j; 0 10j 20+10j];
%
i = Z\v                      % Solve for currents
Zeq = 1/i(1);                % Solve for Zeq
Zeq = [abs(Zeq) angle(Zeq)*360/(2*pi)]
```

Again note that the product symbol, *, can be omitted if the j follows the number. The output from this program is: Zeq = 10.75 ∠59.3.

This answer is the same as that found by standard reduction in Example 7.5. Again, this approach could be used to reduce any passive network of any complexity to an equivalent impedance between any two terminals. The latter portion of this chapter shows how to reduce, and think about, networks that also contain sources. Before proceeding to the next section, a final example will show how to reduce a passive network when the two terminals of interest are in more complicated positions (for example, separated by more than a single element). In this example, we will also solve for the impedance over a range of frequencies.

Example 7.7: Find the equivalent impedance of the circuit shown in Figure 7.6 between terminals A and B. Use MATLAB to find Z_{eq} over a range of frequencies,

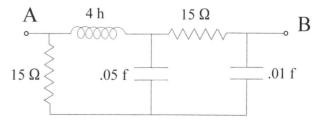

Figure 7.6 Network used in Example 7.7. The goal is to find the equivalent impedance between nodes A and B over a frequency range of 0.01 to 100 rad/sec.

ω, and plot Z_{eq} as a function of log frequency. Use a frequency range of 0.01 to 100 rad/sec. (This frequency range showed the most interesting behavior as determined by trial and error using MATLAB.)

Solution: Apply a source between terminals A and B and solve for the current flowing out of the source. In this problem, the source will have a fixed amplitude (1.0 V), a fixed phase (0.0 degrees), but the frequency must be variable so that Z_{eq} can be determined over the specified range of frequencies. Before applying the source, it is helpful to rearrange the network so that the meshes can be more readily identified. Sometimes this topographical reconfiguration can be the most challenging part of the problem, especially to individuals who are spatially challenged.

In this network, simply rotating the network by 90 degrees makes the mesh arrangement evident as shown in the figure below.

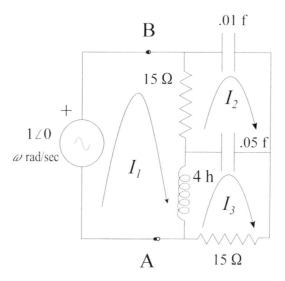

Unfortunately, this network cannot be reduced by simple series-parallel combinations as used previously. While transformations that are more complicated can be used,[1] we can just as easily use the voltage–current method used in the previous example. Again, the problem is solved by applying standard network analysis. Assigning a voltage source and mesh currents as shown above, the total current out of the source will be i_1.

---

[1] Two slightly more complicated transformations exist that allow configurations such as this to be reduced by the first method. These transformations are known as the Π(pi) *to* T transformation and vice versa. However, the voltage–current approach applies to any network and lends itself well to computer analysis as shown here.

After converting all elements to phasor notation, the matrix equation can be written as:

$$\begin{vmatrix} 1 \\ 0 \\ 0 \end{vmatrix} = \begin{vmatrix} 15+j4\omega & -15 & -j4\omega \\ -15 & 15-j100/\omega-j20/\omega & j20/\omega \\ -j4\omega & j20/\omega & 15+j4\omega-j20/\omega \end{vmatrix} \begin{vmatrix} I_1 \\ I_2 \\ I_3 \end{vmatrix}$$

Note that the frequency, ω, must remain a variable in this equation because we need to find the value of Z_{eq} over a range of frequencies. MATLAB can be used to find specific values of Z_{eq} over the range of frequencies, and these will be plotted:

```
%Example 7.7 To find and plot the values of an equivalent
%impedance between .01 and 100 rad/sec
clear all; close all;
%
% Define frequency, use .01 rad/sec increments
w = .01:.01:100;                    % Define frequency vector
v = [1; 0; 0];                      % Define voltage vector
%
% Loop over all frequencies, solving for Z_eq
for k = 1:length(w)
  % Define impedance vector (Use continuation statement)
  Z = [15+4j*w(k), -15, -4j*w(k); -15, 15-120j/w(k),...
      20j/w(k); -4j*w(k), 20j/w(k),
      15+4j*w(k)-20j/w(k)];
  i = Z\v;                          % Solve for current
  Z_eq (k) = 1/i(1);                % Solve for Z_eq
end
.......plot and label magnitude and phase.
```

The graph produced by this program is shown in Figure 7.7. Both the magnitude and phase of Z_{eq} are functions of frequency.

It is easy to find the maximum and minimum of MATLAB variables using the MATLAB max or min functions. Applying these functions to the Z_{eq}:

```
[max_Z_eq _abs, abs_freq] = max(abs(Zeq));
[max_Zeq_phase, phase_freq] = min(angle(Zeq)*360/(2*pi));
%
%Now display the maximum and minimum values including
 frequency
disp([max_Zeq_abs, w(abs_freq); max_Zeq_phase,...
w(phase_freq)])
```

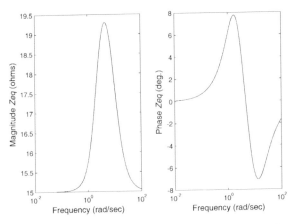

Figure 7.7 The value of Z_{eq} as a function of frequency for the network in Example 7.7. The impedance is plotted in terms of magnitude and phase. These plots were obtained from the MATLAB code used in Example 7.7.

This produces the values:

$$|Z_{eq}|_{max} = 19.29\ \Omega \quad \omega_{max} = 4.5\ \text{rad/sec}$$

$$\angle Z_{eq\_max} = 7.07\ \text{degrees} \quad \omega_{max} = 13.3\ \text{rad/sec}$$

Note that the `max` or `min` functions give the maximum or minimum and the variable index at which these values occur. To convert the index to the appropriate frequency, just take `w(index)`, as in the code above.

The remainder of this chapter examines the characteristics of sources, both real and ideal, and develops methods for reducing networks that contain sources. Many of the principles used here in passive networks apply equally well to networks that are more general.

7.2 IDEAL AND REAL SOURCES

Before developing methods to reduce networks that contain sources, it is helpful to revisit the properties of ideal sources described in Section 4.2.3 and to examine how ideal and real sources differ. In this discussion, only constant output sources (i.e., DC sources) are considered, but the arguments presented generalize without only minor modification to time varying sources as shown in the next section.

7.2.1 The Voltage–Current or *v-i* Plot

Essentially an ideal source can supply any amount of energy that is required by whatever is connected to that source. Discussions of real and ideal sources often use plots of voltage against current (or force against velocity), which provide a visual representation of the source characteristics. Such '*v-i*' plots are particularly effective at demonstrating the equivalent resistance of an element.

Recall that the equation for a straight line is $y = mx + b$, where m is the slope of the line and b is the intercept. From Ohm's law, $v = Ri$, so the voltage–current relationship for all resistors will plot as straight lines with a slope equal to the value of the resistance and an intercept of 0.0. The *v-i* plots of five different resistors are shown in Figure 7.8: 0, 10, 100, 1,000, and infinity Ω.

Taking the reverse argument, an element that plots as a straight line on a *v-i* plot is either a resistor or contains a resistance, and the slope of the line indicates the value of the resistance. The steeper the slope, the greater the resistance: a vertical line that has a slope of infinity indicates the presence of an infinite resistance and a horizontal line that has a slope of zero indicates the presence of a 0.0-Ω resistance.

The *v-i* plot of an ideal DC voltage source follows directly from the definition: a source of voltage that is constant regardless of the current though it. For a time-varying source, such as a sinusoidal source, the voltage would vary as a function of time, but *not* as a function of the current. An ideal voltage source cares naught about the current through it. Hence, the *v-i* plot of an ideal DC voltage source, V_S,

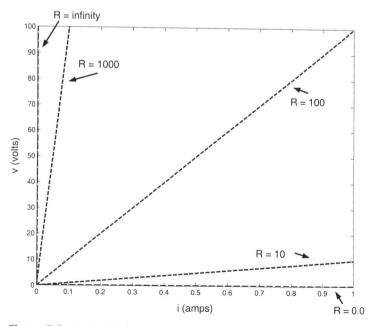

Figure 7.8 A *v-i* plot, showing voltage against current, for resistors having 5 different values between 0.0 Ω and infinity Ω. This shows that the *v-i* plot of a resistor is a straight line passing through the origin and having a slope equal to the resistance.

would be a flat line intersecting the horizontal (voltage) axis at $v = V_S$ (Figure 7.9). If the voltage source were time varying, the *v-i* plot would look essentially the same except that the height of the horizontal line would vary in some specific manner with time.

This straight-line plot with zero slope clearly demonstrates that the resistive component of an ideal voltage source is zero. In other words, the equivalent resistance of an ideal voltage source is 0.0 Ω. With regard to the *v-i* plot, an ideal source looks like a resistor of 0.0 Ω with an offset of V_S. It may seem strange to talk about the equivalent resistance of a voltage source, especially when the resistance is zero, but the concept of equivalent resistance, the resistance of a component ignoring its other electrical properties, becomes important in real sources where the equivalent resistance is no longer zero. The concept of equivalent resistance, or more generally equivalent impedance, is a useful concept in network reduction and has important implications in transducer analysis and design.

Resistive elements having either zero or infinite resistance have special significance and have their own terminology. Resistances of zero will produce no voltage drop regardless of the current running through them. As stated in Chapter 4, an element that has zero voltage for any current is termed a *short circuit* because current flows freely through such elements. Resistors with infinite resistance have the opposite voltage–current relationship, they allow no current regardless of the voltage (assuming that it is finite). Again in Chapter 4, devices that have zero current for any voltage are termed *open circuits* because they do not provide a path for current. Hence, elements having infinite resistance are open circuit. To summarize, short circuits have zero resistance and zero voltage drop for any current while open circuits have infinite resistance and zero current for any voltage.

The *v-i* plot of an ideal current source is also evident from its definition: an element that produces a specified current regardless of the voltage across it. This leads to the *v-i* plot shown in Figure 7.10 of a vertical line that intersects the current

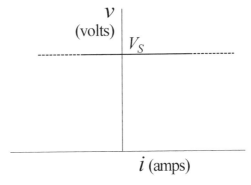

Figure 7.9   A *v-i* plot of an ideal voltage source. This plot shows that the resistor-like properties of a voltage source have a zero value.

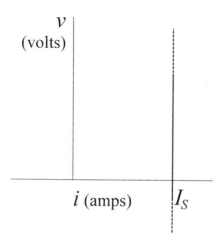

Figure 7.10 A *v-i* plot of an ideal current source. This plot shows that the resistor-like properties of a current source have an infinite value.

axis at $i = I_S$. By the arguments above, the equivalent resistance of an ideal current source is infinite.

The concepts of ideal voltage and current sources are somewhat counterintuitive. An ideal voltage source has the resistive properties of a short circuit, but also somehow maintains a nonzero voltage. The trick is to understand that an ideal voltage source is a short circuit with respect to current, but not with respect to voltage. Understanding this apparent contradiction is critical to understanding the properties of ideal voltage (or force) sources. A similar apparent contradiction applies to current sources: They are open circuits with respect to voltage yet produce a specified current.

In this section, voltage and current sources are described in terms of fixed values (i.e., DC sources), but the basic arguments do not change if V_S or I_S are time varying. This generalization will hold true for the other arguments presented below.

7.2.2 Real Voltage Sources: The Thévenin Source

Unlike ideal sources, real voltage sources are not immune to the current flowing through them, nor are real current sources immune to the voltage falling across them. In real voltage sources, the source voltage drops as more current is drawn from the source. This gives rise to a *v-i* plot such as shown in Figure 7.11, where the line is no longer horizontal but decreases with increasing current. The decrease indicates the presence of an internal, nonzero, resistance having a value equal to the negative slope of the line. The slope is negative because the voltage drop across the internal resistance is opposite in sign to V_s and hence subtracts from the values of V_s.

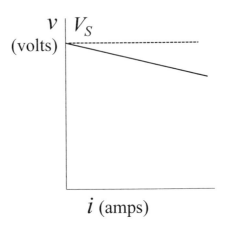

Figure 7.11   The *v-i* plot of a real voltage source (*solid line*). The nonzero slope of this line shows that the source contains an internal resistor. The *slope* indicates the value of this internal resistance.

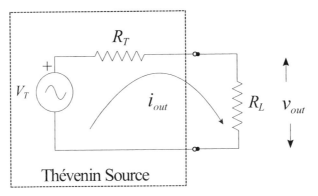

Figure 7.12   Representation of a real source also known as a Thévenin circuit shown with a load resistor, R_L that is used draw current from the source. This circuit is used in Example 7.8.

Real sources, then, are simply ideal sources with some nonzero resistance. They can be represented as an ideal source in series with a resistor as shown in Figure 7.12. This configuration is also known as a Thévenin source, named after an engineer who developed a network reduction theory described below. Finding values for V_T and R_T given a physical (and therefore *real*) source is straightforward. The value of the internal ideal source, V_T, is just the voltage that would be measured at the output, v_{out}, if no current was flowing through the circuit; that is, if the output was an open circuit. For this reason, V_T is described as equivalent to the *open circuit voltage* v_{oc}. To find the value of the internal resistance, R_T, we need to draw current from the circuit and measure how much the output voltage decreases. A resistor placed across the output of the Thévenin source will do the trick. This resister, R_L, is often referred to as a *load resistor* or just the *load* because it makes the Thévenin source do work by drawing current from the source ($P = v_{out}^2/R_L$). The smaller the resistor, the more current that will be drawn for the source, and the more work required by the source. For this reason, decreasing R_L is referred to as *increasing the loading on the source*. Assuming a current i_{out} is being drawn from the source and the voltage measured is v_{out}, the difference in voltage between the no-current voltage and current conditions is $v_D = V_T - v_{out}$. Because the voltage difference v_D is due entirely to R_T, the value of R_T can be determined as:

$$R_T = \frac{v_D}{i_{out}} = \frac{(V_T - v_{out})}{i_{out}} \qquad \text{[Eq. 7.11]}$$

The maximum current out of the source will occur when the source is connected to a short circuit, and the current that will flow, the *short-circuit current*, i_{sc}, will be:

$$i_{sc} = \frac{V_T}{R_T} \qquad \text{[Eq. 7.12]}$$

Remembering that V_T is the voltage that would be measured under open circuit conditions and defining the open circuit voltage as v_{oc}, we can write:

$$V_T = v_{oc} = R_T i_{sc}$$

$$R_T = \frac{v_{oc}}{i_{sc}} \qquad \text{[Eq. 7.13]}$$

In words, the internal resistance is equal to the open circuit voltage (v_{oc}) divided by the short-circuit current (i_{sc}). Eq. 7.13 is a viable method for determining R_T in theoretical problems, but is not practical in real situations with real sources because shorting a real source may draw excessive current and damage the source. When dealing with real sources, it is safer to draw only a small amount of current out of the source by placing a large resistor across the source, not a short-circuit. In this case, Eq. 7.11 can be used to find to find R_T since v_{out} and i_{out} can be measured and V_T can be determined by a measurement of open-circuit voltage. Example 7.8 takes this approach to determine the internal resistance of a voltage source.

Example 7.8: In the laboratory, the voltage of a real source is measured using a device that draws negligible current from the source; hence the voltage recorded can be taken as the open-circuit voltage. This voltage is measured as 9.0 V. A resistor, R_L, is placed across the output terminals of the device as in Figure 7.12 and a current of $i_{out} = 5$ mA (5×10^{-3} A) flows from the source. (Assume this current is measured using an ideal current measurement device, although it could also be calculated from v_{out} if the value of R_L is known.) Under this load condition, the output voltage falls to 8.6 V. What is the internal resistance of the source, R_T? What is the resistance of the load, R_T, that produced this current?

Solution: When there is no load resistor (i.e., $R_L = \infty$), then $i_{out} = 0$ and $v_{out} = V_T$. When the load resistor is attached to the output, $i_{out} = 5$ mA, and $v_{out} = 8.6$ V. Applying Eq. 7. 11:

$$R_T = \frac{v_D}{i_{out}} = \frac{(V_T - v_{out})}{i_{out}} = \frac{(9.0 - 8.6)}{0.005} = \frac{0.4}{0.005} = 80 \ \Omega$$

To find the load resistor, R_L, use Ohm's law:

$$R_L = \frac{v_{out}}{i_{out}} = \frac{8.6}{0.005} = 1,720 \ \Omega$$

In summary, a real source can be represented by an ideal source with a series resistance. In the examples shown here, the sources were DC and the series element a pure resistor. In the more general case, the source could generate sinusoids or other waveforms (but would still be ideal) and the series element would be an impedance. In the case where the source was sinusoidal, the equations above still hold, but would require phasor analysis for their solution.

7.2.3   Real Current Sources: The Norton Source

As mentioned in Chapter 4, current sources are somewhat counterintuitive since current is really an affect of voltage not a causal process. Current sources actually adjust the voltage to produce the desired current. For an ideal current source, the larger the load resistor, the more work they have to do since they must generate a larger voltage to get the desired current. Current sources prefer small load resistors, the opposite for voltage sources. For a real current source, as the load resistor increases, the voltage requirement increases, and limits on the ability to generate this voltage cause a reduction of the output current. This is reflected by the *v-i* plot circuit in Figure 7.13, and can be represented by an internal resistor.

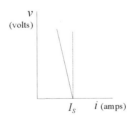

Figure 7.13   The *v-i* plot of a real current source (*solid line*). The noninfinite slope of this line shows that the current cannot keep up with the voltage required when high voltages are necessary to produce the desired current so the current decreases. The reduction in current output with increased voltage can be represented by an internal resistor and the slope is indicative of its value.

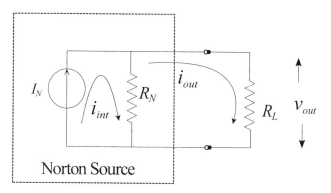

Figure 7.14 A circuit diagram of a real current source connected to a resistor load. The current source circuit is often referred to as a Norton equivalent circuit.

For current sources, the negative slope of the line in the v-i plot is equal to the internal resistance just as for voltage sources. The circuit diagram of a real current source is shown in Figure 7.14 as an ideal current source in parallel with an internal resistance. This configuration is often referred to as a *Norton equivalent* circuit. Inspection of this circuit shows how this circuit will produce a falloff in current output at higher voltages. As the voltage at the source output increases more current flows though the internal resistor, R_N, and less comes out of the source. If there were no internal resistor, all of the current would have to flow out of the source regardless of the output voltage and the source would be ideal.

Referring to Figure 7.14, when the output is a short circuit (i.e., $R_L = 0.0\ \Omega$), then no current will flows through R_N and all the current will flow though the output. (Note that like the open-circuit condition for a voltage source, the short-circuit condition produces the least work for a current source—in fact, no work at all.) By KCL:

$$I_N - i_{R_N} - i_{out} = 0;$$

$$i_{out} = i_{sc} = I_N \qquad \text{[Eq. 7.14]}$$

Hence I_N equals the short-circuit current, i_{sc}. When R_L is not a short circuit, some of the current will flow through R_N and i_{out} will decrease. Essentially, the internal resistor *steals* current from the current source when the output voltage is anything other than zero. Applying KCL to the upper node of the Norton circuit (Figure 7.14), paying attention to the current directions:

$$I_N - i_{R_N} - i_{out} = 0$$

$$I_N - \frac{v_{out}}{R_N} - i_{out} = 0 \qquad \text{[Eq. 7.15]}$$

Solving for R_N:

$$\frac{v_{out}}{R_N} = I_N - i_{out}; \quad R_N = \frac{v_{out}}{(I_N - i_{out})}$$

$$\text{Since } I_N = isc: \quad R_N = \frac{v_{out}}{(i_{sc} - i_{out})} \qquad \text{[Eq. 7.16]}$$

When the output of the Norton circuit is an open circuit (i.e., $R_L \rightarrow \infty$), all the current flows through the internal resistor, R_N. Hence:

$$v_{oc} = I_N R_N \qquad \text{[Eq. 7.17]}$$

Combining this equation with Eq. 7.14, we can solve for R_N in terms of the open circuit voltage, v_{oc}, and the short-circuit current, i_{sc}.

$$v_{oc} = I_N R_N = i_{sc} R_N$$

$$R_N = \frac{v_{oc}}{i_{sc}} \qquad \text{[Eq. 7.18]}$$

This relationship is the same as for the Thévenin circuit as given in Eq. 7.13 if we make $R_T = R_N$.

Example 7.9:   A real current source produces a current of 500 mA under short-circuit conditions and a current of 490 mA when the short is replaced by a 20-Ω resistor. Find the internal resistance.

Solution: Given i_{sc} as 500 mA, find v_{out} when the load is 20 Ω, then apply Eq. 7.16 to find R_N.

$$v_{out} = R_L i_{out} = 20(0.49) = 9.8 \text{ V}$$

$$R_N = \frac{v_{out}}{i_{sc} - i_{out}} = \frac{9.8}{0.5 - 0.49} = \frac{9.8}{0.01} = 980 \text{ }\Omega$$

The Thévenin and Norton circuits have been presented in term of sources, but they can also be used to represent entire networks as well as mechanical and other nonelectrical systems. These representations can be especially helpful when two systems are being connected. Imagine you are connecting two systems and you want to know how the interconnection will affect the behavior of the overall system. If you could represent the system serving as a source as a Thévenin or Norton source and determine the effective input impedance of the system serving as load, you would be able to calculate the loss of signal due to the interconnection. The same could be stated for a biological measurement where the biological system is the source and the measurement system is the load. You may not have much control over the nature of the source, but as a biomedical engineer, you will have some control over the effective impedance of the load. These concepts are exploded further in a later section.

7.2.4 Thévenin and Norton Circuit Conversion

It is easy to convert between the Thévenin and Norton equivalent circuits, that is, to determine a Norton circuit that has the same voltage–current relationship as a given Thévenin circuit and vice versa. Such conversions would allow you to apply KVL to system with current sources (by converting them to an equivalent voltage source) or to use KCL in system with voltage sources (by converting them to an equivalent current). Consider the voltage current relationship shown in the v-i plot of Figure 7.15. Because the curve is a straight line, it is uniquely determined by any two points. The horizontal and vertical intercepts, v_{oc} and i_{sc}, are particularly convenient points to use.

The equations for equivalence are easy to derive based on previous definitions:

$$\text{For a Thévenin circuit: } v_{ocT} = V_T; \quad \text{and since } R_T = \frac{v_{ocT}}{i_{scT}}; \quad i_{scT} = \frac{v_{ocT}}{R_T}$$

$$\text{For a Norton circuit: } I_N = i_{scN}; \quad \text{and since } R_N = \frac{v_{ocN}}{i_{scN}}; \quad v_{ocN} = R_N i_{scN}$$

For a Norton circuit to have the same $v - i$ relationship as a Thévenin, $i_{scN} = i_{scT}$ and $v_{ocN} = v_{ocT}$. Equating these terms in the equations above:

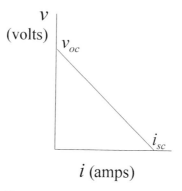

Figure 7.15 The *v-i* plot of the output of some device to be represented as either a Thévenin or Norton circuit. The voltage–current relationship plots as a *straight line* and can be uniquely represented by two points: v_{oc} and i_{sc}, for example.

$$I_N = i_{scN} = i_{scT} = \frac{v_{ocT}}{R_T}; \quad \text{but } v_{ocT} = V_T$$

$$I_N = \frac{V_T}{R_T} \tag{Eq. 7.19}$$

$$R_N = \frac{v_{ocN}}{i_{scN}} = \frac{v_{ocT}}{i_{scT}}; \quad \text{but } \frac{v_{ocT}}{i_{scT}} = R_T$$

$$R_N = R_T \tag{Eq. 7.20}$$

To go the other way and convert from a Norton to an equivalent Thévenin:

$$V_T = v_{ocT} = v_{ocN}; \quad \text{but } v_{ocN} = I_N R_N$$

$$V_T = I_N R_N \tag{Eq. 7.21}$$

$$R_T = \frac{v_{ocT}}{i_{scT}} = \frac{v_{ocN}}{i_{scN}}; \quad \text{but } \frac{v_{ocN}}{i_{scN}} = R_N$$

$$R_T = R_N \tag{Eq. 7.22}$$

These four equations allow for easy conversion between Thévenin and Norton circuits. Note that the internal resistance, R_N or R_T, is the same for either configuration. This is reasonable, since the internal resistance defines the slope of the v-i curve so to achieve the same v-i relationship you need the same sloped curve and hence the same resistor.

The ability to represent any linear v-i relationship by either a Thévenin or a Norton circuit implies that, given a real source, it is impossible to determine whether it is, in reality, a current or voltage source based solely on external measurement of voltage and current. If the v-i relationship of a source is more-or-less a vertical line as in Figure 7.13, indicating a large internal resistance, we might guess that the source is probably a current source, or at least more appropriately represented as such. In fact, one simple technique for constructing a current source in practice is to place a voltage source in series with a large resistor. Alternatively, if the v-i relationship is approximately horizontal as in Figure 7.11, a nonideal voltage source would be a better guess. However, if the v-i curve is neither particularly vertical nor horizontal, as in Figure 7.15, it is anyone's guess as to whether it is a current or voltage source and either would be an equally appropriate representation unless other information was available.

Conversion between Thévenin and Norton circuits can be used in order to apply nodal analysis in circuits that contain voltage sources or to use mesh analysis in circuits that contain current sources. This application of Thévenin-Norton conversion is shown in the example below.

Example 7.10: Find the voltage, V_A, in the circuit shown below using nodal analysis.

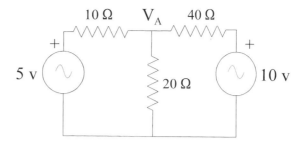

Solution: This circuit can be viewed as containing two Thévenin circuits: a 5-V source and 10-Ω resistor, and a 10-V source and 40-Ω resistor. After converting these two Thévenin circuits to equivalent Norton circuits using Eq. 7.19 and Eq. 7.20, apply standard nodal analysis.

After Thévenin-Norton conversion, the circuit becomes:

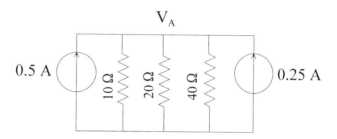

Writing the KCL equation around node A.

$$i_{s1} + i_{s2} + i_{R1} + i_{R2} + i_{R3} = 0$$

$$0.5 + 0.25 - \frac{V_A}{R_1} - \frac{V_A}{R_2} - \frac{V_A}{R_3} = 0$$

$$0.5 + 0.25 - V_A\left(\frac{1}{10} + \frac{1}{20} + \frac{1}{40}\right) = 0$$

$$V_A = \frac{0.5 + 0.25}{1/10 + 1/20 + 1/40} = \frac{0.75}{0.1 + 0.05 + 0.025} = \frac{0.75}{0.175} = 4.29 \text{ V}$$

The Thévenin and Norton circuits and their inter-conversions are also of value in reducing or simplifying networks that contain sources as shown in the next section. These concepts also apply to mechanical systems, with appropriate modifications, as shown in Section 7.5.

7.3   THÉVENIN AND NORTON THEOREMS: NETWORK REDUCTION WITH SOURCES

The Thévenin theorem states that any network of passive elements and sources can be reduced to a single voltage source and series impedance. Thus the reduced network would look like a Thévenin circuit such as that shown in Figure 7.12, except that the internal resistance, R_T, would be replaced by a generalized impedance, $Z\angle\theta$. The Norton theorem makes the same claim for Norton circuits, which is reasonable because Thévenin circuits can easily be converted to Norton circuits via Eq. 7.19 and Eq. 7.20.

There are a few constraints on these theorems. The elements in the network being reduced must be linear, and if there are multiple sources in the network, they must be at the same frequency. As has been done in the past, the techniques for network reduction will be developed using phasor representation and hence limited to networks with sinusoidal sources. However, the approach for generalizing this and other techniques involving phasor analysis to a wide range of signals is presented in the next chapter.

There are two approaches to finding the Thévenin or Norton Equivalent of a circuit. One is based on the strategy used above to find the equivalence between Thévenin and Norton circuits: find the open-circuit voltage, v_{oc}, and the short-circuit current, i_{sc}. The other method evaluates only the open-circuit voltage, v_{oc}, then determines R_T (or R_N) through network reduction. During network reduction, a source is replaced by its equivalent resistance, that is, open circuits ($R \to \infty \Omega$) substitute for voltage sources, and short circuits ($R = 0.0 \Omega$) substitute for current sources. Both these approaches are straightforward to implement and are best shown though examples.

Example 7.11:   Find the Thévenin equivalent of the circuit below using both the v_{oc}-i_{sc} method and the v_{oc}-network reduction technique.

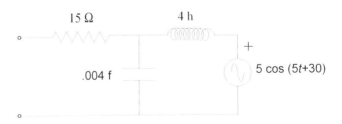

Solution: v_{oc}-Reduction method: First find the open-circuit voltage, v_{oc}, using standard network analysis. Convert all network elements to their phasor representation as shown below.

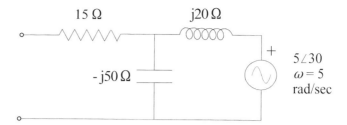

Because in the open-circuit case no current flows through the 15-Ω resistor, there is no voltage drop across this resistor. Therefore, the open-circuit voltage is the same as the voltage across the capacitor. The 15 Ω resistor is essentially *not there* with respect to open-circuit voltage. (The resistor is not totally useless: it does play a role in determining the equivalent impedance and also the short-circuit current.) The open-circuit voltage, the voltage across the capacitor, can be determined by writing the mesh equation around the loop consisting of the capacitor, inductor, and source. Using the usual directional conventions, defining the mesh current as clockwise and going around the loop in a clockwise direction, note that the voltage source will be negative since there is a voltage drop going around in the clockwise direction.

$$-5 \angle 30 - i(-j50 + j20) = 0$$

$$I = \frac{-5 \angle 30}{-j50 + j20} = \frac{-5 \angle 30}{-j30} = \frac{-5 \angle 30}{30 \angle -90} = -0.167 \angle 120 \text{ A}$$

$$v_{oc} = IZ_C = -0.167 \angle 120(-j50) = -0.167 \angle 120(50 \angle -90)$$

$$v_{oc} = -8.35 \angle 30 \text{ V}$$

Note that the Thévenin equivalent voltage is actually larger than the source voltage. This is the result of a partial resonance between the inductor and capacitor as discussed in Chapter 9.

Next, find the equivalent impedance by reduction. To reduce the network, essentially turn off the sources and apply network reduction techniques to what is left. Turning off a source does not mean you remove it from the circuit, rather to turn off a source, you replace it by its *equivalent resistance*. For an ideal voltage source $R_T \rightarrow 0.0$ Ω, so the equivalent resistance of an ideal voltage source is 0.0 Ω. To turn off a voltage source you replace it by a short circuit. After replacing the source by a short circuit, we are left with the network shown below on the left-hand side. Series-parallel reduction techniques will work for this network.

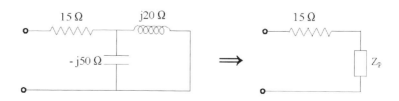

After replacing the voltage source by a short circuit, we are left with a parallel combination of inductor and capacitor. This combines to a single impedance, Z_P:

$$Z_P = \frac{Z_C Z_L}{Z_C + Z_L} = \frac{-j50(j20)}{-j50 + j20} = \frac{1,000}{-j30} = \frac{1,000}{30 \angle - 90}$$

$$Z_P = 33.33 \angle 90 \; \Omega$$

We are left with the series combination of Z_P and the 15-Ω resistor:

$$Z_T = 15 + 33.33 \angle 90 = 15 + j33.33 = 36.55 \angle 65.8 \; \Omega$$

Hence, the original circuit can be equivalently represented by the Thévenin circuit shown below.

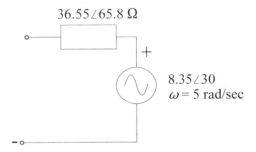

Solution: v_{oc}-i_{sc}: In this method, we solve for the open-circuit voltage and short-circuit current. We have already found the open-circuit voltage above so it is only necessary to find the short-circuit current. After shorting out the output and converting to phasor notation, the circuit is as shown.

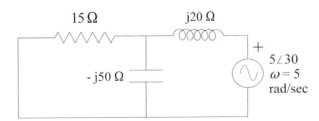

If we solve this using mesh analysis, it is a two-mesh circuit, but if we convert the inductor-source combination to a Norton equivalent, it becomes a single-node equation. To implement the conversion, use Eq. 7.19 and Eq. 7.20:

$$I_N = \frac{V_T}{R_T} = \frac{5 \angle 30}{j20} = \frac{5 \angle 30}{20 \angle 90} = 0.25 \angle - 60 \text{ volts}; \quad R_N = R_T = j20 \; \Omega$$

The network becomes:

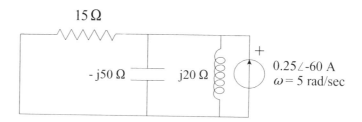

15 Ω

− j50 Ω   j20 Ω   0.25∠-60 A
$\omega = 5$ rad/sec

Writing KCL about the single node:

$$0.25\angle -60 - v\left(\frac{1}{15} + \frac{1}{-j50} + \frac{1}{j20}\right) = 0$$

$$v = \frac{0.25\angle -60}{\dfrac{1}{15} + \dfrac{1}{-j50} + \dfrac{1}{j20}} = \frac{0.25\angle -60}{0.0667 + j.02 - j.05} = \frac{0.25\angle -60}{0.0667 - j.03}$$

$$v = \frac{0.25\angle -60}{0.073\angle -24.2} = 3.42\angle -35.8 \text{ V}$$

$$i_{sc} = \frac{v}{15} = \frac{3.42\angle -35.8}{15} = 0.228\angle -35.8 \text{ A}$$

Now solve for R_T:

$$R_T = \frac{v_{oc}}{i_{sc}} = \frac{8.35\angle 30}{0.228\angle -35.8} = 36.6\angle 65.8 \ \Omega$$

This is the same value for R_T as found by network reduction above.

Networks that are more complicated can be reduced using MATLAB as shown in the example below.

Example 7.12:   Find the Norton Equivalent of the circuit below with the aid of MATLAB.

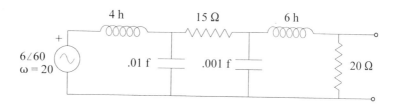

4 h   15 Ω   6 h

6∠60
$\omega = 20$ .01 f   .001 f   20 Ω

Solution: The open-circuit voltage and short-circuit current can be solved directly using mesh analysis in conjunction with MATLAB. In fact, the mesh equations in both cases (solving for v_{oc} and i_{sc}) will be similar. The only difference is that when solving for the short-circuit current, the 20-Ω resistor will be short-circuited and not appear in the equation. First convert to phasor notation, then encode the network directly into MATLAB. Because we are using MATLAB and the computational load is reduced, we will keep ω as a variable in case we want to find the Norton equivalent for other frequencies.

After converting to phasor notation and assigning the mesh current, the circuit is as shown below. Note that the open-circuit voltage, v_{oc}, is the voltage across the 20-Ω resistor and the short-circuit current, i_{sc}, is just i_3, when the resistor is shorted.

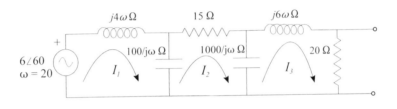

Writing the KVL equations for the open-circuit condition:

$$\begin{vmatrix} 6 \angle 60 \\ 0 \\ 0 \end{vmatrix} = \begin{vmatrix} j4\omega + \dfrac{100}{j\omega} & -\dfrac{100}{j\omega} & 0 \\ -\dfrac{100}{j\omega} & 15 + \dfrac{100 + 1{,}000}{j\omega} & -\dfrac{1{,}000}{j\omega} \\ 0 & -\dfrac{1{,}000}{j\omega} & 20 + j6\omega + \dfrac{1{,}000}{j\omega} \end{vmatrix} \begin{vmatrix} i_1 \\ i_2 \\ i_3 \end{vmatrix}$$

where $v_{oc} = 20i_3$.

The mesh equation for the short-circuit condition is quite similar:

$$\begin{vmatrix} 6 \angle 60 \\ 0 \\ 0 \end{vmatrix} = \begin{vmatrix} j4\omega + \dfrac{100}{j\omega} & -\dfrac{100}{j\omega} & 0 \\ -\dfrac{100}{j\omega} & 15 + \dfrac{100 + 1{,}000}{j\omega} & -\dfrac{1{,}000}{j\omega} \\ 0 & -\dfrac{1{,}000}{j\omega} & j6\omega + \dfrac{1{,}000}{j\omega} \end{vmatrix} \begin{vmatrix} i_1 \\ i_2 \\ i_3 \end{vmatrix}$$

where $i_{sc} = i_3$

The program to solve these equations and find I_N and R_N is shown below:

```
%Example 7.12 To find the Norton equivalent of a three mesh
%circuit
clear all; close all;
```

```
w = 20;                              % Define frequency
theta = 60*2*pi/360;
VS = 6*cos(theta) + ...              % Define Vs as rectangular
  6*sin(theta)*j;
v = [VS 0 0]';                       % Define voltage vector
%
% Define open-circuit impedance matrix
Zoc = [4j*w+100/(j*w), -100/(j*w), 0; -100/(j*w),...
      15+1100/(j*w), -1000/(j*w),...
      0, -1000/(j*w), 20+6j*w+1000/(j*w)];
ioc = Zoc\v;                         % Solve for currents
voc = 20*ioc(3);                     % and open-circuit voltage
%
% Define short-circuit impedance matrix and solve for
  short-circuit current
Zsc = [4j*w+100/(j*w), -100/(j*w), 0; -100/(j*w),...
      15+1100/(j*w), -1000/(j*w);...
      0, -1000/(1j*w), 6j*w+1000/(j*w)];
i = Zsc\v                            % Solve for currents
isc = i(3);                          % Find isc
Zn = voc/isc;                        % Solve for Z_N (Eq. 7.18)
%
% Output magnitude and phase of I_N and R_N
IN_mag = abs(isc)
IN_phase = angle(isc)*360/(2*pi)
ZN_mag = abs(Zn)
ZN_phase = angle(Zn)*360/(2*pi)
```

This program produces the following outputs:

$$I_N = 0.0031 \angle 20.6 \text{ A}$$

$$Z_N = 19.36 \angle 9.8 \ \Omega$$

It would be easy to modify this program to find, and perhaps plot, the Norton element values over a range of frequencies. A similar exercise is given in Problem 16 the end of the chapter.

7.4 MEASUREMENT LOADING

We now have the tools to analyze the situation when two systems are connected together. For Bioengineers not involved in electronic design, this situation most frequently occurs when making a measurement so we will analyze the problem in that context. However, the approach followed here applies to any situation when two processes are connected together.

7.4.1 Ideal and Real Measurement Devices

One of the important tasks of biomedical engineers is to make measurements, usually on some physiological system. Any measurement requires withdrawing some energy from the system of interest and that, in turn, will alter the state of the system and the value of the measurement. This alteration is referred to here as *measurement loading*. The word *load* is applied to any device or system that is attached to the system of interest and *loading* is the influence the attached device has on the system. Hence, measurement loading is the influence a measurement device has on the system being measured. This well-known phenomenon extends down to the smallest systems and has a significant impact on fundamental principles of particle physics. The concepts developed above can be used to analyze the effect of a measurement device on a given system. In fact, the evaluation of the influence of a measurement device is one of the major applications of the Thévenin and Norton equivalent circuits.

Just as there are ideal and real sources, there are ideal and real measurement devices or loads. An ideal source is one that can supply any amount of energy or power required by its load. For voltage and current sources, $P = vi$, so an ideal voltage source supplies a given voltage at any current and an ideal current source supplies a given current at any voltage. Both these devices can supply infinite power (vi) if need be. Ideal measurement devices have the opposite characteristics: they can make a measurement without drawing any energy or power from the system being measured. Of course, we know from basic physics that such an idealization is impossible, but some measurement devices can provide nearly ideal measurements, at least for all practical purposes.

The goal in practical situations is to be able to make a measurement without significantly altering the system being measured. The ability to attain this goal will depend on the characteristics of the source as well as the load; a given device might have little effect on one system providing a reliable measurement, yet significantly alter another system giving a measurement that does not reflect the underlying conditions. It is not just a matter of how much energy a measurement device requires, but how much energy the system being measured can supply without significant change.

Just as ideal voltage and current sources have quite different properties, ideal measurement devices for voltage and current differ significantly. A device that measures voltage is termed a *voltmeter*. An ideal voltmeter would draw no power from the circuit being measured. Since $P = vi$, and v cannot be zero (that is what is being measured), an ideal voltmeter must draw no current while making the measurement. The current will be zero for any voltage only if the equivalent resistance of the voltmeter is infinite. An ideal voltmeter is effectively an open circuit, and the v-i plot is a vertical line. Practical voltmeters do not have infinite resistances, but they do have very large impedances, of the order of hundreds of megaohms ($100 \times 10^6 \ \Omega$), and this can be considered ideal for all but the most challenging conditions. The characteristics of ideal sources and loads are summarized in Table 7.1.

TABLE 7.1 Electrical Characteristics of Ideal Sources and Loads

| Characteristic | Source | | Measurement Device (Load) | |
	Voltage	Current	Voltage	Current
Impedance	$0.0\ \Omega$	$\infty\ \Omega$	$\infty\ \Omega$	$0.0\ \Omega$
Voltage	V_S	up to ∞ v	$V_{measured}$	0.0 v
Current	up to ∞ A	I_S	0.0 A	$I_{measured}$

Current measuring devices are termed *ammeters*. An ideal ammeter also needs no power from the circuit to make its measurement. Again, since $P = vi$, and i cannot be zero in an ammeter, voltage must be zero if no energy is to be drawn from the system being measured. This means that an ideal ammeter is effectively a short circuit having an equivalent resistance of $0.0\ \Omega$. The *v-i* plot of an ideal ammeter would be a horizontal line. Practical ammeters are generally not very ideal having resistances approaching a tenth of an ohm or more. However, current measurements are rarely made in practice because that usually involves breaking a circuit connection to make the measurement.

An illustration of the error caused by a less-than-ideal ammeter is given in the example below.

Example 7.13: A practical ammeter having an internal resistance of $2\ \Omega$ is used to measure the short-circuit current of the three-mesh network used in Example 7.12. How large is the error—that is, how much does the measurement differ from the true short-circuit current?

Solution: As with all issues of measurement loading, it is easiest to use the Thévenin or Norton representation of the system being loaded. The Norton equivalent of the three-mesh circuit was determined in Example 7.12 and is shown below loaded by the ammeter. From the Norton circuit, we know the true short-circuit current is: $i_{sc} = I_N = 0.0031\angle\|20.6$ A. The measured short-circuit current can be determined by applying nodal analysis to the circuit.

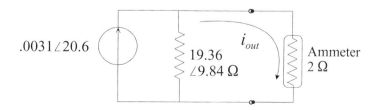

Applying KCL:

$$0.0031 \angle 20.6 - v\left(\frac{1}{19.36 \angle 9.8} + \frac{1}{2}\right) = 0$$

$$v = \frac{0.0031 \angle 20.6}{\dfrac{1}{19.36 \angle 9.8} + \dfrac{1}{2}} = \frac{0.0031 \angle 20.6}{0.052 \angle - 9.8 + 0.5} = \frac{0.0031 \angle 20.6}{0.051 - j.009 + 0.5} = \frac{0.0031 \angle 20.6}{0.55 \angle 0.9}$$

$$v = 0.0056 \angle 19.7 \text{ volts}$$

$$i_{sc\_measured} = \frac{v}{R} = \frac{0.0056 \angle 19.7}{2} = 0.0028 \angle 19.7 \text{ amps}$$

The short-circuit current actually measured is slightly less that the actual short-circuit current which is equal to I_N. The error is:

$$Error = \frac{(0.0031 - 0.0028)}{0.0031} 100 = 9.7\%$$

The difference in the measured current and the actual short-circuit current represents the current flowing through the internal resistor. Since the external load is much less (an order of magnitude) than the internal resistor, the current taking the internal pathway is small. So the smaller the external impedance with respect to the internal impedance, the more ideal the measurement. In this case, the load resistor as approximately one-tenth the value of internal resistor, leading to an error of approximately 10%. For some measurements, this may be sufficiently accurate. Usually a ratio of one to 100 (i.e., two orders of magnitude) between the internal and load resistor is adequate for the type of accuracy required in biomedical engineering measurements.

The same rule of thumb can be used for voltage measurements, except now the load resistance should be much greater than the internal resistance (or impedance). In voltage measurements, if the load resistor is 100 or more times the internal resistance, the loading can usually be considered negligible and the measurement sufficiently accurate.

These general rules also apply whenever one network is attached to another. If voltages carry the signal (usually the case), the influence of the second network on the first can be ignored if the effective input impedance of the second network is much greater than the effective output impedance of the first network (Figure 7.16). In other words, if $Z_L \gg Z_T$, the transfer functions derived for each of the two networks independently, can be taken as valid when the two are interconnected. Recall that the two basic assumptions in deriving the transfer function were that the input was an ideal source and the output was an ideal load ($Z_L \rightarrow \infty$). If $Z_L \gg Z_T$ (say, 100 times) these assumptions are reasonably met, at least with respect to these two networks. (The input to network 1 and the load on network 2 could still present problems, but with respect to the interconnection shown, the two assump-

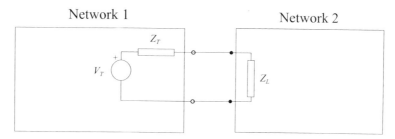

Figure 7.16 The input of network 2 is connected to the output of network 1. If the equivalent input impedance of network 2, Z_L, is much greater than the output impedance of network 1, Z_T, the alternation of the networks by the connection can be ignored.

tions would hold reasonably well.) An analysis of voltage loading is found in the problems.

If the signal is carried as a current, then the opposite would be true. The output impedance of network 1 should be much greater than the input impedance of network 2: $Z_L \ll Z_T$. In this situation the current loading of network 2 can be considered negligible with respect to network 1. Signals are rarely carried as current except for the output of certain transducers, particularly those that respond to light and in these cases the signal is converted to a voltage by a special amplifier circuit (see Chapter 10).

What if these conditions are not met; for example, if Z_L were approximately equal to Z_T? In this case, you have two choices: Calculate the transfer function of the two-network combination or estimate the error that will occur due to the interconnection as in the last example. Sometimes you actually want to increase the load on a system, purposely making Z_L, close to the value of Z_T. The motivation for such a strategy is explained in the next section.

7.4.2 Maximum Power Transfer

The goal in most measurement applications is to extract minimum energy from the system. This is also the usual goal when one system is connected to another. In these situations, the load resistance (or impedance) should be either much less than the internal resistance if the output is current or much greater than the internal resistance if the output is a voltage. However, what if the goal is to extract maximum energy from the system? To determine the conditions for maximum power out of the system, we consider the Thévenin circuit with its load resistor in Figure 7.17.

To address this question, we assume that R_T is part of a source and cannot be adjusted; that is, it is an internal (and therefore inaccessible) property of the system from which we are trying to extract the maximum power. If only R_L can be adjusted, the question becomes, what is the value of R_L that will extract maximum power from the system? The power out of the system is $P = v_{out}i_{out}$ or $P = R_L i_{out}^2$. To find the value of the load resistor R_L that will deliver the maximum power to itself, we

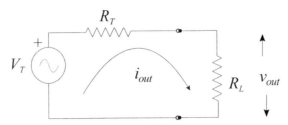

Figure 7.17 A Thévenin circuit is shown with a load resistor, R_L. For minimum power out of the system R_L should be much greater than R_T. In this section we seek to determine the value of R_L that will extract *maximum* power from the Thévenin source.

use the standard calculus trick for maximizing a function: Solve for power in terms of R_L, then take dP/dR_L and set it to zero:

$$P = R_L i^2; \quad i = \frac{V_T}{(R_L + R_T)}; \quad \text{hence, } P = \frac{R_L V_T^2}{(R_L + R_T)^2}$$

$$\text{Solving for } \frac{dP}{dR_L} \text{ by parts:}$$

$$\frac{dP}{dR_L} = \frac{V_T^2(R_L + R_T)^2 - 2V_T^2 R_L(R_L + R_T)}{(R_L + R_T)^4} = 0$$

$$V_T^2(R_L + R_T)^2 = 2V_T^2 R_L(R_L + R_T)$$

$$R_L + R_T = 2R_L;$$

$$R_L = R_T \qquad \qquad \text{[Eq. 7.23]}$$

So for maximum power out of the system, R_L should equal R_T (or, more generally, $Z_L = Z_T$), a condition known as *impedance matching*. This is known as the *maximum power transfer theorem*. Using this theorem, it is possible to find the value of load resistance that extracts maximum power from any network. Just convert the network to a Thévenin equivalent and set R_L equal to R_T. Recall that the maximum power theorem applies when R_T is fixed and R_L is varied. If R_T *can* be adjusted, then just by looking at Figure 7.17 we can see that maximum power will be extracted from the circuit when $R_T = 0$.

When sinusoidal signals are involved, Eq. 7.23 can be modified to include impedances in the Thévenin equivalent circuit and the load. When impleances are involved, maximum power transfer occurs when the resistors are matched ($R_L = R_T$) and the reactive components are also equal but opposite in sign:

$$Z_L = R_L + jX_L = R_T - jX_T \qquad \qquad \text{[Eq. 7.25]}$$

where R_L and R_T are the same as in Eq. 7.25 and X_L and X_T are the *reactive* or none-resistive components of Z_L. Complex variables with the same real part but oppositely signed imaginary parts are called *complex conjugates*. Putting the maximum power transfer theorem in more general context, for maximum power transfer the load impedance should be the complex conjugate of the Thévenin impedance.

7.5 MECHANICAL SYSTEMS

All of the concepts described in this chapter are applicable to mechanical systems with only minor modifications. The concepts of equivalent impedances and impedance matching are often used in mechanical systems, particularly in acoustic applications. Of particular value are the concepts of real and ideal sources and real and ideal loads or measurement devices. As mentioned in Chapter 4, an ideal force generator produces a specific force regardless of the velocity, just as an ideal voltage source produces the required voltage at any current. An ideal force generator would generate the same force at 0.0 velocity, or 10 miles per hour (mph), or 10,000 mph, or even beyond the speed of light (clearly impossible), if necessary. The force produced by a real force generator would decrease with velocity. This can be expressed in a force–velocity plot (analogous to the *v-i* plot) as shown in Figure 7.18A. An ideal velocity (or displacement) generator would produce a specific velocity against any force, be it 1 oz. or 1 ton, but the velocity produced by a real velocity generator would decrease as the force against it increased (Figure 7.18B). Again, there can be an ambiguity between real force and velocity generators. The device producing the force–velocity curve shown in Figure 7.18C could be interpreted as a nonideal force generator or a nonideal velocity generator, and it is not possible to determine its true nature from only the force–velocity plot.

Ideal measurement devices follow the same guiding principle in mechanical systems: they should extract no energy from the system being measured. For a

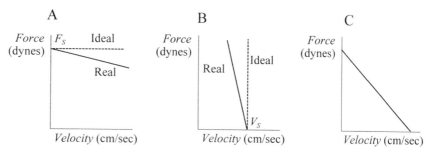

Figure 7.18 **A:** The force–velocity plot of an ideal (*dashed line*) and a real (*solid line*) force generator. The force produced by a real generator decreases the faster it must move to generate that force. **B:** The force–velocity plot of an ideal (*dashed line*) and a real (*solid line*) velocity generator. The velocity produced by a real generator decreases as the strength of the force that opposes its movement increases. **C:** The force–velocity plot of a generator that may be interpreted as either a nonideal force or velocity transducer.

TABLE 7.2 Mechanical Characteristics of Ideal Sources and Loads

Characteristic	Source		Measurement Device (Load)	
	Force	Velocity	Force	Velocity
Impedance	$0.0\ \Omega$	$\infty\ \Omega$	$\infty\ \Omega$	$0.0\ \Omega$
Force	F_S	up to ∞ dynes	$F_{measured}$	0.0 dynes
Velocity	up to ∞ cm/sec	v_S	0.0 cm/sec	$v_{measured}$

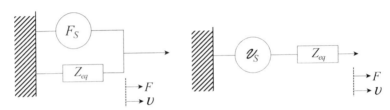

Figure 7.19 A mechanical analog of a Thévenin (*left side*) and Norton (*right side*) equivalent system. These configurations can be used to determine the effect of loading by a measurement device or by another mechanical system.

force-measuring device, a *force transducer*, velocity should be zero and position a constant (the *isometric* condition). An ideal force transducer would require no movement to make its measurement—it would appear as a solid, immobile wall to the system being measured. This corresponds to infinite mechanical impedance. For an ideal velocity transducer, the force required to make a measurement would be zero. This is equivalent to a mechanical impedance of zero. The characteristics of ideal mechanical sources and loads are given in Table 7.2 in a fashion analogous to the electrical characteristics on Table 7.1.

The concept of equivalent impedances and sources is useful in determining the alteration produced by a nonideal measurement device or load on a mechanical system. The analog of Thévenin and Norton equivalent circuits can also be constructed for mechanical systems. The mechanical analog of a Thévenin circuit is a force generator with an effective impedance in parallel (the configuration is reversed in a mechanical system), while the mechanical equivalent analog of a Norton circuit is a velocity (or displacement) generator in series with the equivalent impedance (Figure 7.19). To find the values for either of the two equivalent mechanical systems in Figure 7.19, we will use the analog of the $v_{oc} - i_{sc}$ method: find *isometric* force, the force when velocity is zero (position is constant), and the unloaded velocity, the velocity when no force is applied to the system.

Most practical lumped-parameter mechanical systems are not so complicated and do not usually require reduction to a Thévenin or Norton-like equivalent circuit. Nonetheless, all of the concepts regarding source and load impedances developed previously apply to mechanical systems. If the output of the mechanical system is a force, the minimum load, and most accurate measurement, is produced by a system with a large equivalent mechanical impedance, one that tends to produce a large opposing force and allows little movement. Conversely, if the output of the mechanical system is a velocity, the minimum load is produced by a system having very small equivalent mechanical impedance, one that produces very little force. Finally, if the goal is to transfer maximum power from the source to the load, the mechanical impedances of each should be equal. These principles are explored in the examples below.

Example 7.14: The mechanical elements of a real force generator are shown on the left side of the figure below. The mechanical elements of a real force transducer are shown on the right-hand side.

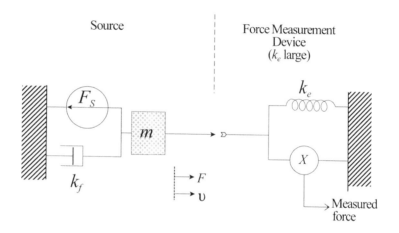

The left side shows a real force generator consisting of an ideal force generator (F_S) in parallel with friction and mass. The right side is a real force transducer consisting of a displacement transducer (marked X) with a parallel spring. This transducer actually measures the displacement of the spring that is proportional to force ($F = k_e X$).

First find the force that would be measured by an ideal force generator, that is, the force produced by the force generator when the velocity is zero. The force transducer actually measures displacement of the spring (the usual construction of a force transducer). What is the force that would be measured by this transducer? How could this transducer be improved to make a measurement with less error? The system parameters are:

$$k_f = 20 \, \text{dyne-sec/cm}; \quad m = 5 \, \text{gm}; \quad k_e = 2,400 \, \text{dyne/cm}; \quad F_S = 10 \cos(4t).$$

Solution: To find the force measured by an ideal force transducer, write the equations for the force generator and set velocity to zero as would be the case for an ideal force transducer. Note that F_S is negative based on its defined direction (the arrow pointing to the left).

$$F_{measured} = -F_s - v(j\omega m + k_f) \quad \text{if } v = 0$$

$$F_{measured} = -F_s = 10 \angle 0 \text{ dynes}$$

When the measurement transducer is attached to the nonideal force generator, the velocity will no longer be zero. To measure the force out of the force generator under this load write the equations for the combined system. Because there is the equivalent of only one node in the combined system, only a single equation will be necessary, but we need to pay attention to the signs.

$$-F_S - v\left(j\omega m + k_f + \frac{k_e}{j\omega}\right) = 0$$

$$-10 - v\left(j4(5) + 20 + \frac{2,400}{j4}\right) = -10 - v(j20 + 20 - j600) = 0$$

$$v = \frac{-10}{20 - j580} = \frac{-10}{580.3 \angle -88} = -17.2 \times 10^{-3} \angle 88 \text{ cm/sec}$$

$$F_{measured} = vZ_{k_e} = -17.2 \times 10^{-3} \angle 88(-j600) = 10.3 \angle -2 \text{ dynes}$$

Hence, the measured force is 3% larger than the actual force. The measured force is larger because of a very small resonance between the mass in the force generator and the elastic element in the transducer. Resonance is explored in Chapter 9. Note that the elastic element is very stiff ($k_e = 2,400$ dyne/cm) in order to make the transducer a good force transducer: the stiffer the elastic element, the closer the velocity will be to zero. To improve the measurement further, this element could be made even stiffer. For example, if the elasticity, ke, were increased to 9,600 dyne/cm, the force measured would be $10.08 \angle -0.5$ reducing the error to less than a percent. This evaluation is presented in Problem 17 at the end of the chapter.

The next example involves the measurement of a velocity generator.

Example 7.15:   The mechanical elements of a real velocity generator are shown on the left side of the figure below.

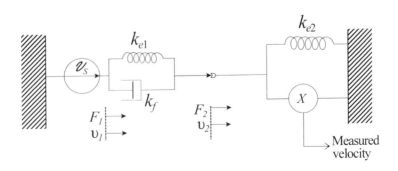

The left side shows real velocity generator consisting of an ideal velocity generator ($\mathcal{V}_s$) in series with friction and elastic element. The right side is a real velocity transducer consisting of an ideal velocity transducer (marked X, its output is proportional to v_2) with a parallel spring. Assume $\mathcal{V}_s$ is a sinusoid with $\omega = 4$ rad/sec and: $k_f = 20$ dyne sec/cm; $k_{e1} = 20$ dyne/cm; $k_{e2} = 2$ dyne/cm.

Find the velocity that would be measured by the velocity transducer on the right side of the figure. Note that this transducer is the same as the force transducer except the elasticity is very much lower. Improvement in the accuracy of this transducer is given in Problem 18 at the end of the chapter.

Solution: The solution proceeds in exactly the same manner as in the previous problem except that now there are two velocity points. However, because velocity v_1 is equal to $\mathcal{V}_s$, it is not independent and only one equation need be solved.

Writing the sum of forces equation about the point indicated by v_2 point:

$$-\left(\frac{k_{e1}}{j\omega} + k_f\right)(v_2 - v_1) - \left(\frac{k_{e2}}{j\omega}\right)v_2 = 0 \quad \text{Substituting } v_1 = \mathcal{V}_s \text{ and other variables}$$

$$-\left(\frac{20}{j4} + 20\right)(v_2 - \mathcal{V}_s) - \left(\frac{2}{j4}\right)v_2 = (j5 - 20)(v_2 - \mathcal{V}_s) + j0.5(v_2) = 0$$

$$(-20 + j5.5)v_2 - (-20 + j5)\mathcal{V}_s = 0 \quad \text{Solving for } v_2$$

$$v_2 = \frac{(-20 + j5)\mathcal{V}_s}{-20 + j5.5} = \frac{(20.6\angle 165.9)\mathcal{V}_s}{20.7\angle 164.6} = (0.995\angle 1.3)\mathcal{V}_s \text{ cm/sec}$$

The measured value of v_2 is very close to that of $\mathcal{V}_s$, i.e., the value that would be measured by an ideal velocity transducer, one that produced no resistance to movement. This is because the impedance of the transducer is much less than that of the source ($k_{e2} = 0.1 k_{e1}$). If the elasticity of the transducer, which provides the only resistance to movement, is increased the error increases, and if it is reduced the error is reduced. The influence of transducer impedance is demonstrated further in several of the problems.

Matching mechanical impedances is particularly important in ultrasound imaging. In this imaging technology, a high-frequency (1 MHz and up) pressure pulse wave is introduced into the body and reflects off various internal surfaces. The time-of-flight for the return signal is used to estimate the depth of a given surface. Using a scanning technique, a large number of individual pulses are directed into the body at different directions and a two-dimensional image is constructed. Because the return signals can be very small, it is important that the maximum energy be obtained from the return signals. The following example illustrates the advantage of matching acoustic (i.e., mechanical) impedances in ultrasound imaging.

Example 7.16:   An ultrasound transducer that uses a barium titinate piezoelectric device is applied to the skin. Calculate the percent power returned to the transducer. The transducer is round with a diameter of 2.5 cm. Use an acoustic impedance of

24.58 × 10⁶ kg-sec/m² for barium titinate and an acoustic impedance of 1.63 × 10⁶ kg-sec/m² for the skin. What is the maximum percent power that could be returned?

Solution: The interface between the skin and transducer can be represented as two series mechanical impedances as shown below.

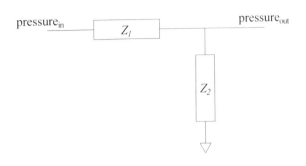

The ratio of pressure in and pressure out (which is the same as force in and force out since the areas are the same) can be determined by a *force divider* equation analogous to the voltage divider equation:

$$\frac{pressure_{out}}{pressure_{in}} = \frac{Z_2}{Z_1 + Z_2} \qquad \text{[Eq. 7.26]}$$

Note that the impedances are given in MKS units not the cgs units used throughout this text. However, since they are used in a ratio, the units would cancel anyway (along with the 10⁶) so there is no need to convert them to cgs.

If the impedances are matched, $Z_1 = Z_2$, then the pressure ratio is 0.5 as given by the maximum power transfer theorem. Under the unmatched conditions, the pressure ratio can be calculated as the following:

$$\frac{pressure_{out}}{pressure_{in}} = \frac{Z_2}{Z_1 + Z_2} = \frac{1.63}{24.58 + 1.63} = 0.062$$

The power ratio would actually be equal to the pressure ratio squared because power is proportional to pressure, or force, squared. Hence, the power ratio in the unmatched case becomes 0.0039 as opposed to 0.25 in the matched case. This shows the importance of matching impedances to improve power transfer. By matching impedances, the power transferred into the body would be 64 times larger (0.25/0.0039) than without impedance matching. Because we have no control over the impedance of the skin, we must adjust the impedance of the ultrasound transducer. In fact, special coatings can be applied to the active side of the transducer that will match tissue impedance. In addition, a gel having the same acoustic (i.e., mechanical) impedance is used to improve coupling and insure impedance matching between the transducer and skin.

7.6   **MULTIPLE SOURCES: REVISITED**

In Chapter 6, we applied multiple sources in the form of sinusoids at different frequencies to a single input to find the frequency characteristics of a system. Superposition allows us to compute the transfer function for variable frequencies with
the assurance that if multiple frequencies were applied to the system, the response
would just be the summation of the responses to individual frequencies. However,
what if the sources have both different frequencies and different locations? (If the
sources are at the same frequency, but different locations, we have no problem as
standard analysis techniques apply, see Example 5.3.) Even if the sources have different locations and different frequencies, superposition, still can be used to analyze
the network. We can solve the problem for each source separately knowing that the
total solution will be the algebraic summation of all the partial solutions. We simply
"turn off" all sources but one, solve the problem using standard techniques, and
repeat until the problem has been solved for all sources. Then we add all the partial
solutions for a final solution. As stated previously, turning off a source does not
mean removing it from the system; rather, it means replacing it by its equivalent
impedance. Hence, *voltage sources are replaced by short circuits* and *current sources
by open circuits* (so current sources are essentially removed from the circuit). Similarly, force sources become straight connections and velocity sources essentially dis-
appear. The example below uses superposition in conjunction with source equivalent
impedance to solve a circuit problem with two sources having different frequencies.

Example 7.17:   Find the voltage across the 30-Ω resistor in the circuit below.

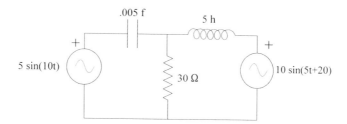

Solution: First turn off the right-hand source by replacing it with a short circuit
(its internal resistance), solve for the currents through the 30-Ω resistor, and then
solve for the voltage across it. Then turn off the left-hand source and repeat the
process. Note that the impedances of the inductor and capacitor will be different
since the frequency is different. Add the two partial solutions to get the final voltage
across the resistor.

   Turning off the right-hand source and converting to the phasor domain gives
circuit shown below.

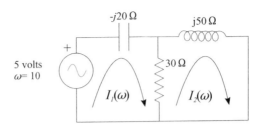

Applying KVL leads to a solution for the voltage across the center resistor:

$$\begin{vmatrix} 5 \\ 0 \end{vmatrix} = \begin{vmatrix} 30 - j20 & -30 \\ -30 & 30 + j50 \end{vmatrix} \begin{vmatrix} I_1(\omega) \\ I_2(\omega) \end{vmatrix}$$

Solving: $I_1 \ (\omega = 10) = 0.217 \ \angle 17$; $I_2 \ (\omega = 10) = 0.112 \angle - 42$:

$$V_R(\omega = 10) = [I_1(\omega) - I_2(\omega)]R = 5.57 \ \angle \ 48 \ \text{V}$$

Now turning off the left-hand source and converting to the phasor domain leads to the circuit below. Note that the impedances are different since the new source has a different frequency.

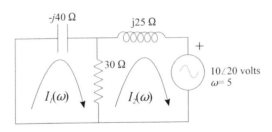

Again, apply the standard analysis.

$$\begin{vmatrix} 0 \\ -10 \angle 20 \end{vmatrix} = \begin{vmatrix} 30 - j40 & -30 \\ -30 & 30 + j25 \end{vmatrix} \begin{vmatrix} I_1(\omega) \\ I_2(\omega) \end{vmatrix}$$

Solving: $I_1(\omega = 5) = 0.274 \ \angle - 175$; $I_2(\omega = 5) = 0.457 \ \angle \ 131$;

$$V_R(\omega = 5) = (I_1(\omega) - I_2(\omega))R = 10.98 \ \angle - 85 \ \text{V}$$

The total solution is just the sum of the two partial solutions:

$$v_R(t) = 5.57 \cos(10t + 48) + 10.98 \cos(5t - 85) \text{V}$$

This approach extends directly to any number of sources. It applies equally well to current sources as shown in Problem 7.13.

7.7 SUMMARY

Even very complicated circuits can be reduced using the rules of network reduction. These rules allow networks containing one or more sources (at the same frequency) and any number of passive elements to be reduced to a single source and a single impedance. This single source–impedance combination could be either a voltage source in series with the impedance, a so-called Thévenin source, or a current source in parallel with the impedance, a Norton source. Conversion between the two representations is straightforward.

One of the major applications of network reduction is to evaluate the performance of the system when circuits are combined together. The transfer function of each isolated network was determined based on the assumption that the circuit was driven by an ideal source and connected to an ideal load. This can be taken as true *if* the impedance of the source driving the network is much less than the network's input impedance, and the equivalent impedance of the load is much greater than the network's output impedance. Network reduction techniques provide a method for determining these input and output impedances.

The ratio of input to output impedance is particularly important when making physiological measurements. Usually the goal is to make the input impedance of the measurement device as high as possible to minimally load the process being measured; that is, to draw minimum energy from the process. Sometimes it is desirable to transfer a maximum amount of energy between the process being measured and the measurement system. This is often true if the measurement device must also inject energy into the process in order to make its measurement. In such situations, an impedance matching strategy is used where the input impedance of the measuring device is adjusted to equal the output impedance of the process being measured.

All of the network reduction tools apply equally to mechanical systems. Indeed, one of the major applications of impedance matching in bioengineering is in ultrasound imaging where the ultrasound transducer mechanical impedance must be matched the impedance of the tissue.

This chapter concluded with the analysis of networks containing multiple sources at different frequencies. To solve these problems, the effect of each source on the network must be determined separately and to the solutions for each source added together, an approach supported by the principle of superposition. To solve for the influence of each source on the network, the other sources are turned off by replacing them by their equivalent impedances: Voltage sources are replaced by short circuits and current sources are replaced by open circuits.

PROBLEMS

1. Find the combined values of the elements below.

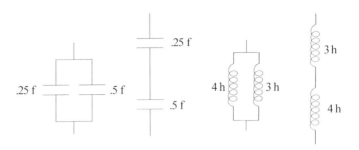

2. Find the value of R so the resistor combination equals 10 Ω. (*Hint*: Use Eq. 7.9.)

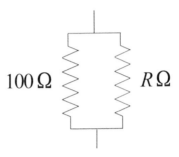

3. A two-terminal element has an impedance of $100 \angle -30$ Ω at $\omega = 2$ rad/sec. The element consists of two components in series. What are they: Two resistors, two capacitors, resistor capacitor, resistor inductor, or two inductors? What are their values?
4. An impedance, Z, has a value of $60 \angle 25$ Ω at $\omega = 10$ rad/sec. What type of element should be added in series to make the combination look like a pure resistance at this frequency? What is the value of the added element at $\omega = 10$ rad/sec? What is the value of the combined element at $\omega = 10$ rad/sec?
5. Find the equivalent impedance of the network below between terminals A and B. Assume $\omega = 5$ rad/sec.

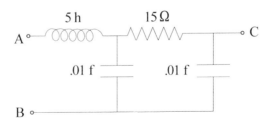

6. Find the equivalent impedance of the network in Problem 7.5 between terminals A and C.

7. Find the equivalent impedance of the network below between terminals A and B as a function of frequency.

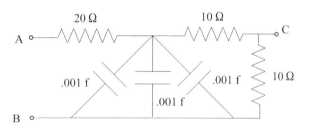

8. The following *v-i* characteristics were measured on a two-terminal device. Model the device by a Thévenin circuit and a Norton circuit.

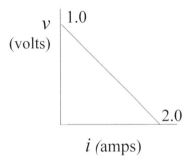

9. Plot the *v-i* characteristics of this network at two different frequencies $\omega = 2$ and $\omega = 200$ rad/sec. Note: the *v-i* plot is a plot of the voltage *magnitude* against current *magnitude*.

10. Two different resistors were placed across a two-terminal source (a battery) known to contain an ideal voltage source in series with a resistor. The voltages measured using the two load resistors are $V = 8.5$ V when $R = 1,000\ \Omega$; and $V = 8.2$ V when $R = 100\ \Omega$. Find the load resistor, R_L, that will extract the maximum power from this source. What is the power extracted?

11. The following magnitude v-i plot was found for a two-terminal device at the two frequencies shown. Model the device as a Thévenin equivalent. (*Hint:* to find Z_T, use Eq. 7.13 generalized for impedances.)

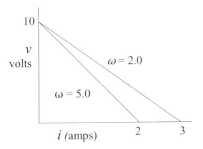

12. The voltage of a nonideal voltage source shown below is measured with two voltmeters. One has an internal resistance of 10 MΩ whereas the other, a "cheapie," has an internal resistance of only 100 KΩ. What are the two voltages read? How do they compare with the true Thévenin voltage? (Assume the voltmeter reads peak-to-peak voltage, although voltmeters usually read in rms.) Note: The Thévenin source is composed of more realistic values normally encountered in electronics.

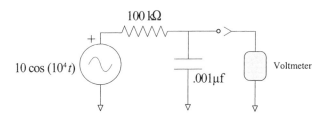

13. Find the voltage across the 0.01-f capacitor in the circuit below. Note that the two sources are at different frequencies.

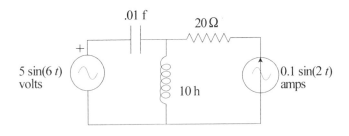

MATLAB Problems

14. Find the Thévenin equivalent circuit of the network below. (*Hint:* Modify the code in Example 7.12.)

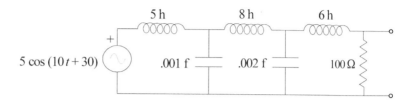

15. Plot the magnitude and phase of the impedance, Z_{eq}, in Example 7.3 over a range of frequencies from $\omega = 0.01$ to 1,000 rad/sec. Plot both magnitude and phase in log frequency and plot the phase in degrees.

16. For the circuit in Problem 7.14, further modify the code in Example 7.12 to determine and plot the Norton current source and Norton impedance as a function of frequency. Plot both magnitude and phase of both variables for a range of frequencies between 0.1 and 100 Hz.

17. For the mechanical system shown in Example 7.14, find the difference between the measured force $F_{measured}$, and F_S (the ideal source) if the friction in the source, k_f, increased from 20 dyne-sec/cm to 60 dyne cm/sec. Find the difference between $F_{measured}$, and F_S with the increased friction in the source if the elasticity of the transducer were also increased by a factor of 3.

18. For the mechanical system shown in Example 7.15, find the difference between the measured velocity, v_2, and $\mathcal{V}_S$ if the elastic coefficient of the transducer, k_{e2} were doubled, quadrupled, or halved.

19. For the mechanical system shown in Example 7.15, find the difference between the measured velocity, v_2, and $\mathcal{V}_S$ if the velocity transducer contained a mass of 4 gm.

20. For the network shown below, what should the value of Z_L be to extract maximum power from the network? (*Hint:* Use the same approach as in Example 7.12, but remember Z_L equals the complex conjugate of Z_{equv}.) If the voltage source were increased to $15 \cos(20t)$, what should the value of Z_L be to extract maximum power from the network?

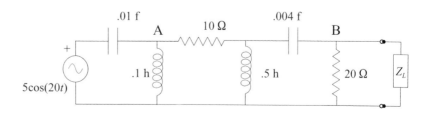

21. In the network shown in Problem 7.20, what is the *impedance* between nodes A and B? Use the approach shown in Example 7.7. Apply a hypothetical 1-V source between these two points and use MATLAB-aided mesh analysis to find the current. In addition, you need to remove the influence of the 5-V source by turning it off; that is, replacing it by a short circuit. (Note: This becomes a four-mesh problem, but still requires only three lines of MATLAB code to solve!)

22. The Thévenin equivalent of a circuit consists of a 1 volt source and a Thévenin impedance of $10 + j10\ \Omega$. Using MATLAB plot the power transferred to the load impedance of $Z_L = R_L + jX_L$ where R_L varies between 1 and 100 Ω in 1 Ω increments. Plot P transferred to R_L as a function of R_L for five values of X_L: $-50, -10, -5, 0.0,$ and $+10\ \Omega$. Plot the five curves superimposed and label each curve. (Note: The power transferred to the load is only to the load resistor and is equal to the load resistance times the absolute value of the current squared.)

8 THE ANALYSIS OF TRANSIENTS: THE LAPLACE TRANSFORM

8.1 THE LAPLACE TRANSFORM

Phasor analysis allows us to analyze any linear system using algebraic operations, provided all sources are sinusoidal and in steady-state. With the aid of the Fourier series analysis, any periodic signal can be broken down into sinusoidal components. Using phasor analysis and the principle of superposition, we can analyze each component separately, then add all the individual contributions to get the total response (see Example 6.11). This approach can be extended to aperiodic functions through the application of the continuous Fourier transform. What phasor analysis cannot handle are waveforms that suddenly change and never return to a baseline level. A classic example is the *step function*, which changes from one value to another at one point in time (often taken as $t = 0$) and remains at the new value for all eternity (Figure 8.1).

One-time changes, or changes that do not return to some baseline level, are common in nature and even in one's life. For example, this text should leave the reader with a lasting change in his or her knowledge of biosystems and biosignals, a knowledge that will not return to the baseline, pre-course level. (However, this change is not expected to occur instantaneously as with the step function.) A signal or input could also change any number of times in any manner, but if it is not periodic, or does not return to a baseline level so it could be considered aperiodic, we do not have the tools to analyze a system's response to these signals. Of course, we could always return to calculus, but the analysis would soon become very complicated particularly for large systems. It would be extraordinarily useful to have a transformation such as phasors that would enable an algebraic treatment of processes driven by signals that do not return to a baseline level. With such a tool, it would be possible to extend all of the techniques developed with phasors (mesh and nodal analysis, network reduction, Thévenin and Norton equivalent circuits) to systems exposed to this wider class of signals. Fortunately, a technique based on the Laplace transform allows us to analyze, using only algebra, processes exposed to so-called *transient signals*, those such as the step function that do not return to a baseline. (As all signals vary in time they could be called transient signals, but this

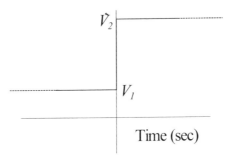

Figure 8.1 The time plot of a step function which changes from value V_1 to value V_2 at $t = 0.0$ seconds. It will remain at this value for all time.

term is often reserved for signals that have one-time, or steplike, changes. This is another term where the context can vary its meaning.)

8.1.1   Definition of the Laplace Transform

The reason that phasor analysis cannot be used when signals change in a steplike manner is simply that functions such as the step function do not have a Fourier series or Fourier transform representation. If they did, they could be decomposed in sinusoids and phasor techniques could be applied. Consider a function similar to that shown in Figure 8.1, a function that is 0.0 for a $t \leq 0$ and 1.0 for all $t > 0$:

$$x(t) = \begin{cases} 0 & t \leq 0 \\ 1 & 5 > 0 \end{cases} \qquad \text{[Eq. 8.1]}$$

This special case of the function in Figure 8.1 is known as the *unit step function* because it begins at zero and jumps to 1.0 at $t = 0$. The Fourier transform of this function would be taken as:

$$FT(\omega) = \int_0^\infty x(t)e^{-j\omega t}dt = \int_0^\infty 1e^{-j\omega t}dt \Rightarrow \infty \qquad \text{[Eq. 8.2]}$$

In the continuous Fourier transform, the function, $x(t)$, has a finite life span: It is nonzero over some period but is zero outside the time period. The Fourier transform integral need only be taken over the time when the function is nonzero. However, if the function never returns to zero, the integration must be carried out to infinity. Because the sinusoidal function $e^{-j\omega t}$ has nonzero values out to infinity, the integral goes to infinity. The integral does not *converge* as t becomes large. The trick that is used to solve this problem of nonconvergence is to modify the exponential function so that it does converge to zero at large values of t even if the signal, $x(t)$, does not. This could be accomplished by multiplying the sinusoidal term, $e^{-j\omega t}$, by a decaying exponential such as $e^{-\sigma t}$ where σ is some positive real variable. In this case, the sinusoid becomes a complex sinusoid, or rather a sinusoid with a complex frequency:

$$e^{-j\omega t}e^{-\sigma t} = e^{-(\sigma+j\omega)t} = e^{-st} \qquad\qquad \text{[Eq. 8.3]}$$

where $s = \sigma + j\omega$ and is termed the *complex frequency* because it is a complex variable having both a real and imaginary part, but fulfills the same role as frequency, ω, in the Fourier transform exponential. The complex variable, s, is also known as the *Laplace variable* because it plays a critical role in the Laplace transform. A modified version of the Fourier transform can now be constructed using complex frequency in place of real frequency; that is, $s = \sigma + j\omega$ instead of just $j\omega$. This modified transform is termed the *Laplace transform*:

$$X(s) = \mathcal{L}x(t) = \int_0^\infty x(t)e^{-st}dt \qquad\qquad \text{[Eq. 8.4]}$$

For a given function, $x(t)$, the product of $x(t)e^{-st} = x(t)e^{(\sigma+j\omega)t}$ may not necessarily converge to zero as $t \to \infty$, in which case the Laplace transform does not exist.[1] Some advanced signal processors spend a lot of time worrying about such functions: Which functions converge, their ranges of convergence, or how to get them to converge. Such matters need not concern us because most common transient functions, including the step function, do converge for some value of σ and do have a Laplace transform. The range of σ's over which a given product of $x(t)$ and $e^{-(\sigma+j\omega)t}$ converge is of interest to mathematicians and some signal processing theoreticians, but also does not concern us here. If a signal has a Laplace transform (as given in Appendix B), obviously the product must converge for that signal.

The Laplace transform has two downsides: It cannot be applied to functions for negative values of t, and it is difficult to evaluate from the definition (Eq. 8.4) for any but the simplest functions. The restriction against representing functions for negative times comes from the fact that e^{-st} becomes a positive exponential and will go to infinity as t goes to large negative values: for negative t, the real part of the exponential becomes $e^{\sigma t}$ and does exactly the opposite of what we want it to do; it diverges rather than forces convergence. The only way around this is to limit our analyses to $t > 0$, but we will see that this is not a severe restriction. The other downside, the difficulty in evaluating Eq. 8.4, stems from the fact that s is complex, so although the integral in Eq. 8.4 does not look that complicated, the complex integration becomes very involved for all but a few simple functions of $x(t)$. To get around this problem we use tables that give the Laplace transform of often-used functions. Such a table is given in Appendix B, and a more extensive list can be found in references such as the *Handbook of Physics and Chemistry* (Lide, 2004). The Laplace transform table is used to calculate both the Laplace transform of a signal and the inverse Laplace transform of the output or response. The only

---

[1] For example, the function $x(t) = e^{t^2}$ will not converge when multiplied by e^{st} and hence does not have a Laplace transform.

difficulty in finding the inverse Laplace transform is rearranging the output Laplace function into the same format that is found in the table.

Example 8.1:   Find the Laplace transform of the step function in Eq. 8.1.

Solution: The step function is one of the few functions that can be evaluated easily using the basic defining equation of the Laplace transform, Eq. 8.4.

$$X(s) = \int_0^\infty x(t)e^{-st}dt = \int_0^\infty 1e^{-st}dt = -\frac{e^{-st}}{s}\Big|_0^\infty = 0 - \left(-\frac{1}{s}\right)$$

$$X(s) = \frac{1}{s}$$

8.1.2 Laplace Representation of Elements: Calculus Operations in the Laplace Domain

The Laplace transform of the derivative operation can be determined from the defining equation, Eq. 8.4.

$$\mathcal{L}\frac{dx(t)}{dt} = \int_0^\infty \frac{dx(t)}{dt}e^{-st}dt \quad \text{integrating by parts}$$

$$= x(t)e^{-st}\Big|_0^\infty + s\int_0^\infty x(t)e^{-st}dt$$

But from the definition of the Laplace transform, $\int_0^\infty x(t)e^{-st}dt = X(s)$, so the right term in the summation is just $sX(s)$, and the equation becomes:

$$\mathcal{L}\frac{dx(t)}{dt} = x(\infty)e^{-\infty} - x(0)e^{-0} + sX(s)$$

$$\mathcal{L}\frac{dx(t)}{dt} = sX(s) - x(0-) \qquad\qquad \text{[Eq. 8.5]}$$

Equation 8.5 shows that differentiation in the Laplace domain becomes multiplication by the Laplace variable s with the additional subtraction of the value of the function at $t = 0$. The value of the function at $t = 0$ is known as the *initial condition*. This value can be used, in effect, to account for all negative time history of $x(t)$. In this approach, all of the behavior of $x(t)$ when t was negative is lumped together as a single initial condition at $t = 0$. This trick allows us to include some aspects of the system's behavior over negative values of t even if the Laplace transform does not itself apply to such time values. If the initial condition is zero, as is frequently the case, differentiation in the Laplace domain is simply multiplication by s paralleling the differentiation operation in the phasor domain, multiplication by $j\omega$.

Multiple derivatives can also be taken in the Laplace domain, although this is not such a common operation. Multiple derivatives involve multiplication by s^n times, and taking the derivatives of the initial conditions:

$$\mathcal{L}\frac{d^n x(t)}{dt^n} = s^n X(s) - s^{n-1}x(0-) - s^{n-2}\frac{dx(0-)}{dt} \cdots \frac{d^{n-1}x(0-)}{dt^{n-1}} \qquad \text{[Eq. 8.6]}$$

Again if there are no initial conditions, taking n derivatives becomes just a matter of multiplying $x(t)$ by s^n.

Integration in the Laplace domain is simply a matter of dividing by s, just as integration involved dividing by $j\omega$ in the phasor domain. Again, the initial conditions need be taken into account:

$$\mathcal{L}\left[\int_0^T x(t)dt\right] = \frac{1}{s}X(s) + \frac{1}{s}\int_{-\infty}^0 x(t)dt \qquad \text{[Eq. 8.7]}$$

If there are no initial conditions, then integration is accomplished by simply dividing by s. The direct integral that accounts for the initial conditions is again a way of accounting for the negative time history of the system. Although it looks complicated, we will find that elements whose variable relationships involve integration (the capacitive elements) provide this integral easily.

8.1.3 Initial Conditions

As noted previously, the Laplace transform cannot be used in situations where $t < 0$ since the transform equation (Eq. 8.4) diverges for negative time. By definition, $t < 0$ is the time before the signal transition, and $t(0-)$ is the time immediately before the transition. The transient signal is defined to begin at $t = 0$ (or later) with the initial, and perhaps only, transition usually taking place at this time. However, Laplace analysis can be applied to components that have nonzero variable values at $t = 0$, by making use of the initial conditions feature of the Laplace transform. Using this concept, a system that was active during negative t (i.e., before the onset of the transient) can still be analyzed as long as that prior activity can be summarized down to an initial condition at $t = 0$.

As an example, consider a capacitor that has had various amounts of current flowing, perhaps in and out, during negative time. This entire $t < 0$ history can be summarized into a single voltage at $t = 0$ using the basic equation that defines the voltage–current relationship of a capacitor:

$$V_C(t = 0) = \frac{1}{C}\int_{-\infty}^0 i_C(t)dt \qquad \text{[Eq. 8.8]}$$

This equation has the same form as the right-hand term in Eq. 8.7, the initial condition term. Hence, the initial voltage on a capacitor can be included in the Laplace equation for integration, although the Laplace analysis itself is still limited to times greater than zero. Because it is not possible to change the voltage across a capacitor instantaneously (Eq. 4.23), the voltage on the capacitor when t is on the negative side of zero ($t = 0-$) will be the same as when t is on the positive side of zero ($t = 0+$). Hence, if conditions change at $t = 0$, the voltage on the capacitor just before that time, $V_C(0-)$, should be used as the initial voltage. A similar argument holds for the initial force on a spring.

The application of initial conditions to account for behavior during negative time also applies to an inductor. In an inductor, energy is stored in the form of current flow, so the salient initial condition is the current at $t = 0$. This can be obtained based on the history of the inductor's voltage:

$$I_L(0) = \frac{1}{L} \int_{-\infty}^{0} v_L dt \qquad \text{[Eq. 8.9]}$$

An inductor treats current flowing through it the same way a capacitor treats voltage: The current flowing through an inductor cannot be changed instantaneously and is continuous (Eq. 4.17) so the current just before the $t = 0$ transition, $I_L(0-)$ will be the same after the transition, $I_L(0+)$. Again, a similar argument could be made for the initial velocity for a mass.

8.1.4 Voltage–Current and Force–Velocity Relationships in the Laplace Domain

Combining the derivative and integral operations in the Laplace domain with integration, it is possible to construct the equations for the various passive electrical and mechanical elements. Because the Laplace transform of a constant is just the same constant (by inspection of Eq. 8.4), resistor and friction elements have the same representation in both the time domain and Laplace domain, just as they have the same representation in the time and phasor domains.

To find the v-i relationship for an inductor in the Laplace domain, substitute $i(t)$ for $x(t)$ in Eq. 8.5:

$$v = L\frac{di}{dt} \Rightarrow V(s) = L(sI(s) - i(0)) = sLI(s) - Li(0) \qquad \text{[Eq. 8.10]}$$

A similar relationship holds for the force–velocity relationship of a mass:

$$F = m\frac{dv}{dt} \Rightarrow F(s) = smv(s) - mv(0) \qquad \text{[Eq. 8.11]}$$

If the initial current, $i(0)$, or velocity, $v(0)$ is nonzero, is must be included in the relationship equations. Often these values are known, but in some cases, they must be calculated from the history of the system.

The v-i relationship for a capacitor is obtained using Eq. 8.7 substituting in $i(t)$ for $x(t)$:

$$v(t) = \frac{1}{C}\int i(t)dt \Rightarrow V(s) = \frac{1}{Cs}I(s) + \frac{1}{Cs}\int_{-\infty}^{0} i(t)dt \qquad \text{[Eq. 8.12]}$$

From Eq. 8.8 we see that the second term is just the initial voltage, $V_C(0)$, divided by s, so Eq. 8.12 becomes:

$$v(t) = \frac{1}{C}\int i(t)\,dt \Rightarrow V(s) = \frac{1}{Cs}I(s) + \frac{V_C(0)}{s} \qquad \text{[Eq. 8.13]}$$

Similarly, the force–velocity relationship for a mass in the Laplace domain can be written as:

$$F(t) = k_e \int v(t)\,dt \Rightarrow F(s) = \frac{k_e}{s}v(s) + \frac{k_e x(0)}{s} = \frac{k_e}{s}v(s) + \frac{F(0)}{s} \qquad \text{[Eq. 8.14]}$$

As with phasors, Laplace domain variables are usually given capital letters.

Using these definitions it is also possible to represent these elements as impedances (or admittances), similar to the impedances defined in the phasor domain, but now they will be functions of s instead of $j\omega$. Ignoring for the moment the initial condition term (or assuming the initial conditions are zero), these impedances would be defined for inertial elements as:

$$Z_L(s) = sL\Omega; \qquad \text{[Eq. 8.15]}$$

$$Z_m(s) = sm\Omega; \qquad \text{[Eq. 8.16]}$$

and for capacitive elements as:

$$Z_C(s) = \frac{1}{Cs}\Omega; \qquad \text{[Eq. 8.17]}$$

$$Z_e(s) = \frac{k_e}{s}\Omega; \qquad \text{[Eq. 8.18]}$$

There is a clever way to deal with the initial condition terms and retain the concept of impedance even when the initial conditions are nonzero. Regarding the two terms in the v-i equation for an inductor (Eq. 8.10), the first term is the impedance, sL, and the second term is a constant, $Li(0)$. The constant term can be viewed as a constant voltage source with a value of $Li(0)$. It is a strange voltage source, being dependent on the initial current and the inductance, but is still clearly a voltage source in the Laplace domain. [To verify this, just set the first term to zero, then $V(s) = Li(0)$.] Thus the symbol for an inductor in the Laplace domain would actually be two elements in series: a Laplace impedance representing the inductor impedance (sL), and a peculiar voltage source [$Li(0)$] representing the initial current (Figure 8.2). Because the second term in Eq. 8.10 is negative, the voltage source would have its positive terminal on the side that current was exciting the device.

A similar situation holds for the v-i equation for a capacitor (Eq. 8.13): It is made up of an impedance term, $1/Cs$, and what can be taken as a voltage source of $V_C(0)/s$. This leads to the combined Laplace elements of capacitor impedance in series with a voltage source that is dependent on initial voltage. This voltage source would have the same polarity as the voltage charge on the capacitor. This would lead to the phasor domain symbol for a capacitor shown in Figure 8.3.

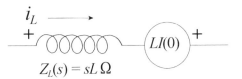

$$Z_L(s) = sL \; \Omega$$

Figure 8.2   The symbol for an inductor in the Laplace domain showing the impedance term as a standard inductor, sL, and the initial condition term as a voltage source, $Li(0)$. Note the polarity of the initial condition voltage source, which is due to the negative sign in Eq. 8.10.

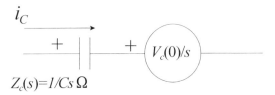

$$Z_c(s) = 1/Cs \; \Omega$$

Figure 8.3   Symbol for a capacitor in the Laplace domain with an impedance element, $1/Cs$, and a voltage source element representing the initial condition $V_C(0)/s$. The initial condition term is positive in Eq. 8.13 so the polarity of the voltage element is in the same direction as the polarity of the impedance element.

In an analogous manner, the mechanical elements mass and elasticity would have force generators added to their Laplace representation if initial conditions were present. These Laplace domain representations are summarized with the other passive elements in Table 8.1.

8.1.5   Sources: Common Signals in the Laplace Domain

Sources are also represented by functions of s and these functions depend on the specific source signal and are the Laplace transforms of that signal. Although it might seem that there could be a large variety of such signals, in practice only a few different signal types commonly occur in Laplace analysis. The signal most frequently encountered in Laplace analysis is the step function shown in Figure 8.1, or its more constrained version, the *unit* step function given in Eq. 8.1 and repeated here:

$$x(t) = u(t) = \begin{cases} 0 & t \le 0 \\ 1 & t > 0 \end{cases} \qquad\qquad \text{[Eq. 8.19]}$$

TABLE 8.1   Representation of Electrical and Mechanical Elements in the Laplace Domain

Element (Symbol)	Laplace Equation	Laplace Impedance Z(s)	Modified Symbol
Resistor (R)	$V(s) = R\ I(s)$	$Z_R(s) = R\ \Omega$	
Inductor (L)	$V(s) = sL\ I(s)$	$Z_L(s) = sL\ \Omega$	
Capacitor (C)	$V(s) = \dfrac{I(s)}{sC}$	$Z_C(s) = \dfrac{1}{sC}\,\Omega$	
Friction (k_f) (dyne-sec/cm)	$F(s) = k_f v(s)$	$Z(s) = k_f$	
Mass (m) (gm)	$F(s) = ms v(s)$	$Z(s) = ms$	
Elasticity (k_e) (spring) (dyne/cm)	$F(s) = \dfrac{k_e v(s)}{s}$	$Z(s) = \dfrac{k_e}{s}$	

The symbol u is frequently used to represent the unit step function. The Laplace transform of the step function was found in Example 8.1:

$$X(s) = U(s) = \frac{1}{s} \qquad\qquad \text{[Eq. 8.20]}$$

Again, it is common to use capital letters to represent the Laplace transform of a function. Two functions closely related to the unit step function are the ramp and impulse function (Figure 8.4).

These functions are related to the step function by differentiation and integration. The unit ramp function is just a straight line with slope of 1.0.

$$r(t) = \begin{cases} t & t > 0 \\ 0 & t \le 0 \end{cases} \qquad\qquad \text{[Eq. 8.21]}$$

Because the unit ramp function is just the integral of the unit step function, its Laplace transform will be that of the step function divided by s:

$$R(s) = \frac{1}{s}\left(\frac{1}{s}\right) = \frac{1}{s^2} \qquad\qquad \text{[Eq. 8.22]}$$

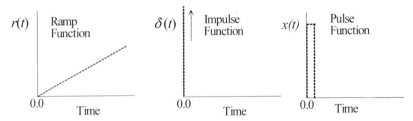

Figure 8.4 The ramp and impulse are two functions related to the step function and commonly encountered in Laplace analysis. The ideal impulse occurs at $t = 0$ and is infinitely narrow and infinitely tall. A real impulse function looks like a short pulse (*right-hand plot*).

The impulse function is the derivative of a unit step that leads to one of those mathematical fantasies: a function that is infinitely short, infinite in amplitude, and yet the area under the function is one as shown in Eq. 8.23. From the defining equation, we see that as $a \rightarrow \infty$, the pulse becomes infinitely tall, infinitely narrow, but retains an area of 1.0.

$$x(t) = \delta(t) = \lim_{a \to 0} \frac{1}{a} \quad \frac{-a}{2} \le t \le \frac{a}{2} \qquad \text{[Eq. 8.23]}$$

In practice, an impulse function would be a pulse with decidedly finite amplitude and finite pulse width, but the pulse width would be short (Figure 8.4, right-side plot). In real situations, it is only necessary that the pulse be much shorter (say, an order of magnitude) than the fastest process in the system that is driven by this signal. The method to determine when a pulse can be taken as a *practical impulse function* is demonstrated in Example 9.3 in the next chapter.

Because the impulse response is the derivative of the unit step function, its Laplace transfer function would be that of a unit step multiplied by s:

$$\Delta(s) = s\left(\frac{1}{s}\right) = 1 \qquad \text{[Eq. 8.24]}$$

Hence, the Laplace transform of an impulse function is a constant, and if it is a unit impulse (the derivative of a unit step), that constant is 1. As you might guess, this fact will have special relevance with regard to transfer functions analyzed in the Laplace domain. The Laplace transform of other common signal functions are given in Appendix B.

8.1.6 Converting the Laplace Transform to the Frequency Domain

Remember that s is complex frequency and equals a real term plus the standard frequency term $j\omega$ ($s = \alpha + j\omega$). To convert a Laplace function to the phasor domain, we simply substitute in $j\omega$ for s. Essentially we are agreeing to restrict ourselves to sinusoidal steady-state signals, so the real component of s is no longer needed. (Recall that the real component was introduced to ensure convergence in the Fourier integral and is not needed for sinusoidal steady-state signals.) Converting the Laplace transform function back to the phasor domain then allows us to determine the frequency characteristics of that function using Bode plot techniques or plotting with MATLAB. This approach is used in the next section to evaluate the frequency characteristics of a time-delay process.

8.1.7 The Time-Delay Element

Many physiological processes have a delay before they even begin to respond to the stimulus. Such delays are termed *reaction time*, or *response latency*, or simply *response delay*. In these processes, there is a period between stimulus onset when nothing happens, before the transient response begins. In neurological control systems, this delay is due to processing delays in the brain. The Laplace transform can be used to represent such delays.

The *time-delay* theorem can be derived from the defining equation of the Laplace transform (Eq. 8.4). Assume a function $x(t)$ that is zero for negative time. If this function is delayed from $t = 0$ by a time of τ seconds, the delayed function would be $x(t-\tau)$, where again τ is the delay. From the defining equation, the Laplace transform of such a delayed function would be:

$$\mathcal{L}[x(t-\tau)] = \int_0^\infty x(t-\tau)e^{-st}dt; \quad \text{Defining a new variable: } \gamma = t - \tau,$$

$$\mathcal{L}[x(t-\tau)] = \int_0^\infty x(\gamma)e^{-s(\tau+\gamma)}d\gamma = e^{-s\tau}\int_0^\infty x(\gamma)e^{-s\gamma}d\gamma$$

But the integral in the right-hand term is just the Laplace transform of the function, $x(t)$, before it was shifted. Hence, the Laplace transform of the shifted function becomes:

$$\mathcal{L}[x(t-\tau)] = e^{-s\tau}\mathcal{L}[x(t)] \qquad \text{[Eq. 8.25]}$$

Equation 8.25 is known as the time-delay theorem. This theorem can also be used to construct an element that represents a pure time delay; specifically an element that has a transfer function:

$$TF(s) = e^{-sT} \qquad \text{[Eq. 8.26]}$$

Where T = the delay usually in seconds. Just as an element having a transfer function of $1/s$ is an integrator, an element of e^{-sT} is a pure time delay of T seconds. Such an element is commonly found in systems models of neurological control processes where it represents neural processing delays.

To determine the frequency characteristics of the transfer function of a process that consists of a pure time delay, substitute $j\omega$ for s in Eq. 8.26.

$$TF(\omega) = e^{-j\omega T}; \quad |e^{-j\omega T}| = 1; \quad \angle e^{-j\omega T} = -\omega T \qquad \text{[Eq. 8.27]}$$

As shown in Eq. 8.27, a pure time delay has a magnitude transfer function of 1. A signal that experiences a pure time-delay process has the same magnitude frequency characteristics as the original signal. The time-delay process does have a frequency-dependent phase term, one that increases linearly with frequency ($-\omega T$). Hence, a pure time-delay process increases the phase component of a signal's frequency characteristics in a linear manner. The larger the delay, T, the greater the linear increase in the signal's phase component. Such a process is explored in Problem 8.16.

8.1.8   The Inverse Laplace Transform

Working in the Laplace domain is essentially the same as working in the phasor domain. First, the system elements are converted to Laplace notation using the conventions shown in Table 8.1. Next, the signals or sources are converted to Laplace functions using the table in Appendix B. Equations are then formulated using the techniques described in Chapters 5 and/or 7, and are solved using algebra for the output or other desired variables. You are left with a solution in the Laplace domain that is a function of s. Sometimes the Laplace representation of the solution or transfer function is sufficient, but if a time domain solution is desired then the inverse Laplace transform must be determined. The equation for the inverse Laplace transform is given as:

$$x(t) = \mathcal{L}^{-1} X(s) = \frac{1}{2\pi} \int_{\sigma-j\infty}^{\sigma+j\infty} X(s) e^{st} ds \qquad \text{[Eq. 8.28]}$$

Unlike the inverse Fourier transform, this equation is quite difficult to solve even for simple functions. So to evaluate the inverse Laplace transform, we use the Laplace transform table in Appendix B in the reverse direction: Find a transform function (on the right side of the table) that matches your output function and convert to the equivalent time domain function. The trick is usually in rearranging your output function to conform to one of the Laplace functions given in the table. Methods for doing this are described next and specific examples given.

8.2   LAPLACE ANALYSIS: THE LAPLACE TRANSFER FUNCTION

The analysis of systems using the Laplace transform is no more difficult than in the phasor domain, except that there may be the added task of accounting for initial conditions. In addition, to go from the Laplace domain back to the time domain may require some algebraic manipulation of the Laplace output function.

The transfer function, introduced for phasors in Chapter 6, is ideally suited to Laplace domain analysis, particularly when there are no initial conditions. In the phasor domain, the transfer function was used primarily to determine the frequency characteristic of a system. It can also be used to determine the system's output or response to any input provided that input can be expressed as a phasor; that is, as a sinusoid or series of sinusoids. In the Laplace domain, the transfer function can be used to determine the output response of a system to a much broader class of inputs. In fact, the Laplace domain transfer function is a widely applied concept that has applications in a number of biomedical engineering specialties and in many other areas of science and engineering. The Laplace domain transfer function is similar to its cousin in the phasor domain, except the frequency variable, ω, is replaced by the Laplace variable, s:

$$Transfer\ Function(s) = \frac{Output(s)}{Input(s)} \qquad \text{[Eq. 8.29]}$$

As mentioned in Section 8.1.1, the Laplace variable s is often called the *complex frequency* as it contains a real term along with the imaginary frequency term ($s = \alpha + j\omega$). Like its phasor domain cousin, the general Laplace domain transfer function will consist of a series of polynomials (see Eq. 6.4) and can be written as:

$$TF(s) = \frac{Bs(s+\omega_1)(s^2 + 2\delta_1\omega_{n1}s + \omega_{n1}^2)\ldots}{s(s+\omega_2)(s^2 + 2\delta_2\omega_{n2}s + \omega_{n2}^2)\ldots} \qquad \text{[Eq. 8.30]}$$

Again, the assumption is that any higher order polynomial (third-order or above) can be factored into the first- and second-order terms shown.

A comparison of the general Laplace transform equation above with its phasor domain equivalent (Eq. 6.5) shows two major differences: It is a function of s instead of ω (well, of course), and the polynomials have all been normalized so that the *highest power* of s has a coefficient of 1. This is opposite to the normalization strategy used in the phasor transfer function where the *lowest power* of s, usually the constant term, is set to 1. However, in all other respects the Laplace domain transfer function is quite similar to its phasor cousin. The coefficients terms that are used, ω_1, ω_2, ω_n, δ, and so forth, have the same definitions as in the phasor domain, and techniques used to determine a system's transfer function are the same. Even the terminology used to describe the elements is the same as those given for the phasor domain in Table 6.1. For example an ($s + \omega_1$) in the numerator is called a *zero*, or a first-order term, and an ($s^2 + 2\delta\omega s + \omega_{n1}^2$) is known as a second-order term. It is also possible to go from the Laplace domain transfer function to the phasor representation simply by substituting $j\omega$ for s.

In analyzing the inverse Laplace transform, it makes sense to discuss the various components of the transfer function individually as we did in our discussion of Bode plots.

The isolated s terms have already been discussed. When the isolated s appears in the numerator, it is a derivative operation and when it is in the denominator, it is a pure integrator. Physical systems consisting only of a pure differentiator [i.e., $TF(s)$ = s] or integrator [i.e., $TF(s) = 1/s$] are rare and relatively simple to understand: Their outputs are the same as the inputs either integrated or differentiated. Of much greater interest are the ($s + \omega_1$)—type terms, or first-order terms. We begin describing the behavior of systems that can be represented by these first-order terms, then continue to the intriguing behavior of systems that are represented by second-order terms.

8.2.1   First-Order Processes

First-order processes will contain an $(s + \omega_1)$ or similar term in the denominator. As noted in Chapter 6, first-order systems contain a single energy-storage device, either a capacitor or inductor in electric circuits, or a mass or elastic element in mechanical circuits. The RC and RL circuits shown in Figure 8.5 are examples of first-order processes.

From Example 6.5, we know that the phasor domain transfer function of the RC circuit is:

$$TF(\omega) = \frac{1}{1+j\omega RC} = \frac{1}{1+j\omega/\omega_1} \quad \text{where } \omega_1 = 1/RC \qquad \text{[Eq. 8.31]}$$

To determine the Laplace transfer function, we apply the same analysis techniques, except using the Laplace representation of the elements given in Table 8.1 and assuming zero initial conditions as shown in Figure 8.6.

A KVL (Kirchhoff's voltage law) analysis of this system gives rise to the equations below:

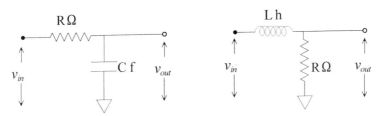

Figure 8.5 Two first-order systems that give rise to similar transfer functions.

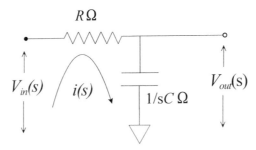

Figure 8.6 The RC circuit after assigning the mesh current and converting the elements and variables to Laplace notation.

$$V_{in}(s) - i(s)\left(R + \frac{1}{Cs}\right) = 0; \quad i(s) = \frac{V_{in}(s)}{R + \dfrac{1}{Cs}} = \frac{V_{in}(s)Cs}{1 + RCs}$$

$$V_{out}(s) = i(s)\left(\frac{1}{Cs}\right) = \frac{V_{in}(s)}{1 + RCs}$$

$$TF(s) = \frac{V_{out}(s)}{V_{in}(s)} = \frac{1}{1 + RCs} = \frac{1/RC}{s + 1/RC} \qquad \text{[Eq. 8.32]}$$

To standardize the first-order Laplace transfer function, a new variable is introduced so that the first-order transfer function has the form:

$$TF(s) = \frac{1/\tau}{s + 1/\tau} \qquad \text{[Eq. 8.33]}$$

Equating coefficients with the transfer function for an RC circuit (Eq. 8.32), it is evident that the new variable, τ is equal to the product of RC. The new variable, τ, is termed the *time constant* for reasons that will become apparent. Comparing the Laplace transfer function in Eq. 8.33 with the phasor transfer function in Eq. 8.31, we note that the time constant, τ, is just the inverse of the cutoff frequency, ω_1:

$$\tau = \frac{1}{\omega_1} \qquad \text{[Eq. 8.34]}$$

A similar analysis applied to the circuit on the right-hand side of Figure 8.5 would produce the transfer function:

$$TF(s) = \frac{V_{out}(s)}{V_{in}(s)} = \frac{R/L}{s + R/L} = \frac{1/\tau}{s + 1/\tau} \quad \text{where } \tau = L/R \qquad \text{[Eq. 8.35]}$$

The two circuits will have similar transfer functions, and identical responses to the same stimulus if the component values are adjusted so that:

$$\tau = \frac{R}{L} = RC \qquad \text{[Eq. 8.36]}$$

From Example 6.5, we know that these two systems have the frequency characteristics of a lowpass filter, a first-order lowpass filter. These frequency characteristics are shown in Figure 6.20. The cutoff frequency and bandwidth of the LR (inductor-resistor) filter is just $\omega_1 = 1/\tau = L/R$.

Now that the transfer function is in the Laplace domain, we can explore the response of these two similarly behaving circuits using input signals other than sinusoids. Two popular signals that are used are the step function, Eq. 8.19 (also shown in Figure 8.1), and the impulse function, Eq. 8.23. Typical impulse and step responses of a first-order system are shown in Figure 8.7 for two time constant values. Both the impulse and step response are exponential with time constants of $\tau = RC$. When the time variable equals the time constant (i.e., $t = \tau$), the value of

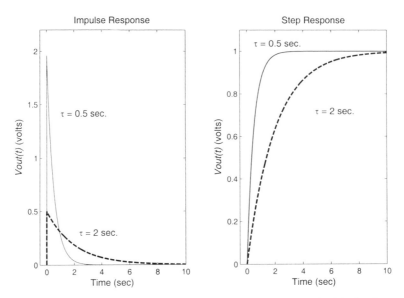

Figure 8.7   The impulse response (*left*) and unit step response (*right*) of a first-order process, in this case the RC circuit shown in Figure 8.6. The responses are shown for two different time constants: $\tau = 0.5$ seconds and $\tau = 2$ seconds.

the exponential becomes: $e^{-t/\tau} = e^{-1} = 0.37$. Hence at time $t = \tau$, the exponential is within 37% of its final value. An alternative, equivalent statement is that the exponential has attained 63% of its final value in one time constant. The time constant makes a good measure of the relative "speed" of an exponential. As a rule-of-thumb, an exponential is considered to have reached its final value when $t > 5\tau$, although in theory the final value is reached only then $t = \infty$. The determination of the impulse and step response of the circuit in Figure 8.6 is given in the next example and was plotted with the aid of MATLAB.

Example 8.2:   Find the output of the RC circuit to a unit step function input. This is sometimes referred to as the *step response* of the system. Also find the output of the RC circuit to an impulse function. Not surprisingly, this is also known as the *impulse response* of the system.

Solution: Step Response: Often the first step in these types of problems is to determine the transfer function; however, in this example we already have the transfer function. Accordingly, we need only use the transfer function equation to solve for the output, $V_{out}(s)$:

$$V_{out}(s) = V_{in}(s)TF(s) \qquad \text{[Eq. 8.37]}$$

In this example, $V_{in}(s)$ is the step function:

$$V_{out}(s) = V_{in}(s)TF(s); \quad \text{where } V_{in}(s) = \frac{1}{s} \text{ (Step function)}$$

After we insert $V_{in}(s)$ into the transfer function equation, we need to rearrange the resulting Laplace function so that it has the same form as one of the functions in the Laplace transform table (Appendix B). Often, this can be the most difficult part of the problem. Solving for $V_{out}(s)$:

$$V_{out}(s) = \frac{1/RC}{(s+1/RC)}\left(\frac{1}{s}\right) = \frac{1/RC}{s(s+1/RC)} = \frac{1/\tau}{s(s+1/\tau)} \qquad \text{[Eq. 8.38]}$$

Referring to the Laplace transform table in Appendix B, we see that the form of the equation for $V_{out}(s)$ matches the Laplace function in entry no. 4. Thus, the inverse Laplace transform for $V_{out}(s)$ in Eq. 8.38 can be obtained as

$$\frac{1/\tau}{s(s+1/\tau)} \text{ which has the form: } \frac{\alpha}{s(s+\alpha)} \Leftrightarrow 1-e^{-\alpha t}$$

Where $\alpha = 1/\tau = 1/RC$. Hence, the step response in the time domain for this system is a simple exponential:

$$V_{out}(t) = 1 - e^{-t/\tau} = 1 - e^{-t/RC}\,\text{V}$$

Solution: Impulse Response: Solving for the impulse response is even easier since for the impulse function, $V_{in}(s) = 1$. Therefore, the response of a system to an impulse function in the Laplace domain is the transfer function itself, and the impulse response in the time domain is the inverse Laplace transform of the transfer function:

$$\text{For the impulse response: } V_{out}(s) = TF(s) \text{ and } v_{out}(t) = \mathcal{L}^{-1}(TF(s))$$

$$V_{out}(s) = \frac{1/RC}{s+1/RC} = \frac{1}{RC}\frac{1}{s+1/RC}\,\text{V}$$

The Laplace function is matched by entry no. 3 in the Laplace transform table except for the constant term. From the definition of the Laplace transform (Eq. 8.4), any constant term can be removed from the integral, so the Laplace transform of a constant times a function is just the constant times the Laplace transform of the function. Similarly, the inverse Laplace transform of a constant times a Laplace function is just the constant times the inverse Laplace transform. Stating these two characteristics formally:

$$\mathcal{L}[kx(t)] = k\mathcal{L}x(t) \qquad \text{[Eq. 8.39]}$$

$$\mathcal{L}^{-1}[kX(s)] = k\mathcal{L}[X(s)] \qquad \text{[Eq. 8.40]}$$

So the time domain solution for $V_{out}(s)$ is obtained:

$$V_{out}(s) = \left(\frac{1}{\tau}\right)\left(\frac{1}{s + 1/\tau}\right) \text{ which has the same form as: } k\frac{1}{s + \alpha} \Leftrightarrow ke^{-\alpha t}$$

$$V_{out}(s) = \left(\frac{1}{\tau}\right)e^{-t/\tau} = \left(\frac{1}{\tau}\right)e^{-t/RC}\text{V}$$

There are many other possible input waveforms in addition to the step and impulse function. As long as they have a Laplace representation, the response to any signal can be found using Eq. 8.37. However, the response of the system to a step and/or impulse usually provides the most insight into the general behavior of the system.

There are also other configurations for first-order circuits in addition to those shown in Figure 8.5. For example, reversing the position of the resistor and capacitor (or the resistor and inductor) gives rise to a highpass filter having the same cutoff frequency (i.e., $\omega_1 = 1/RC = L/R$ as was shown for RC circuits in Problem 6.15). Of course, the transfer function will be slightly different for this configuration; specifically, it will contain an extra s in the numerator. This transfer function and the related time response are evaluated in Problem 8.4 at the end of this chapter.

What bonds all first-order systems is the denominator term, $(s + 1/\tau)$. All first-order systems will contain this term in the denominator, although they might have different numerator terms. In the next section, we will find that all second-order systems contain a quadratic polynomial of s in the denominator of the transfer function. The close link between the denominator term and the system order explains why the denominator of the transfer function is sometimes called the *characteristic equation*: The characteristic equation defines the type of system associated with the transfer function. The characteristic equation also tells us something about the behavior of the system even without evaluating the time domain solution. For example, a characteristic equation such as $(s + 3)$, tells us: (a) the system is a first-order system, and (b) the system's response will include an exponential having a time constant of 1/3 or 0.33 seconds. Second-order characteristic equations are even more informative as shown in the next section.

8.2.2　Second-Order Processes

Second-order processes will have quadratic polynomials of s in the denominator of the transfer functions:

$$s^2 + 2\delta\omega_n s + \omega_n^2 = as^2 + bs + c \qquad \text{[Eq. 8.41]}$$

The right-hand term is the familiar notation for a standard quadratic equation and can be factored using Eq. 8.42 below. As noted in Section 6.6.3, second-order systems contain two energy storage devices. In electrical systems, this could be two independent inductors or capacitors, or both an inductor and capacitor. For mechanical systems, there must be two independent masses or elasticities, but often it is a mass and an elastic element.

One method for dealing with second-order terms would be to factor them into two first-order terms. We could then treat these two factors, which would have the form $(s + \alpha)$, as two first-order terms. Indeed this method is perfectly satisfactory if the factors, the roots of the quadratic equation, are real. Examination of the classic quadratic equation demonstrates when this approach will work. Since the coefficient of the s^2 term is always normalized to 1.0, the a coefficient is always one and the roots of the quadratic equation become:

$$r_1, r_2 = \frac{-b}{2} \pm \frac{1}{2}\sqrt{b^2 - 4c} \qquad \text{[Eq. 8.42]}$$

If $b^2 \geq 4c$, the roots will be real and the quadratic can be factored into two first-order terms. However, if $b^2 < 4c$, both roots will be complex and have real and imaginary parts:

$$r_1 = \frac{-b}{2} + j\frac{1}{2}\sqrt{4c - b^2} \text{ and } r_2 = \frac{-b}{2} - j\frac{1}{2}\sqrt{4c - b^2} \qquad \text{[Eq. 8.43]}$$

If the roots are complex, they will both have the same real part ($b/2$) while the imaginary parts also have the same values, but opposite signs. Complex number pairs that feature this relationship, the same real part, but oppositely signed imaginary parts, are called *complex conjugates*. Whether the roots of a second-order characteristic equation are real or complex has important consequences on the behavior of the system. Sometimes all we want to know about a second-order system is the type of roots in the characteristic equation. This saves the effort of finding the inverse Laplace transform.

The RLC circuit shown in Figure 8.8 is a familiar example of a second-order process.

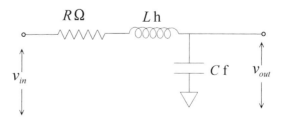

Figure 8.8 A second-order RLC circuit.

We can determine the transfer function by writing applying KVL for the single mesh. In this example we will assume the initial conditions are zero:

$$V_{in}(s) - i(s)(R + Ls + 1/Cs) = 0; \quad i(s) = \frac{V_{in}(s)}{R + Ls + 1/Cs} = \frac{V_{in}(s)Cs}{R + LCs^2 + 1}$$

$$V_{out}(s) = \left(\frac{1}{Cs}\right)\frac{V_{in}(s)Cs}{RCs + LCs^2 + 1} = \frac{V_{in}(s)/LC}{s^2 + \frac{R}{L}s + \frac{1}{LC}}$$

$$TF(s) = \frac{V_{out}(s)}{V_{in}(s)} = \frac{1/LC}{s^2 + \frac{R}{L}s + \frac{1}{LC}} \qquad \text{[Eq. 8.44]}$$

To *standardize* the second-order transfer function, the variables δ and ω_n were first introduced in Section 6.2.1.4 and reintroduced in Eq. 8.30. The names of these terms were given, but not explained, in Chapter 6. The parameter δ is called the *damping factor* and ω_n is called the *undamped natural frequency*. As with the name *time constant*, these names relate to the step and/or impulse response behavior of the second-order transfer function. As shown later, second-order systems with low damping factors will respond with an exponentially decaying oscillation and the smaller the damping factor, the longer it will take for the oscillation to damp out to zero. The rate of oscillation is related to ω_n. Specifically, the rate of oscillation, ω_d, when it occurs is:

$$\omega_d = \omega_n\sqrt{1 - \delta^2} \qquad \text{[Eq. 8.45]}$$

So as δ becomes smaller and smaller, the oscillation frequency ω_d approaches ω_n. When δ equals 0.0, the system is *undamped* and the oscillation frequency is ω_n. Hence, the term *undamped natural frequency* for ω_n, the frequency at which the system oscillates if the damping factor was zero and the system was undamped. Using these two terms, the transfer function for a second-order system is:

$$TF(s) = \frac{\omega_n^2}{s^2 + 2\delta\omega_n s + \omega_n^2} \qquad \text{[Eq. 8.46]}$$

To determine the values of ω_n and δ in terms of R, L, and C, we equate coefficients between Eq. 8.44 and Eq. 8.46:

$$\omega_n^2 = \frac{1}{LC}; \quad \omega_n = \frac{1}{\sqrt{LC}} \qquad \text{[Eq. 8.47]}$$

$$2\delta\omega_n = \frac{R}{L}; \quad \delta = \frac{R}{2\omega_n L} = \frac{R\sqrt{LC}}{2L} = \frac{R}{2}\sqrt{\frac{C}{L}} \qquad \text{[Eq. 8.48]}$$

Equating the variables ω_n and δ to variables a and b in Eq. 8.42 and inserting them into the solution to the quadratic equation:

$$r_1, r_2 = \frac{-2\delta\omega_n}{2} \pm \frac{\sqrt{4\delta^2\omega_n^2 - 4\omega_n^2}}{2} = -\delta\omega_n \pm \omega_n\sqrt{\delta^2 - 1} \qquad \text{[Eq. 8.49]}$$

From Eq. 8.49 we see that the damping factor, δ, alone determines if the roots will be real or complex. Specifically, if $\delta \geq 1$, the constant under the square root is positive and the roots will be real. Conversely, if $\delta < 1$ the square root will contain a negative number and the roots will be complex. If $\delta = 1$, the two roots are also real and are the same: both roots are equal to $-\omega_n$.

As mentioned, the behavior of the system is quite different if the roots are real or imaginary and the form of inverse Laplace transform used is different. Accordingly, it is best to examine the behavior of a second-order system with real roots separately from that of a system with complex roots, acting as if they were two different animals.

8.2.2.1 Second-Order Processes with Real Roots

After constructing the transfer function and finding that it is second-order (i.e., having the same characteristic equation as in Eq. 8.46), the next step is to evaluate the value of δ. If $\delta \geq 1$, the roots are real, and the procedure outlined here should be used. Second-order systems that have a $\delta \geq 1$ and real roots are said to be *overdamped* because they do not exhibit oscillatory behavior in response to a step or impulse input. Such systems have responses that consist of two exponentials.

The best way to analyze overdamped second-order systems is to factor the quadratic equation into two first-order terms each with its own time constant, τ_1 and τ_2:

$$TF(s) = \frac{1}{s^2 + 2\delta\omega_n s + \omega_n^2} = \frac{1}{(s + 1/\tau_1)(s + 1/\tau_2)} \qquad \text{[Eq. 8.50]}$$

These two time constants can be determined from the roots of the quadratic equation, Eq. 8.42. Note that the actual equation could have numerator terms (here shown only as 1), but the evaluation strategy would be the same. Typical overdamped impulse and step responses are shown in Figure 8.9 for different values of τ_1 and τ_2.

After factoring the quadratic equation, the next step is to either separate this function into two individual first-order terms, $\dfrac{k_1}{s + 1/\tau_1} + \dfrac{k_2}{s + 1/\tau_2}$ using partial fraction expansion (see below), or, if available, find a Laplace transform that matches the form in Eq. 8.50. If the numerator is a constant (i.e., not a function of s), then entry no. 9 in the Laplace transform table (Appendix B) matches this form. If the numerator is more complicated, a match may not be found and partial fraction expansion will be necessary. Both these strategies are presented as examples.

Example 8.3: Find the impulse response and step response of the RLC circuit shown in Figure 8.8. Component values are: $R = 50\ \Omega$; $L = 4$ h; $C = 0.01$ f.

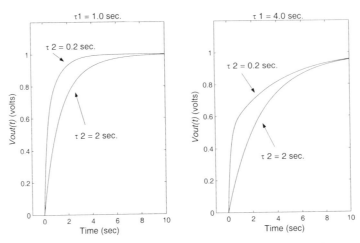

Figure 8.9 Typical response of a second-order system with real roots. These responses are termed *overdamped* because of the exponential-like behavior of the response. Four different combinations of τ_1 and τ_2 are shown. The speed of the response depends on both time constants.

Solution: Impulse Response: As in the last example, we already have the transfer function. Inserting the values of R, L, and C into Eq. 8.44:

$$TF(s) = \frac{25}{s^2 + 12.5s + 25}$$

We can find the values of δ and ω_n by equating coefficients with Eq. 8.46 or by using Eq. 8.47 and Eq. 8.48 entering the values of R, L, and C. Equating coefficients:

$$\omega_n^2 = 25; \quad \omega_n = 5$$

$$2\delta\omega_n = 12.5; \quad \delta = \frac{12.5}{2\omega_n} = \frac{12.5}{10} = 1.25$$

Alternatively we can evaluate δ directly from Eq. 8.48:

$$\delta = \frac{R}{2}\sqrt{\frac{C}{L}} = \frac{50}{2}\sqrt{\frac{0.01}{4}} = 1.25$$

Encouragingly, both approaches give the same results. Because $\delta > 1$, the roots will be real. In this case, the next step is to factor the denominator using the quadratic equation, Eq. 8.42:

$$r_1, r_2 = \frac{-b}{2} \pm \frac{1}{2}\sqrt{b^2 - 4c} = \frac{-12.5}{2} \pm \frac{1}{2}\sqrt{12.5^2 - 4(25)} = -6.25 \pm 3.75$$

$$r_1, r_2 = -10.0 \text{ and} - 2.5$$

For an impulse function input, the output $V_{out}(s)$ is the same as the transfer function because $V_{in}(s) = 1$. The output function can be rearranged to match entry no. 9 in the Laplace transform table:

$$V_{out}(s) = \frac{25}{(s+10)(s+2.5)} = \left(\frac{25}{7.5}\right)\frac{7.5}{(s+10)(s+2.5)};$$

This matches:

$$\frac{\gamma - \alpha}{(s+\alpha)(s+\gamma)} \Leftrightarrow e^{-\alpha t} - e^{-\gamma t}$$

where $\alpha = 2.5$ and $\gamma = 10$.
So $V_{out}(t)$ becomes:

$$v_{out}(t) = \left(\frac{25}{7.5}\right)(e^{-2.5t} - e^{-10t}) = 3.33(e^{-2.5t} - e^{-10t})V$$

The time response of this function is shown in Figure 8.10 (dashed line).

Solution: Step Response: The same initial steps apply: construct the transfer function, and if it is second-order, find δ. If $\delta > 1$, factor the denominator into two first-order terms. These steps have all been performed in the solutions above. To find $V_{out}(s)$, multiply the transfer function by the Laplace transform of the step function:

$$V_{out}(t) = \left(\frac{1}{s}\right)\frac{25}{(s+2.5)(s+10)} = \frac{25}{s(s+2.5)(s+10)} \qquad \text{[Eq. 8.51]}$$

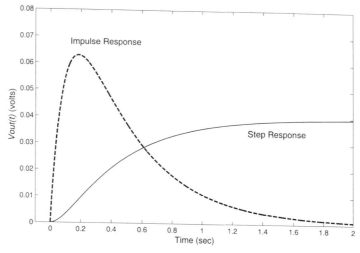

Figure 8.10 The impulse (*dashed curve*) and step response (*solid curve*) of the RLC circuit used in Example 8.3.

With the extra s added to the denominator, this function no longer matches any in the Laplace transform table. However, we can expand this function using partial fraction expansion. The expansion will have the form:

$$V_{out}(t) = \frac{k_1}{s} + \frac{k_2}{s+2.5} + \frac{k_3}{s+10}$$

8.2.2.2   Partial Fraction Expansion

Partial fraction expansion is the opposite of finding the common denominator: Instead of trying to combine fractions into a single fraction, we are trying to break them apart to determine what series of simple fractions will add up to the faction we are trying to decompose. The technique described here is a simplified version of partial fraction expansion that deals with distinct linear factors; that is, denominator components of the form $(s - p)$. Moreover, this analysis will be concerned only with single poles, not multiple poles such as $(s - p)^2$.

Under these restrictions, the partial fraction expansion can be defined as:

$$TF(s) = \frac{N(s)}{(s-p_1)(s-p_2)(s-p_3)\ldots} = \frac{k_1}{s-p_1} + \frac{k_2}{s-p_2} + \frac{k_3}{s-p_3} + \ldots \quad \text{[Eq. 8.52]}$$

where:

$$k_n = (s-p_n)TF(s)\big|_{s=p_n} \quad\quad\quad \text{[Eq. 8.53]}$$

Because the constants in the Laplace equation denominator will always be positive, the values of p will always be negative. Applying partial fraction expansion to Laplace function of Eq. 8.51, the values for p_1, p_2, and p_3 are, respectively, -0, -2.5, and -10, which produces the numerator terms:

$$k_1 = (s+0)\frac{25}{s(s+2.5)(s+10)}\bigg|_{s=-0} = \frac{25}{2.5(10)} = 1.0$$

$$k_2 = (s+2.5)\frac{25}{s(s+2.5)(s+10)}\bigg|_{s=-2.5} = \frac{25}{-2.5(-2.5+10)} = -1.33$$

$$k_3 = (s+10)\frac{25}{s(s+2.5)(s+10)}\bigg|_{s=-10} = \frac{25}{-10(2.5-10)} = 0.33$$

This gives rise to the expanded version of Eq. 8.51:

$$V_{out}(s) = \frac{25}{s(s+2.5)(s+10)} = \frac{1}{s} - \frac{1.33}{s+2.5} + \frac{0.33}{s+10}$$

Each of the terms in Eq. 8.53 has an entry in the Laplace table. Taking the inverse Laplace transform of each of these terms gives the following:

$$v_{out}(t) = 1.0 - 1.33e^{-2.5t} + 0.33e^{-10t}\,\text{V}$$

This response is shown along with the impulse response is shown Figure 8.10 (solid line).

Example 8.4: Find the velocity of the mass in the mechanical system below to a unit step input $F(S) = u(t)$ (a step function). The parameters for the mass, spring, friction are as follows:

$$k_f = 60 \text{ dyne-sec/cm}; \quad m = 4 \text{ gm}; \quad k_e = 200 \text{ dyne/cm}$$

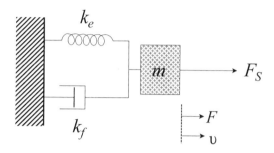

Again, the basic analysis is the same as with phasors, the application of Newton's law, except the variables and elements are represented slightly differently.

$$F_S(s) - v(s)(k_f + ms + k_e/s) = 0; \quad v(s) = \frac{F_s(s)}{k_f + ms + k_e/s}$$

Normalizing the denominator:

$$\frac{v(s)}{F_S(s)} = \frac{s/m}{s^2 + \dfrac{k_f}{m}s + \dfrac{k_e}{m}} = \frac{s\omega_n^2(1/k_e)}{s^2 + 2\delta\omega_n s + \omega_n^2}$$

Where the constants, ω_n and δ are now defined as:

$$\omega_n = \sqrt{\frac{k_e}{m}} \qquad \text{[Eq. 8.54]}$$

$$2\delta\omega_n = \frac{k_f}{m}; \quad \delta = \frac{k_f}{2m\omega_n} = \frac{k_f}{2m}\left(\sqrt{\frac{m}{k_e}}\right) = \frac{k_f}{2\sqrt{k_e m}} \qquad \text{[Eq. 8.55]}$$

From Eq. 8.55, we see that the friction element, k_f, is the only element that can produce an increase in damping by increasing its value. The other two elements decrease the damping when their values increase. In many mechanical systems, damping values much below 0.707 are detrimental since they produce an oscillatory response.

Inserting the values for k_f, k_e, and m:

$$\frac{v(s)}{F(s)} = \frac{s/4}{s^2 + 15s + 50}; \quad \text{since } F(s) = \frac{1}{s}, \text{ the output becomes:}$$

$$v(s) = \left(\frac{1}{s}\right)\frac{s/4}{s^2 + 15s + 50} = \frac{0.25}{s^2 + 15s + 50}$$

Next, we evaluate the value of δ by equating coefficients:

$$\omega_n = \sqrt{50}; \quad 2\delta\omega_n = 15; \quad \delta = \frac{15}{2\omega_n} = \frac{15}{2\sqrt{50}} = 1.06$$

Of course δ could have also been obtained from the component values using Eq. 8.55. Because $\delta = 1.06$, the roots will be real and the system will be overdamped. Accordingly, the next step is to factor the roots using the quadratic equation (Eq. 8.42):

$$r_1, r_2 = \frac{-15}{2} \pm \frac{1}{2}\sqrt{15^2 - 4(50)} = -7.5 \pm 2.5 = -10.0, \quad -5.0$$

and the output becomes: $$v(s) = \frac{0.25}{(s+10)(s+5)}$$

The inverse Laplace transform for this equation can be found in Appendix B (no. 9) where $\gamma = 10$ and $\alpha = 5$. Then multiplying top and bottom by $10 - 5 = 5$, the equation becomes:

$$v(s) = \left(\frac{0.25}{5}\right)\frac{5}{(s+10)(s+5)} = (0.05)\frac{5}{(s+10)(s+5)}$$

This leads to the time function:

$$v(t) = 0.05(e^{-5t} - e^{-10t})\,\text{cm/sec}$$

Alternatively, we could decompose this second-order polynomial in the denominator into two first-order terms using partial fraction expansion.

$$k_1 = (s+10)\frac{0.25}{(s+10)(s+5)}\bigg|_{s=-10} = \frac{0.25}{(-10+5)} = -0.05$$

$$k_2 = (s+5)\frac{0.25}{(s+10)(s+5)}\bigg|_{s=-5} = \frac{0.25}{(-5+10)} = 0.05$$

$$v(s) = \frac{0.05}{s+5} - \frac{0.05}{s+10}$$

Taking the inverse Laplace transform of the two first-order terms leads to the same time function found directly. Figure 8.11 compares this response (Figure 8.11A) with an underdamped response from the same system (Figure 8.11B).

If the friction were reduced in the mechanical system (WD-40!) to less than approximately 56 dyne-sec/cm, δ would be less than 1 and the roots would become complex. This would result in quite different behavior as shown in Figure 8.11. In the next section, we will look at these underdamped systems.

8.2.2.3 Second-Order Processes with Complex Roots

If after determining the transfer function of a second-order system the damping factor, δ, is found to be less than 1, the roots of the characteristic equation will be

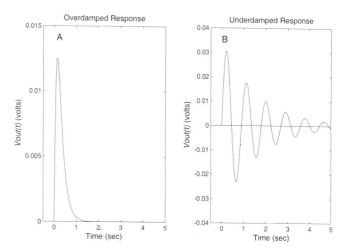

Figure 8.11 Two velocity step responses of the second-order mechanical system analyzed in Example 8.4. The only difference is the coefficient of friction: **(A)** $k_f = 60$ dyne-sec/cm; **(B)** $k_f = 5$ dyne-sec/cm.

complex and the impulse response will have the behavior of a damped sinusoid–a sinusoid that decreases in amplitude exponentially with time (i.e., of the general form $e^{-\alpha t} \sin \omega_n t$). Second-order systems that have complex roots are said to be *underdamped*, a term relating to the oscillatory behavior of such systems. (This terminology can be a bit confusing. Underdamped systems have damped sinusoidal responses Overdamped responses have double exponential, not sinusoidal, behavior.) If the roots are complex, the quadratic equation is not factored. Rather, an inverse Laplace transform can usually be determined directly. Inverse Laplace *transforms* for second-order, underdamped responses are provided in the table both in terms of ω_n and δ and in terms of general coefficients (transform nos. 13–16). Usually the only difficulty in finding the inverse Laplace transform to these systems is in matching coefficients and scaling the transfer function to match the constant term in the table. The next example demonstrates the solution of a second-order underdamped system.

Example 8.5:   In the mechanical system of Example 8.4, the friction coefficient is lowered to $k_f = 5$. Find the step response for $v(t)$ under these conditions.

Solution: After inserting the component values, the transfer function becomes:

$$\frac{v(s)}{F(s)} = \frac{s/4}{s^2 + 1.25s + 50}; \quad \text{and } \delta \text{ is:} \quad 2\delta\omega_n = 1.25; \quad \delta = \frac{1.25}{2\omega_n} = \frac{1.25}{2\sqrt{50}} = 0.088$$

For a step input, $F(s) = 1/s$, so $v(s)$ becomes:

$$v(s) = \left(\frac{1}{s}\right)\frac{s/4}{s^2 + 1.25s + 50} = \frac{0.25}{s^2 + 1.25s + 50}$$

Two of the transforms given in the Laplace tables will work (no. 13 and no. 15). Entry no. 13 will be used here:

$$e^{-\alpha t}\left(\frac{c-b\alpha}{\beta}\sin(\beta t)+b\cos(\beta t)\right) \iff \frac{bs+c}{s^2+2\alpha s+\alpha^2+\beta^2}$$

$$\text{where } b=0; \quad c=0.25; \quad \alpha=\frac{1.25}{2}=0.625;$$

$$\alpha^2+\beta^2=50; \quad \beta=\sqrt{50-0.39}; \quad \beta=7.04$$

Substituting these values into the time domain equivalent on the left side gives $v(t)$:

$$v(t)=e^{-.625t}\left(\frac{0.25-0}{7.04}\sin(7.04t)\right)=0.036e^{-.625t}(\sin(7.04t))\,\text{cm}/\text{sec}$$

This response is plotted in Figure 8.11B. Other examples of second-order under-damped responses will be presented in the next section, which explores systems that have nonzero initial conditions.

8.3 NONZERO INITIAL CONDITIONS

Problems with initial conditions can be divided into two classes: those in which the initial conditions are given, and those in which they have to be determined by analysis of the system's responses over past time. Analysis of problems that have initial conditions is straightforward: It is only a matter of including the initial condition sources (given in Table 8.1) in the transformed network. When initial conditions are known to exist, but not given, there is the extra burden of determining the initial condition values. An example with known, given initial conditions is presented first.

Example 8.6: Find the voltage, $V_{out}(t)$, of the circuit in Figure 8.12. The capacitor is initially charged to 15 V with the positive side indicated. The switch closes at $t = 0$ attaching the resistor to the capacitor.

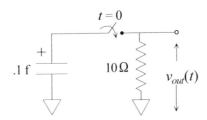

Figure 8.12 The circuit used in Example 8.6. The resistor is attached (instantaneously) to the capacitor at $t = 0$. The capacitor is initially charged to 15 V.

This circuit does not have a separate voltage source; however, the initial charge on the capacitor acts as a voltage source. As with all system's analyses described thus far, the first step is to transform the system into the appropriate domain for analysis, in this case the Laplace domain with initial conditions. This involves all the transformations described previously (variables to Laplace variables, elements to Laplace impedances) along with the addition of sources to account for the initial conditions. In the case of a charged capacitor, the appropriate source is a voltage source of V_C/s (Table 8.1). The transformed circuit is shown in Figure 8.13.

The remainder of the problem is straightforward following the lines of other circuit problems. Applying KVL:

$$\frac{15}{s} - I(s)\left(\frac{10}{s} + 10\right) = 0; \quad I(s) = \frac{15/s}{10/s + 10} = \frac{1.5}{s+1}$$

$$V_{out}(s) = I(s)10 = \frac{10(1.5)}{s+1} = \frac{15}{s+1}; \quad v_{out}(t) = 15e^{-t}\,\text{V}$$

The next example is considerably more challenging. It includes a mass and a spring, both of which have nonzero initial conditions. Moreover, the initial conditions are not given, but must be determined, in this case using phasor analysis. Hence, this problem uses both phasor and Laplace analysis.

Example 8.7: In the mechanical system used in Example 8.4, F_S is now a sinusoidal force generator that applies a force of $10\cos(2t)$ dyne. It has been applied for a long time, so we can assume the system is in sinusoidal steady state. At $t = 0$ the connection to the source breaks, so the force is no longer applied to the mechanical system. Find $x(t)$ for $t > 0$. For this problem, the values for the elements are: $k_f = 3$ dyne/cm; $k_e = 20$ dyne-sec/cm; $m = 10$ gm.

Solution: Even though we are only interested in the output for $t > 0$, we need to find the initial conditions and this necessitates analyzing the problem for negative times to find the initial state at $t = 0$. Because we have both a mass and a spring in the problem, we will need to find the initial position for the spring and the initial velocity for the mass. During negative times, the source is sinusoidal and it has been

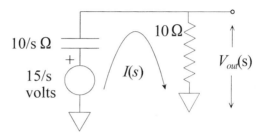

Figure 8.13 The circuit in Figure 8.12 after transformation to the Laplace domain.

connected for long time so we can use phasor analysis for this portion of the problem.

For $t < 0$, we first convert the elements and variables to the phasor domain to obtain the system shown in Figure 8.14.

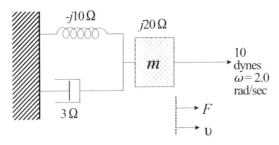

Figure 8.14   Phasor representation of the mechanical system in Figure 8.10.

Writing the basic conservation equation:

$$10 - v(3 + j20 - j10) = 0; \quad v = \frac{10}{3 + j10} = \frac{10}{10.4 \angle 73} = 0.96 \angle -73 \, \text{cm/sec}$$

$$v(t) = 0.96 \cos(2t - 73) \, \text{cm/sec}$$

$$x = \frac{v}{j\omega} = \frac{0.96 \angle -73}{2 \angle 90} = 0.48 \angle -163; \quad x(t) = 0.48 \cos(2t - 163) \text{cm}$$

Substituting in $t = 0$ and solving for $v(0)$ and $x(0)$:

$$v(t) = 0.96 \cos(2t - 73); \quad v(0) = 0.96 \cos(-73) = 0.28 \, \text{cm/sec}$$

$$x(t) = 0.48 \cos(2t - 163); \quad x(0) = 0.48 \cos(-163) = -0.46 \, \text{cm}$$

With these initial conditions, we can now transform the mechanical system into the Laplace domain for $t > 0$. For $t > 0$ there is no force applied to the system (F_S has broken off); only the initial conditions drive the response. Care must be taken with the direction of the initial condition force generators. Since the spring is actually compressed (x is negative), the force produced by the initial condition will be in the positive (rightward) direction. The value of the force is $k_e x(0)/s = -0.46(20)/s = -9.2/s$ dyne. The velocity is positive (i.e., moving to the right), so the mass will also generate a rightward force with a value of $mv(0) = 10(0.28) = 2.8$ dyne. The transformed system is shown in Figure 8.15.

From here on, the problem is similar to any of the others we have solved. Applying Newton's law about the point indicated by the dashed vertical line:

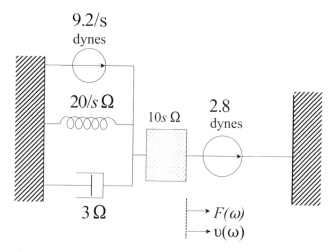

Figure 8.15 Laplace representation of the mechanical system in Example 8.4 showing additional force generators to account for the nonzero initial conditions in both velocity and position.

$$\frac{9.2}{s} + 2.8 - \upsilon(s)\left(3 + 20s + \frac{10}{s}\right) = 0; \quad \upsilon(s) = \frac{9.2/s + 2.8}{3 + 20s + 10/s}$$

$$\upsilon(s) = \frac{(2.8s + 9.2)/20}{s^2 + 3/20^2 + 10/20} = \frac{0.14s + 0.46}{s^2 + 0.15s + 0.5}$$

$$X(s) = \frac{1}{s}\upsilon(s) = \frac{0.14s + 0.46}{s(s^2 + 0.15s + 0.5)}$$

The damping is:

$$2\delta\omega_n = 0.15; \quad \delta = \frac{0.15}{2\omega_n} = \frac{0.15}{2\sqrt{0.5}} = 0.106 < 1$$

The system is clearly underdamped and the appropriate transfer function should be used for matching $X(s)$ (entry nos. 13–16 in the Laplace transform table). The function $x(s)$ matches entry no. 14 in the Laplace transform table, but will require some rescaling to match the numerator. Considering only the dominator, $\alpha_2 + \beta_2 = 0.5$, but the constant in the numerator is $\alpha_2 + \beta_2$ and must also equal 0.5. To make this match, we need to multiply the numerator by $0.5/0.46 = 1.09$. Multiplying top and bottom by 1.09 the rescaled Laplace function becomes:

$$X(s) = \left(\frac{1}{1.09}\right)\frac{0.15s + 0.5}{s(s^2 + 0.15s + 0.5)} = \frac{0.92(0.15s + 0.5)}{s(s^2 + 0.15s + 0.5)}$$

Now equating coefficients with entry no. 14:

$$b = 0.15; \quad \alpha = \frac{0.15}{2} = 0.075; \quad \alpha^2 + \beta^2 = 0.5; \quad \beta^2 = 0.5 - 0.075^2; \quad \beta = \sqrt{0.49} = 0.7$$

The inverse Laplace transform becomes:

$$x(t) = 0.92\left[1 - e^{-0.075t}\left(\frac{0.075 - 0.15}{0.7}\sin(0.7t) + \cos(0.7t)\right)\right]$$

$$x(t) = 0.92(1 - e^{-0.075t}(-0.14\sin(0.7t) + \cos(0.7t)))\text{cm}$$

This could also be put in terms of a single sinusoid using the Eq. 2.9 and Eq. 2.10 presented in Chapter 2.

8.4   INITIAL AND FINAL VALUE THEOREMS

The time representation of a Laplace function is obtained by taking the inverse Laplace *transform* using tables such as found in Appendix B. Sometimes we are only looking for the value of the function at its very beginning, $x(t = 0)$, or at its very end as $x(t \to \infty)$. Two useful theorems that can supply us with this information are the *initial and final value theorems*. These theorems can give us the function's initial and the final value without the need for taking the inverse Laplace transform. The initial value theorem provides us with the value of the function at $t = 0$, while the final value theorem, as you might expect, gives us the value of the function as $t \to \infty$.

Because the Laplace variable, s, is a form of complex frequency and frequency is inversely related to time, you might expect that the value of a Laplace function at $t = 0$ might be obtained by letting $s \to \infty$. In particular, the initial value theorem states:

$$x(0+) = \lim_{s \to \infty} sX(s) \qquad \text{[Eq. 8.56]}$$

This is easy to demonstrate. Beginning with the definition of the Laplace transform of the time derivative of $x(t)$:

$$\mathcal{L}\,dx(t)/dt = \int_0^\infty \frac{dx(t)}{dt}e^{-st}dt = sX(s) - x(0-) \qquad \text{[Eq. 8.57]}$$

Assuming that $x(t)$ is continuous at $t = 0$, and taking the limit of both sides:

$$\lim_{s \to \infty}\int_0^\infty \frac{dx(t)}{dt}e^{-st}dt = \int_0^\infty \frac{dx(t)}{dt}\left(\lim_{s \to \infty}e^{-st}\right)dt = \lim_{s \to \infty}sX(s) - x(0) = 0$$

$$\lim_{s \to \infty}sX(s) = x(0)$$

The validity of the theorem can also be demonstrated if there is a discontinuity at $t = 0$.

The final value theorem follows the same logic, except $s \to 0$. Specifically, the final value theorem states:

$$\lim_{t \to \infty} x(t) = \lim_{s \to 0} sX(s)$$ [Eq. 8.58]

To validate this theorem, taking the limit as $s \to 0$ of Eq. 8.57:

$$\lim_{s \to 0} \int_0^\infty \frac{dx(t)}{dt} e^{-st} dt = \lim_{s \to 0} sX(s) - x(0)$$

$$\lim_{s \to 0} \int_0^\infty \frac{dx(t)}{dt} e^{-st} dt = \int_0^\infty \frac{dx(t)}{dt} \lim_{s \to 0}[e^{-st}] dt = \int_0^\infty \frac{dx(t)}{dt} dt$$

Evaluating the integral:

$$\lim_{s \to 0} sX(s) - x(0) = \lim_{x \to \infty} x(t) - x(0)$$

Which reduces to Eq. 8.58.

The application of either of the theorems is straightforward as shown in the following example.

Example 8.8:   Use the final value theorem to find the final value of $x(t)$ in the previous example.

Solution: Apply Eq. 8.58 to the Laplace function found for $X(s)$:

$$X(s) = \frac{0.14s + 0.46}{s(s^2 + 0.15s + 0.5)}$$

Substituting into Eq. 8.58:

$$\lim_{s \to 0} sX(s) = \lim_{s \to 0} s \frac{0.14s + 0.5}{s(s^2 + 0.15s + 0.5)} = \lim_{s \to 0} \frac{0.14s + 0.46}{s^2 + 0.15s + 0.5} = \frac{0.46}{0.5} = 0.92$$

This is the same value that is obtained when letting $t \to \infty$ in the time solution $x(t)$.

An example of the application of the initial value theorem is found in the problems.

8.5   THE LAPLACE DOMAIN AND THE FREQUENCY DOMAIN

Because s is a complex frequency variable, there is a relationship between the Laplace domain and the frequency domain. Given a Laplace transfer function, it is easy to find the phasor or frequency equivalent by substituting $s = j\omega$. After renormalizing, the coefficients so the constant term equals 1, the frequency plot can be constructed using the Bode plot techniques described in Chapter 6. Some of the relationships between the Laplace transfer function and the frequency characteristics have already been mentioned, and these depend largely on the characteristic equation. A first-order characteristic equation gives rise to first-order frequency characteristics such as shown in Figures 6.5 and 6.6. A classic example of a first-order circuit is the RC lowpass filter whose frequency characteristics are shown in Figure 6.20.

Second-order frequency characteristics, like second-order time responses, are highly dependent on the value of the damping coefficient, δ. As shown in Figure 6.7, the frequency curve develops a peak of values of $\delta < 1$, and the height of that peak increases as δ decreases. In the time domain, this corresponds to an overshoot response. In the frequency domain, the peak in the frequency curve occurs at frequency ω_d which is close to ω_n, the undamped natural frequency, Eq. 8.45. In the time domain, the response will oscillate at frequency ω_d. As shown in Eq 8.45, the oscillating frequency is not quite the same as the undamped frequency, the resonant frequency at which the system would like to oscillate. This is because the decay in the oscillation due to the damping (i.e., the $e^{-\omega_n t}$ term) lowers the actual resonant frequency to ω_d. As the damping factor, δ, decreases, the resonant frequency, ω_d, approaches the undamped natural frequency, ω_n.

The next example in this chapter uses MATLAB to compare the frequency and time characteristics of a second-order system for various values of damping.

Example 8.9:   In the RLC circuit of Figure 8.8, assume $L = 1$ mh (10^{-3} h) and $C = 0.1$ μf. These are more realistic values for L and C. Find values for R corresponding to damping factors of 0.1, 0.5, and 2.0. Plot the frequency characteristics (i.e., Bode plot) of the transfer function and the step response of the system. Use MATLAB for both calculation and plotting.

Solution: To find the required resistor values use Eq. 8.48:

$$\delta = \frac{R}{2}\sqrt{\frac{C}{L}}; \quad R = 2\delta\sqrt{\frac{L}{C}} = 2\delta\sqrt{\frac{1\times 10^{-3}}{.1\times 10^{-6}}} = \delta(200)\Omega$$

Now solve for the Laplace transfer function leaving R as a variable. The transfer function of the RLC circuit has already been found in a previous example and is given in Eq. 8.44:

$$TF(s) = \frac{1/LC}{s^2 + \frac{R}{L}s + \frac{1}{LC}} = \frac{10^{10}}{s^2 + R(10^3)s + 10^{10}}$$

The step response can be obtained by multiplying the transfer function by $1/s$, then determining the inverse Laplace transform:

$$V_{out}(s) = \frac{10^{10}}{s(s^2 + R(10^3)s + 10^{10})}$$

This matches entry no. 16 for $\delta < 1^*$ with no rescaling required:

$$v(t) = 1 - \frac{e^{-\delta\omega_n t}}{\sqrt{1-\delta^2}}\sin(\omega_n\sqrt{1-\delta^2}\,t + \theta); \quad \theta = \tan^{-1}\left(\frac{\sqrt{1-\delta^2}}{\delta}\right)$$

* This solution also applies when $\delta = 2$.

where:

$$\omega_n^2 = 10^{10}; \quad \omega_n = 10^5; \quad \delta = \frac{R(10^3)}{2\omega_n} = \frac{R10^3}{2 \times 10^5} = \frac{R}{200}$$

$$R = 200\delta$$

The equivalent time response will be programmed directly in to the MATLAB code.

To find the frequency response, convert the Laplace transfer function to phasor by substituting $s = j\omega$ and rearranging:

$$TF(\omega) = \frac{10^{10}}{(j\omega)^2 + R10^3 j\omega + 10^{10}} = \frac{1}{1 - (10^{-10}\omega^2) + j\omega R10^{-7}}$$

This can also be programmed directly into MATLAB. The resulting program and plots are shown below.

```
% Example 8.9 Comparison of Time and Frequency
% characteristics of a second-order system.
%
close all; clear all;
w = 10000:1000:1000000;        % Set frequency vector, w
R = [2 .5 .1 ]*200;            % Set values of R
%
% Calculate and plot frequency characteristics
for k = 1:length(R)
  TF = 1./(1 - 10.^-10*w.^2 + j*R(k)*10.^-7.*w); % Transfer
                                                 % function
  Mag = 20*log10(abs(TF));     % Take magnitude in dB
  semilogx(w,Mag,'k'); hold on;% Plot dB against log freq.
end
%
...... Label axes .......
%
% Calculate and plot time characteristics of step response
figure;
t = 0:10.^-8:2*10.^-4;         % Set time vector, t
for k = 1:length(R)
  wn = 10^5                    % Undamped natural frequency
  del = R(k)/200;              % Delta
  c = sqrt(1-del^2);           % Sqrt of 1- delta squared
  theta = atan(c/del);         % Theta
  % Time function
  xt = 1 - (1/c)*(exp(-del*wn*t)).*(sin(wn*c*t + theta));
  plot(t,xt,'k'); hold on;     % Plot
end
```

```
%
...... Label axes ......
```

The results are shown in Figure 8.16. Comparing Figure 8.16A with Figure 8.16B, the correspondence between the frequency characteristics and the time responses is evident. A frequency response with a peak corresponds to a time domain response with overshoot. Note that the frequency peak associated with a δ of 0.5 is small, but this time domain response still has some overshoot. As shown here and in one of the problems, the larger the peak, the greater the overshoot. The minimum δ for no overshoot in the response is left as an exercise.

The last example is quite challenging but fun and it touches on some real-world concerns of bioengineers. You are given the impulse response to a biological system and want to find the transfer function of that system so that you can predict its response to other inputs. This problem is another form of systems identification introduced in Section 6.4.2, except now you have the impulse response, not the response to a series of sinusoids at different frequencies. Often it is not practical to simulate biological systems with sinusoids while it may be possible to generate an impulse input. For example, it is impractical to stimulate a pharmacological system sinusoidally, but a sudden infusion of a given drug, a drug *bolus*, is a common input. The knee-jerk response is another example of the use of an impulse input.

Example 8.10:   The data file `biologic` contains the impulse response of a biological system. You have reason to believe that the system can be adequately represented as a second-order system, but have no other knowledge about the system. You want to estimate the Transfer function of the system. The impulse response was obtained using a sampling frequency of 100 Hz.

Solution: There are several approaches you might use to estimate the transfer function from the impulse response data. Since you already know that the transfer function is second-order, you only need to find three parameters, the constant amplitude term A, the damped natural frequency, ω_n, and the damping, δ.

Just as the impulse response is the Laplace transform of the transfer function, it is also the inverse Fourier transform of the transfer function in the frequency domain (where $s = j\omega$). Hence, you could take the Fourier transform of the impulse response, plot the magnitude (and possibly phase) and use the Bode plots techniques to estimate A, ω_n, and δ.

Another method would be to try to generate a number of second-order impulse responses with different combinations of A, ω_n and δ and pick the parameter set that produces the best match to the data. This may seem like a more complicated and less elegant approach, but might work better if the data contained noise, as is usually the case. This trial and error approach is an example of an important process known as *optimization*. It could take awhile to find the best match, but MATLAB provides a method for automating the search.

Both methods are tried in the MATLAB program below. The first step in any identification process is to plot the data as is done in the program. Such a plot is

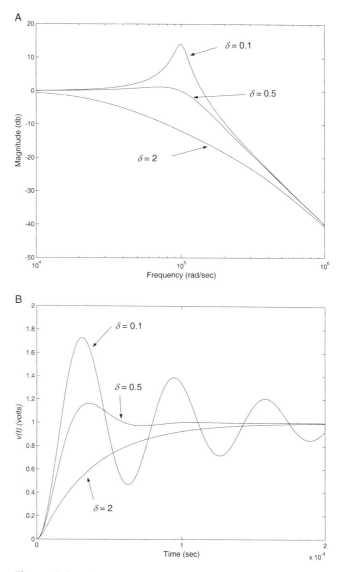

Figure 8.16 **A:** Comparison of Bode plots for a second-order system having three different damping coefficients: $\delta = 0.1$, 0.5, and 2.0. **B:** Comparison of step response for the same second-order system having the three different damping coefficients in **A**.

shown in Figure 8.17 and shows overshoot behavior. The overshoot behavior indicates that system is underdamped. Comparing the general overshoot behavior with that seen for different damping factors in Figure 8.16B suggests the damping factor is somewhere between 0.5 and 0.1. (Figure 8.16B shows step responses while Figure 8.17 is an impulse response, but the decay of the oscillation will be similarly

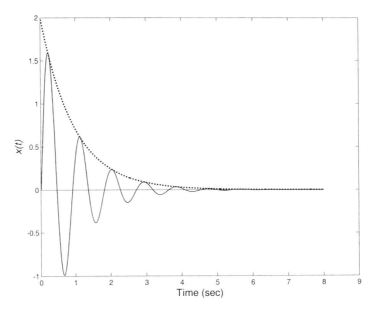

Figure 8.17 A plot of the unknown system's response to an impulse input indicates the system is underdamped (i.e., $\delta < 1$). An envelope of the response, a curve that traces the peaks is shown (*dotted curve*) is helpful in estimating the amplitude.

dependent on δ.) The program to calculate and plot the Fourier transform of the response and then to try to match the time response follows.

```
% Example 8.10 Identify a biological system from impulse
% response data.
% The data are in file 'biologic' and where sampled at 100
% Hz.
%
clear all; close all;
fs = 100;                          % Sampling frequency
load biologic;                     % Get the data
ln = length(x);                    % Data Length
t = (1:ln)/100;                    % Construct time vector
plot(t,x,'k');                     % Plot time data
...... label.......
%
% Take fft of data and plot
figure;
X = 20*log10(abs(fft(x)));
f = (1:ln)*fs/ln;                  % Generate frequency vector
                                     for plotting
```

```
semilogx(f,X,'k');                        % Plot frequency
.......label......
%
% Estimate parameters
param0 = [1 1 1];                         % Initial parameter
                                          % estimates (all ones)
param = fminsearch('second_order',param0,[],x,t);
disp(param)
%
% Plot time estimate and original impulse superimposed
x_trial = param(1)*(exp(-param(2)*param(3)*t)...
  .*sin(param(3)*t));
figure;
plot(t,x,'k',t,x_trial,':k');
....... label .......
```

Analysis: After getting the data from file 'biologic,' the program calculates and plots the Fourier transform in a straightforward manner. The frequency characteristics are shown in Figure 8.18.

The plot indicates an underdamped system with an undamped natural frequency of approximately 1.2 Hz and a peak of approximately 10 dB. This corresponds to an ω_n of:

$$\omega_n = 2\pi f_n = 2\pi 1.2 = 7.5 \text{ rad/sec}$$

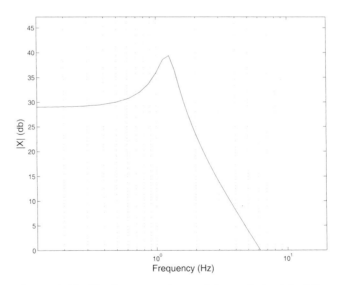

Figure 8.18 The frequency characteristics or Bode plot of the unknown system. This plot will be used to estimate the second-order parameters ω_n and δ.

To find δ, we note that the peak is the negative of $20 \log(2\delta)$. Solving for δ:

$$\delta = \frac{10^{-peak/20}}{2} = \frac{10^{-10/20}}{2} = 0.158$$

It is difficult to estimate the constant gain term A from the frequency plot, but this is not a problem since the constant term can be estimated directly from the time plot. Drawing an exponential envelope over the impulse response curve in Figure 8.17 suggests that the peak amplitude was around 2.0.

The next section of the program estimates the parameters of the impulse response by a trial an error method also known as optimization. The three unknown parameters are initially set to 1.0. The program uses MATLAB's fminsearch to find the parameters that produce a second-order curve that best matches the impulse response. The optimization routine fminsearch calls a function, second-order, that generates the trial response, compares it to the actual response, and outputs the difference as the RMS (root-mean-squared) error. The function second-order is given below.

```
function rms = second_order(param,x,t);
% Function used by fminsearch to compare trial second-
% order response with that of data.
% The output is the RMS error between the trial function
% and the actual data.
% Generate trial second-order system
x_trial = param(1)*(exp(-param(2)*param(3)*t).*...
sin(param(3)*t));
%
% Calculate RMS error
rms = sqrt(sum((x - x_trial).^2));
```

Note that the function second-order is quite short consisting of only two lines of code: One to generate a second-order damped sinusoid and the other to calculate the RMS error which is the output of the function. The equation for the damped sinusoid makes the assumption that $\omega_d = \omega_n$, which is true for small values of δ (see Eq. 8.45). The optimization routine fminsearch calls this routine repeatedly passing to the routine its estimates of the three parameters. The initial values of these parameters were in the first calling argument of fminsearch (param) and the size of the vectors tell fminsearch how many parameters to optimize. It also has the optional capability of passing other arguments, in this case the data variable (x) and the time variable (t), both useful to the function. The routine fminsearch adjusts the parameters until finds it can no longer reduce the RMS error below a certain criterion (see help fminsearch for details). It then outputs the parameters (param) to the main program. The main program displays these parameters and

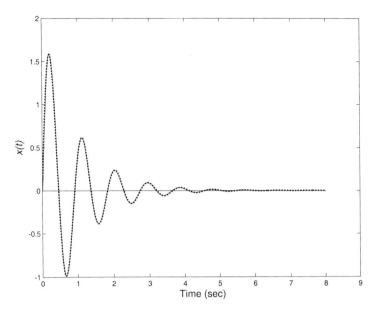

Figure 8.19   The impulse response of the unknown second-order system (*solid curve*) and the response found by optimization (*dashed curve*). The two curves are so similar that it is difficult to tell them apart.

TABLE 8.2   Identification of a Second-Order System

	Actual	**Bode Plot**	**Optimization**
A	2.0	2.0	1.98
ω_n	7.0	7.5	6.85
δ	0.15	0.158	0.15

uses them to construct an impulse response, which it plots along with the original data. As shown in Figure 8.19, the estimated response (dotted line) is very close to the system's actual impulse response (solid line). Table 8.2 compares the values found by the two methods with the actual parameters.

The values found by both approaches are very close to the actual values, but this was a comparatively simple system with noiseless data. One of the problems applies this approach to step response.

8.6 SUMMARY

With the Laplace transform, all of the analysis tools developed in Chapters 5 through 7 can be applied to systems exposed to a broader class of signals. Transfer functions written in terms of the Laplace variable s (i.e., complex frequency) play the same role as transfer functions that use frequency (ω), but to other types of signals. Here, only the responses to the step and impulse signals were explored because these are the two most commonly used stimuli in practice. Their popularity stems from the fact that they provide a great deal of insight into system behavior and they are usually easy to generate in practical situations. However, responses to other signals such as ramps or exponentials, or any signal that has a Laplace transform, can be analyzed using these techniques. Laplace transform methods can also be extended to systems with nonzero initial conditions, a particularly useful feature.

The Laplace transform can be viewed as an extension of the Fourier transform where complex frequency, s, is used instead of imaginary frequency, $j\omega$. With this in mind, it is easy to covert from the Laplace domain back to the frequency domain by substituting $j\omega$ for s in the Laplace transfer functions. Bode plot techniques can then be applied to these converted transforms. Thus, the Laplace transform serves as a gateway into both the frequency domain (through Bode plots) and the time domain (through the inverse Laplace transform).

PROBLEMS

1. Find the Laplace transform of the following time functions:
   A.

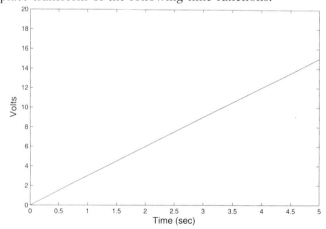

B.

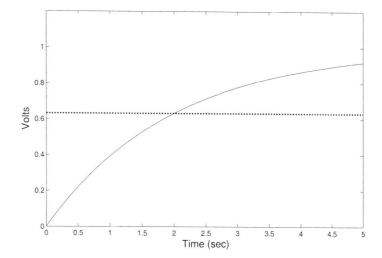

C. $(e^{-2t} - e^{-5t})$
D. $2e^{-3t} - 4e^{-6t}$
E. $5 + 3e^{-10t}$

2. Find the inverse Laplace transform of the following Laplace functions:

A. $\dfrac{10}{s+5}$

B. $\dfrac{10}{s(s+5)}$

C. $\dfrac{5s+4}{s^2+5s+20}$ (*Hint:* Check roots.)

D. $\dfrac{5s+4}{s(s^2+5s+20)}$

3. Use partial fraction expansion to find the inverse Laplace transform of the following functions:

A. $\dfrac{s+4}{s^2+10s+10}$

B. $\dfrac{10}{s(s^2+4s+3)}$

4. Find the transfer function and unit step response of the circuit below.

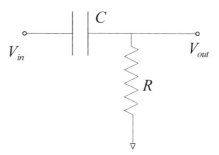

5. Find the transfer function and unit step response of the circuit below.

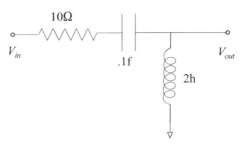

6. Find the transfer function and unit step response of the circuit in Problem 8.5 if the resistance is lowered to 5 Ω.
7. Find the transfer function and *ramp response* of the network below.

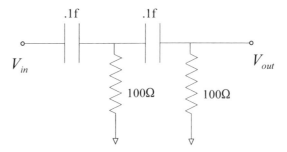

8. An RLC circuit contains a 0.02-f capacitor. Find R and L such that the damping factor, δ, is 0.5 and the response oscillates at 10 rad/sec. (*Hint:* The response oscillates at ω_d not ω_n. See Eq. 8.45.)
9. In the circuit below, the capacitor is charged to 6 V. At $t = 0$ the switch closes. Find the voltage across the resistor for $t > 0$.

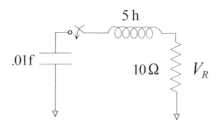

10. The mass shown is supported by two springs, each with coefficients, $k_e = 1,000$ and is pulled doward by gravity. The mass is 100 gm and is initially at rest. At

$t = 0$, one of the springs breaks. Find $x(t)$ for $t > 0$. Assume the only friction is due to the resistance of the air and has a value of $k_f = 0.01$ dyne-sec/cm.

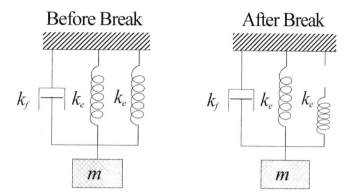

11. The inductor has an initial current of 2.0 A and the capacitor has an initial charge of 10 V when the switch opens at $t = 0$. Find the value of the voltage across the capacitor for $t > 0$. (*Hint:* This becomes a two-mesh problem.)

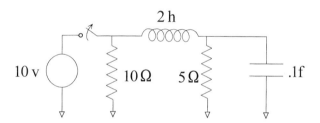

12. At $t = 0$, the mass in the system below is moving at 2 cm/sec to the left, but the spring is uncompressed [$x_e(0) = 0$]. At $t = 0$, a force of 4 dyne is applied in the direction shown (to the right). $k_e = 100$ dynes/cm; $k_f = 5$ dyne-sec/cm; $m = 10$ gm.
    A. Find $v(t)$ for $t > 0$.
    B. Find the force produced by the spring (only the spring for $t > 0$). (*Remember:* At $t = 0$ it is uncompressed, so its initial force is zero.)

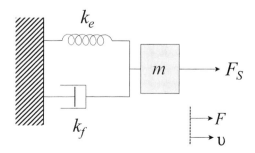

13. In the mechanical system of Problem 8.12, a force is applied that pulls the mass-spring to the right at a steady 2 cm/sec. At 5 seconds, the force is disconnected and the system no longer has an outside force pulling on it. Find the velocity from that time on.

$$m = 10 \text{ gm}; \ k_e = 2 \text{ dyne/cm}; \ k_f = 3 \text{ dyne-sec/cm}.$$

14. The impulse response of a first-order RC filter in the Laplace domain was found in Example 8.2 to be:

$$V_{out}(s) = \left(\frac{1}{\tau}\right)\left(\frac{1}{s + 1/\tau}\right)$$

Use the initial value theorem to find the filter output's value at $t = 0$ [i.e., $V_{out}(t = 0)$] for the filter.

15. The transfer function of an electronic system has been determined as:

$$TF(s) = \frac{5s + 4}{s^2 + 5s + 20}$$

Use the final value theorem to find the value of this system's output for $t \to \infty$ if the input is a 5.0-V step function.

MATLAB Problems

16. Demonstrate the effect of a 0.2-second delay on the frequency characteristics of a second-order system. The system should have a ω_n of 10 rad/sec and a δ of 0.7. Plot the magnitude and phase with and without the delay. (*Hint:* Plot the Bode plot of a standard second-order system by substituting $j\omega$ for s. Then add an $e^{-0.25}$ [$= e^{-j0.2\omega}$] to the transfer function and replot.) Plot for a frequency range of 1 to 100 rad/sec. Note the phase curve with the delay will exceed -180 degrees and will *wraparound*. [You can use the MATLAB command unwrap to correct for this phase wrapping.]

17. For the mechanical system of Problem 8.12, plot the velocity in response to a unit step input for values of mass that produce damping factors of 0.9, 0.5, and 0.2. Assume $k_e = 5$ dyne/cm and $k_f = 10$ dyne-cm/sec. (*Hint:* Modify the code in Example 8.9.)

    If the mass is such that the damping factor is 0.1, how and by how much, should the friction be changed to produce a damping of 0.707?

18. The file 'step_response' contains a data variable x, which is the step response of an unknown system. Use MATLAB's fminsearch in conjunction with a modified version of the function second_order to find the variables that best match the curve. Assume the unknown system can be represented by a second-order process and that it is overdamped which you can easily tell from the time plot of the response. (*Hint:* You can model the step response of a second-order overdamped system as $x = A(e^{-\alpha t} - e^{-\gamma t})$.)

9 SYSTEM MODELS AND BEHAVIOR

9.1 THE SYSTEM MODEL

Most physiological and biological systems are quite complex and lead to complicated analog models, if they can be represented by analog models at all. Moreover, while analog models often provide a good quantitative representation of a physiological system, they are not very effective in describing the general flow of information through the system. For example, when a resistor is connected to a Thévenin source, the source has an influence on the resistor and the resistor affects the source, but these mutual influences are not explicit in the analog representation. In an analog model, every component can interact with every other component and that interaction may not be evident. Many physiological components act this way with distributed, mutually interacting influences, but a model should clarify as well as represent.

The systems model is a process-oriented representation that emphasizes the influences, or flow of information, between modules. A systems model describes how processes interact and what operations these processes perform but does not go into details as to how these processes are implemented. The basic module of the system model is the *black box* process, so-called because the module describes an input–output relationship but is not concerned with the internal mechanism that achieves that relationship: systems modules only define the input-output relationship. Since transfer functions are mathematical correlates of systems modules; they only describe input–output relationships; it is only natural that systems modules be represented by transfer functions. Ignoring the details of mechanism and emphasizing module interactions are both the greatest strengths and greatest weaknesses of systems models.

The graphic symbols used to represent system model elements are straightforward with little artistic merit. Since the basic element of a system model is so general, it is usually represented by a simple rectangle. This element may contain the related Laplace transfer function or simply a symbol for the transfer function. Exceptions are made for a few specific processes such as those that perform arithmetic operations on two or more signals. These are usually represented as circles containing the appropriate arithmetic symbol: Σ for addition or subtraction; $\times$ for multiplication; $\div$ for division. (The latter two are rare in systems models since they are nonlinear processes.) An example of such elements is shown in Figure 9.1 The summation device subtracts x_2 from x_1 as indicated by the plus or minus signs next to the inputs. As shown in Figure 9.1, a system model also contains lines with arrows. These indicate the flow of information, or influence, between the various model elements.

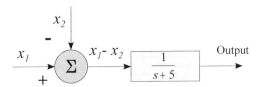

Figure 9.1 Typical systems model elements. The left-hand element is a subtractor whose output signal is the difference between two inputs: $x_1 - x_2$. The output of the subtractor feeds a process represented by a transfer function.

Most system elements are represented by transfer functions, although a module's impulse response would constitute an equivalent representation since in the last chapter we found that the impulse response is just the inverse Laplace transform of the transfer function. The geometry of the systems model shows how these transfer functions interact. Inherent in the representation of a systems module are the assumptions required for the development of the transfer function itself: The source is ideal and there is no effective load on the module's output (the output supplies no energy). In context with the concepts developed in Section 7.4.1, the second assumption could also be stated as: "the output is connected to an ideal load." These assumptions are usually assumed to be met as long as the input impedance of a process is much greater (say, by two orders of magnitude) than the output impedance of the process connected to it. Verifying these assumptions in order to validate a specific systems model is one of the major applications of the source-load concepts presented in Chapter 7. Equation 9.1 restates the definition of a transfer function as a quantitative description of an input–output relationship:

$$TF(s) = G(s) = \frac{Output(s)}{Input(s)} \qquad \text{[Eq. 9.1]}$$

and

$$Output(s) = Input(s)TF(s) \qquad \text{[Eq. 9.2]}$$

In systems models, all interactions are explicitly shown by lines and arrows. For example, in the systems model of Figure 9.2, the process, $G_1(s)$, influences the process, $G_2(s)$, but $G_2(s)$ has no influence on $G_1(s)$.

Figure 9.2 In this model, the process described by transfer function $G_1(s)$ sends its output to process $G_2(s)$, but $G_2(s)$ has no influence on $G_1(s)$.

When two processes are connected together, the overall transfer function is simply the product of the individual transfer functions:

$$V_{out1}(s) = V_{in}(s)G_1(s) \quad \text{and} \quad V_{out}(s) = V_{out1}(s)G_2(s); \quad \text{substituting}$$

$$V_{out}(s) = (V_{in}(s)G_1(s))G_2(s)$$

$$\frac{V_{out}(s)}{V_{in}(s)} = G_1(s)G_2(s) \qquad \text{[Eq. 9.3]}$$

This generalizes to any number of sequential processes:

$$G_{Total}(s) = \prod_{i=1}^{N} G_i(s) \qquad \text{[Eq. 9.4]}$$

If a second process also influences the first process, that influence must be explicitly shown through an additional pathway. Figure 9.3 presents a model in which process $G(s)$ influences process $H(s)$, but process $H(s)$ also influences, along with another signal, process $G(s)$. This is an example of a classic feedback system. As will be shown, feedback can be used to great advantage in certain circumstances.

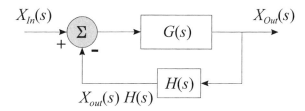

Figure 9.3 A systems model in which the output of process $G(s)$ is fedback to its input after being modified by another process $H(s)$. This is known as a feedback system where G is the *feedforward* process and H is the feedback process. G and H are commonly used symbols for the feedforward and feedback processes.

9.1.1 Feedback

In a feedback system, $G(s)$ is referred to as the *feedforward* processes or Transfer Function, and the upper pathway is known as the *feedforward pathway*. The letter G is commonly used for feedforward processes. If the feedback in the model of Figure 9.3 were missing, the output, $X_{out}(s)$, would be equal to the input times the feedforward transfer function; i.e., $G(s)X_{in}(s)$. The lower pathway is termed the *feedback pathway*, and the letter H is also commonly used to denote the feedback process. The output of the feedback process is $H(s)X_{out}(s)$. In Example 1.4, we solved for the output of a feedback system. This solution gave rise to the classic feedback equation given in Eq. 1.7 and restated here in terms of Laplace variables:

$$TF(s) = \frac{X_{Out}(s)}{X_{In}(s)} = \frac{G(s)}{1 + H(s)G(s)} \qquad \text{[Eq. 9.5]}$$

In some systems, $H(s) = 1$ and such systems are referred to as *unity gain feedback systems*.

Transforming an analog model into a system element is only a matter of solving for the analog model transfer function as has been done for both mechanical and electrical systems in numerous previous examples. Transforming a systems model back into an analog model is not so easy, since the whole point of the systems model is to hide the internal processes represented by an analog model. If something is known about the internal components, it may be possible, but the analog model constructed may not be unique; that is, it may not be the only analog model capable of generating the transfer function and/or meeting other known conditions. In some cases, it may be desirable to combine the two modeling approaches and use analog models for detailed representation of specific processes, and then construct an overall model in which the various processes are represented by system elements derived from the analog models.

Example 9.1:   The RC circuit shown below is the feedforward element in a unity gain feedback control system. Compare the frequency characteristics of the feedforward element along with that of a unity gain feedback system that has this element in the feedforward path.

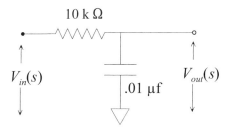

Solution: Find the transfer function for this analog circuit and then use that as $G(s)$ in the feedback equation. Convert both $G(s)$ and the feedback transfer function into the frequency domain ($s \rightarrow j\omega$) and plot the Bode plots.

From several previous examples, we know the transfer function for this simple circuit in the Laplace domain is:

$$G(s) = \frac{1/RC}{S + 1/RC} = \frac{10^4}{s + 10^4}$$

and in the frequency domain:

$$G(\omega) = \frac{10^4}{j\omega + 10^4} = \frac{1}{1 + j\omega/10^4}$$

From experience we know that the frequency characteristics of this transfer function are those of a first-order lowpass filter with a cutoff frequency of $\omega_1 = 10^4$ rad/sec. This feedforward process is placed in a unity gain feedback configuration as shown below. Since this is a unity gain feedback system $H(s) = 1$.

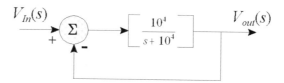

Using the feedback equation (Eq. 9.5), the overall system transfer function becomes:

$$\frac{V_{out}(s)}{V_{in}(s)} = \frac{G(s)}{1 + G(s)} = \frac{\dfrac{10^4}{s + 10^4}}{1 + \dfrac{10^4}{s + 10^4}} = \frac{10^4}{s + 10^4 + 10^4} = \frac{10^4}{s + 2(10^4)}$$

$$\frac{V_{out}(s)}{V_{in}(s)} = 0.5 \frac{2(10^4)}{s + 2(10^4)}$$

and in the frequency domain:

$$G(\omega) = 0.5 \frac{2(10^4)}{j\omega + 2(10^4)} = \frac{0.5}{1 + j\omega/2(10^4)}$$

The new, *closed-loop* transfer function differs in two ways from the original transfer function: Its gain is now half that of the open-loop function and its cutoff frequency has now doubled to 2×10^4 rad/sec. This is demonstrated in the frequency plot of the two transfer functions shown (next page). Thus, we have made a trade-off between gain and bandwidth, increasing the latter at the expense of the former. Indeed, electronic feedback was initially invented to implement this tradeoff. We will see feedback used again in the next chapter on electronic circuits to exactly the same end. Other examples of the use of feedback are found in the problems.

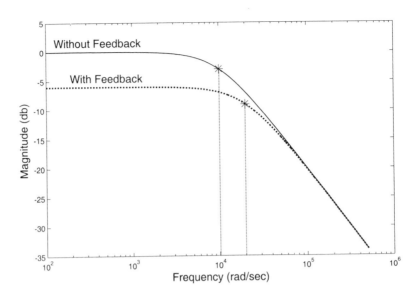

9.2   THE CONVOLUTION INTEGRAL

One of the powerful features of the transfer function is that it allows us to calculate the output to any input (Eq. 9.2) as long as the input has a Laplace transform. Alternatively, if the input can be represented by sinusoids, the transfer function can be rewritten in phasor notation and the output computed. It would be convenient to have a time domain equivalent to the transfer function that would allow the same determination of output to any input, but obtained using only time functions. Such a procedure would eliminate going back and forth into the Laplace or frequency domain, but we should not expect the procedure, whatever it is, to be as simple as the multiplication procedure in the Laplace or frequency domain.

To find such an operation, assume that the input to a system is $x_{in}(t)$ and that the system has a transfer function $G(s)$. From Eq. 9.2 we know that the output of the system in the Laplace domain will be:

$$X_{out}(s) = X_{in}(s)G(s); \quad \text{so in time domain the output would be:}$$

$$x_{out}(t) = \mathcal{L}^{-1}[X_{in}(s)G(s)] \qquad \text{[Eq. 9.6]}$$

Taking the defining equation of the Laplace transform (Eq. 8.4), but using the alternative symbol τ for time, $X_{in}(s)$ can be written as:

$$X_{in}(s) = \int_0^\infty x_{in}(\tau)e^{-s\tau}d\tau \qquad \text{[Eq. 9.7]}$$

When Eq. 9.7 is multiplied by $G(s)$, the equation becomes:

$$X_{in}(s)G(s) = \int_0^\infty x(\tau)[G(s)e^{-s\tau}]d\tau \qquad \text{[Eq. 9.8]}$$

Using time delay equation presented in the last chapter (Eq. 8.25), the term inside the brackets can also be written as:

$$e^{-s\tau}G(s) = \mathcal{L}[g(t-\tau)] = \int_0^\infty g(t-\tau)e^{-st}dt$$

where $g(t) = \mathcal{L}^{-1}G(s)$; and Eq. 9.8 becomes:

$$X_{in}(s)G(s) = \int_0^\infty x(\tau)\left(\int_0^\infty g(t-\tau)e^{-st}dt\right)d\tau \qquad \text{[Eq. 9.9]}$$

Interchanging the order of integration, Eq. 9.9 becomes:

$$X_{in}(s)G(s) = \int_0^\infty e^{-st}\left(\int_0^\infty x(t)g(t-\tau)d\tau\right)dt \qquad \text{[Eq. 9.10]}$$

Referring back to the defining equation of the Laplace transform (Eq. 8.4), the right-hand side of Eq. 9.10 is just the Laplace transform of the second integral in that equation:

$$X_{in}(s)G(s) = \mathcal{L}\left[\int_0^\infty x_{in}(t)g(t-\tau)d\tau\right] \text{ so Eq. 9.6 becomes:}$$

$$x_{out}(t) = \mathcal{L}^{-1}[X_{in}(s)G(s)] = \mathcal{L}^{-1}\left\{\mathcal{L}\left[\int_0^\infty x_{in}(t)g(t-\tau)d\tau\right]\right\}$$

$$x_{out}(t) = \int_0^\infty x_{in}(t)g(t-\tau)dt \qquad \text{[Eq. 9.11]}$$

Since the roles of $G(s)$ and $X_{in}(s)$ could have been reversed in the derivation above, the roles of $x_{in}(t)$ and $g(t)$ can also be reversed in Eq. 9.11, giving an equivalent equation:

$$x_{out}(t) = \int_0^\infty g(t)x(t-\tau)d\tau \qquad \text{[Eq. 9.12]}$$

These equations (Eq. 9.11 and Eq. 9.12) are known as the *convolution equations* and the operation they perform is known as *convolution*. The convolution operation may also be abbreviated by using an * to indicate the convolution:

$$x_{out}(t) = \int_0^\infty g(t)x(t-\tau)d\tau \equiv g(t) * x(t) \qquad \text{[Eq. 9.13]}$$

There are a number of ways of looking at convolution. The one most compatible with our use of this operation is that it is the integration of a weighting function, $g(t)$ moving over the signal. If the weighting function has unit area, convolution is nothing more than a moving weighted average. If the weighting function happens to be the impulse response of a system, convolution gives the *time domain output* of that system to any input signal.

As mentioned several times in the last chapter, the transfer function is the Laplace transform of the impulse response. Going the other way around, the impulse response, $g(t)$, is the inverse Laplace transform of the transfer function. Given the transfer function, the impulse response can be determined directly. The impulse response can also be obtained empirically from a system by monitoring its response to an impulse input.

The basic concept behind convolution is superposition. The impulse function is an infinitely short signal so the impulse response describes how the system responds to an infinitely short input. Using the concepts of basic calculus, any input can be viewed as an infinite string of such infinitesimal segments. Each of these infinitesimal input signal segments will generate its own little impulse response. All of these little impulse responses will look the same except their amplitude will be scaled by the amplitude of the infinitesimal signal segment they represent. These impulse responses will also be shifted in time to correspond with the infinitesimal segment that generated them. If the input signal is large at a given instant, it will give rise to a large impulse response, beginning at that instant. If the infinitesimal segment is negative, is associated impulse response will be negative. If superposition holds, then the output can be determined by summing, or integrating, the impulse responses from all the input signal segments.

An example of the convolution process is given in Figures 9.4 through 9.7. Figure 9.4 shows the impulse response of an example system in the left plot and the input to that system in the right plot. Note the different time scales: the impulse response is much shorter than the input signal as is generally the case.

In Figure 9.5, the impulse responses to four infinitesimal signal segments at 2, 4, 6, and 8 seconds are shown. Each segment produces an impulse response that begins when the segment occurs and is scaled by the amplitude of the input signal at that time. Some responses are larger, some negative, but they all have the basic shape shown in Figure 9.4 (left side).

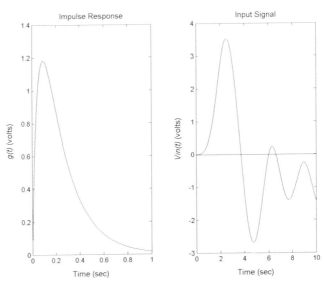

Figure 9.4   The signal on the right is the input to a system having the impulse response shown on the left.

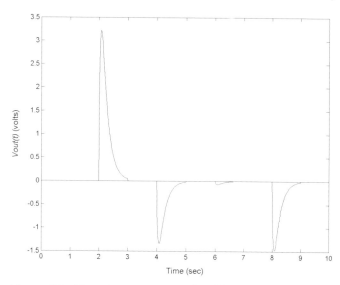

Figure 9.5 The output response produced by four infinitesimal segments of the input signal seen in Figure 9.4 (*right plot*). Impulse
responses from input segments at 2, 4, 6, and 8 seconds are shown in
this figure.

We begin to get the picture in Figure 9.6 (next page) where the impulse responses
to 50 infinitesimal segments are shown. The summation of all these responses is also
shown as a dashed line. This summation begins to look like the actual output.

In Figure 9.7, 150 impulse responses are used and the summation of all these
impulse responses (dotted line) now looks quite similar to the actual output signal
(solid line). (The output signal is scaled down slightly to aid comparison.) Convolution of a continuous signal would require and infinite number of segments, but
the digital version is limited to one for each data point in the input signal. (The
input signal used in the figures above had 1,000 points.)

9.2.1 MATLAB Implementation

The convolution integral can quickly become tedious for more complicated inputs
or input response functions, but it is very easy to implement on a computer. One
important application of convolution is in digital signal processing where it is
frequently used to apply filtering to signals. Example 9.2 explores just such an
application. For discrete signals, the integration becomes a summation and the
convolution equation becomes:

$$x_{out}[n] = \sum_{k=1}^{N} h[n-k]x_{in}[k] \qquad \text{[Eq. 9.14]}$$

where N is the length of the shorter function, usually $h[n]$. (In digital systems it is
common to use $h[n]$ to represent the impulse response, and this term should not be

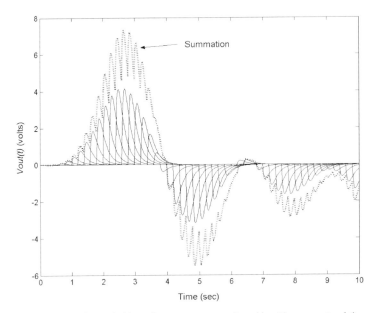

Figure 9.6   The scaled impulse responses produced by 50 segments of the input signal are shown (*solid curves*) along with the summation of those signals. The summation begins to approximate the actual output signal.

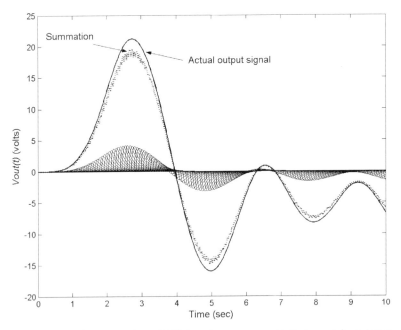

Figure 9.7   The summation of 150 impulse responses (*dotted curve*) from input segments spaced evenly over the input signal now closely resembles the actual output curve (*solid line*).

confused with $H(s)$ used to denote feedback gain in a feedback system.) It does not really matter whether $h[n]$ or $x_{in}[n]$ is shifted, the net effect will be the same, so Eq. 9.14 can also be written as:

$$x_{out}[n] = \sum_{k=1}^{N} h[n]x_{in}[k-n] \equiv h[n] * x_{in}[n] \qquad \text{[Eq. 9.15]}$$

For discrete data, both $h[n]$ and $x_{in}[n]$ must be finite (they are stored in finite memory), so the summation is also finite. It is not difficult to program Eq. 9.14 or 9.15, but in MATLAB it is not necessary since a function to compute convolution already exists. To implement convolution in MATLAB use the conv function:

```
xout = conv(xin,h)
```

where xin and h are vectors containing the waveforms to be convolved, and xout is the output signal. The length of the output waveform is equal to the length of xin plus the length of h minus 1. This will produce additional data points (the output will be longer than the input) and often these additional data points are simply discarded. An example of the use of this routine is given in the next two examples.

An alternative routine termed filter can be used to implement convolution and this routine does not generate any extra points. When used for convolution, the calling structure is:

```
xout = filter(h,1,xin);
```

where the variables are the same as described above. In the examples, the conv route is used and extra points are simply truncated, but the reader is encouraged to experiment using filter where appropriate in the problems.

Example 9.2: Find the output of the electroencephalogram (EEG) signal after filtering by an RC circuit such as in Example 9.1 where $R = 40\ \Omega$ and $C = 0.025$ f. Use convolution in conjunction with the impulse response of the RC circuit to implement the RC filter. Plot the time domain signal before and after filtering and plot the frequency characteristics of these signals. Finally, calculate and plot the Bode plot of the RC circuit (magnitude only) from its impulse response, $h(t)$.

Solution: First construct the Laplace transform for the RC circuit and then compute the impulse response by taking the inverse Laplace transform of the transfer function. Generate that impulse response function in MATLAB, load the EEG signal, then convolve the two signals together to get the output, $v_{out}[n]$. Plot the EEG signal along with $v_{out}[n]$.

Next, take the Fourier transform of these two signals (using fft), and plot. To plot the frequency characteristics of the RC circuit, note that the Fourier transform of the impulse response, $h[t]$, is the transfer function in the frequency domain. Plot the magnitude of the Fourier transform in decibels.

From previous examples, we know that the inverse Laplace transform of the RC circuit is:

$$TF(s) = \frac{1/RC}{s+1/RC} \Leftrightarrow \frac{1}{RC}e^{-t/RC}$$

In this case $RC = 1$ so the impulse response is just $h(t) = e^{-t}$. This is used in the program below.

```
% Example 9.2 Find the output of the EEG signal after
% filtering by an RC circuit.
% Use convolution in conjunction with the impulse response
% of the RC circuit to implement the RC filter.
%
clear all; close all;
%
fs = 50;                          % Sample frequency 50 Hz
t1 = 0:1/fs:10;                   % Time vector from 1 to
                                  %   10 sec .02 sec.
                                  %   intervals
h = exp(-t1);                     % Define h(t) based on
                                  %   inverse Laplace
                                  % Transform
load eeg_data;                    % Get EEG data
%
out = conv(eeg,h);                % Perform convolution
out = out(1:length(eeg));         % Truncate extra points
t = (1:length(eeg))/fs;           % Construct time vector
                                  %   for plotting
%   ...Plot the input and output signals    ....
%
figure;
Vin = abs(fft(eeg));              % Determine Fourier
                                  %   Transform of EEG
Vout = abs(fft(out));             %   and Vout
nf = fix(length(Vout)/2);         % Plot only non-
                                  %   redundant points
f = (1:nf)*fs/(2*nf);             % Construct frequency
                                  %   vector for plotting
% ....... plot and label frequency curves .......
%
% Calculate and plot Bode plot of H(w)
H = 20*log10(abs(fft(h,1024)));   % Zero pad the impulse
                                  %   response
```

```
nf = fix(length(h)/2);          % plot only non-
                                %   redundant points
f = (1:nf)*fs/(2*nf);           % Construct frequency
                                %   vector for plotting
semilogx(f,H(1:nf));            % Plot filter frequency
                                %   curve
....... labels .......
```

Analysis: Two aspects of this program deserve special notice. First, in plotting the frequency plots, only the first half of the Fourier transform is plotted since the second half contains redundant points. Second, when taking the Fourier transform of the impulse response, the response was zero-padded to 1,024 points to improve the apparent resolution of the plot (see Section 3.6.3 and Figure 3.18). The frequency spectrum of the transfer function was then plotted in Bode plot format (decibels against log ω). The plots produced by this program are shown in the next three figures.

As introduced in the last chapter, the ideal impulse function is a pulse that is infinitely tall and infinitely narrow. Of course, such an idealization is not compatible with the real world, and what is actually used is a pulse that is narrow with respect to the response time of the system being impulsed. (*Note:* impulse is not a verb,

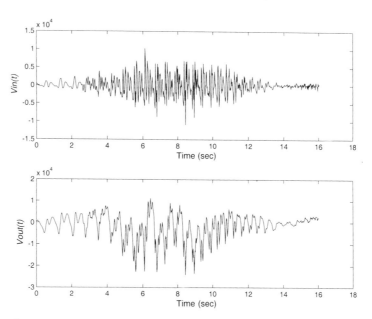

Figure 9.8 An electroencephalogram signal before (*upper*) and after (*lower*) filtering by an RC lowpass filter. Convolution was used to implement the RC filter in the time domain. The filtering process emphasizes the lower frequencies.

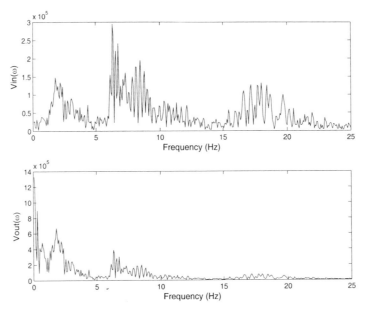

Figure 9.9 Frequency characteristics of the original electroencephalo-gram signal (*upper plot*) and the lowpass filtered electroencephalogram signal (*lower plot*). The action of the lowpass filter is evident when comparing the two curves.

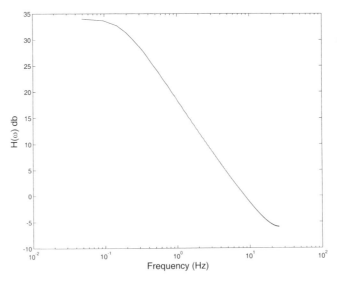

Figure 9.10 Magnitude frequency response of an RC lowpass filter obtained by taking the Fourier transform of its impulse response. As expected, this is the frequency characteristic of a lowpass filter with a cutoff frequency around $1/(2\pi\tau) = 0.16$ Hz.

yet.) A good rule of thumb is that the impulse function should be at least 10 times shorter than the fastest time constant in the system. Unfortunately, you rarely know the system's time constants, but in many cases, an empirical approach can be used. If the pulse being used is more-or-less an impulse with respect to a system's dynamics, decreasing its width should not change the output dynamics. In other words, if the pulse is acting like an impulse, decreasing its width will not make it act any more like an impulse. Finding the right pulse width then is just a matter of decreasing the pulse width until further decreases no longer change the general shape of the response. The amplitude will decrease as pulse width decreases since there is less energy in the pulse (unless there is a compensatory increase in pulse amplitude), but the general shape of the response should stay the same. This approach is explored in the next example.

Example 9.3: Use an empirical approach to find the maximum pulse width that can still be considered an impulse input with respect to a given system. In this example, use convolution to simulate the response of a system to various pulse inputs. Assume the system has a transfer function of:

$$TF(s) = \frac{1}{s^2 + 25s + 70} \qquad \text{[Eq. 9.16]}$$

Solution: This transfer function could be simulated quite easily using MATLAB's Simulink program. However, here we will use convolution in conjunction with the system's impulse response. With this approach, we could find the output of the system to any input, but in this case, we are only interested in pulse inputs. To find the impulse response, we take the inverse Laplace transform of the system's transfer function, and for that we need to evaluate the damping coefficient to determine the proper Laplace transform equation. The damping coefficient can be obtained from the transfer function coefficients:

$$2\delta\omega_n = 25; \quad \delta = \frac{25}{2\omega_n} = \frac{25}{2\sqrt{70}} = 1.49$$

Thus, the system is overdamped and could be factored, or Laplace transform entry no. 11 can be used directly.

$$\left(\frac{b\beta - b\alpha + c}{2\beta}\right)e^{-(\alpha-\beta)t} + \left(\frac{b\beta + b\alpha - c}{2\beta}\right)e^{-(\alpha+\beta)t} \quad \Longleftrightarrow \quad \frac{bs+c}{s^2 + 2\alpha s + \alpha^2 - \beta^2}$$

where $b = 0$; $c = 1$; $2\alpha = 25$; $\alpha = 12.5$; $\alpha^2 - \beta_2 = 70$; $\beta = (156.25 - 70)^{1/2} = 9.29$. Hence, the impulse response becomes:

$$h(t) = \left(\frac{1}{2(9.29)}\right)e^{-(12.5-9.29)t} + \left(\frac{-1}{2(9.29)}\right)e^{-(12.5+9.29)t} = 0.05(e^{-3.21t} - e^{-21.79t})$$

This impulse response is programmed into MATLAB below and the responses to various pulse widths are plotted. To aid comparison the responses are plotted normalized to the same maximum amplitude. (Alternatively, the pulse amplitude could have be changed to keep the pulse area constant.) In addition, some care has to be taken to ensure that the impulse response, $h(t)$, is appropriately represented in the program; that is, the time vector used to generate $h(t)$ is long enough to generate the entire response. This can be done by trial and error, but we note that the slowest time constant is in the first exponential: $\tau_1 = 1/3.21 = 0.31$ seconds. Based on this, a one and a half second period ($5\tau_1$) should be adequate to represent the impulse response. Since the fastest time constant is in the second exponential, $\tau_2 = 1/21.79 = 0.046$ sec, we will experiment with pulses around 0.01 and shorter. Given these numbers, a sampling frequency of 2 kHz should lead to signal representations with a reasonable number of points.

```
% Example 9.3 Example to simulate a second-order Transfer
%    Function and evaluate the responses to pulses
%    of various widths
close all; clear all;
fs = 2000;                              % Sample frequency
PW = [50 25 25 10 10 5 5 2.5]*2;        % Pulse width in msec
t = 0:1/fs:1.5;                         % Generate a time
                                        % vector: 0-1.5 sec.

%
% Now construct the impulse response
h = .05*(exp(-3.21*t) - exp(-21.79*t));
....... plot and label .......
%
figure;
%
for i = 1:length(PW)
  pulse = ones(1,PW(i));                % Generate pulse of
                                        % desired width

  x = conv(pulse,h);                    % Simulate response
  x = x/max(x);                         % Normalize peak to
                                        % 1.0

  subplot(2,2,floor((i+1)/2));          % Plots pairs of
                                        % curves

  plot(t,x(1:length(t)),'k');           % Plot, but not extra
                                        % points

  hold on; axis([0 0.2 0 1.2]);
  xlabel('Time (sec)'); ylabel('\itx(t)');
end
```

Analysis: The program first specified the sampling frequency that will be used in msec. Next the desired pulse widths are specified in pairs (for plotting) and in msec. Because the sampling frequency is assumed to be 2 kHz, the sample interval, the distance between points is 0.5 milliseconds. Thus, the pulse widths are multiplied by two to give the pulse width in number of points. A time vector is then generated ranging between 0 and 1.5 seconds in increments of the sample interval: `1/fs`. The impulse response `h` is then constructed using this time vector. Again, this impulse response is just the inverse Laplace transform of the transfer function given in Eq. 9.16. The impulse response is plotted and shown in the figure below. The 1.5-second data length is adequate to represent the full response.

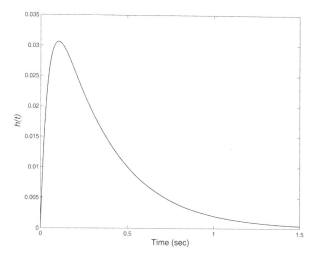

Next, the computer enters a loop in which the pulse is generated by constructing an array containing 1.0's having the desired length specified by the PW vector. The simulated output is generated by convolving this pulse with the impulse response. The output is then plotted in pairs so that pulses of a given width can be compared with shorter pulses. The results are shown in the figure below.

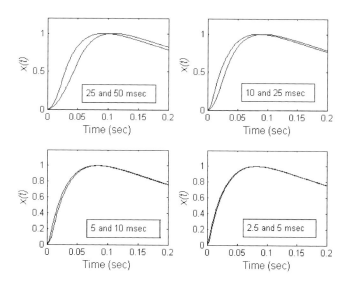

The results show that as pulses reduce in width, the differences in response dynamics decrease. There is very little difference between the responses generated by the 5- and 2.5-millisecond pulses, so that a 5-millisecond pulse is probably close enough to a true impulse for this system. This example is a realistic simulation of the kind of experiment one might do to determine empirically if a given pulse width were short enough to be considered an impulse input.

The next example revisits an empirical approach to system identification developed in Section 6.4.2 where we evaluated the frequency characteristics (i.e., Bode plot) of an unknown system by probing it with sinusoids at different frequencies. Since the frequency characteristics are also the Fourier transform of the impulse response, we can also probe the system with a single impulse and take the Fourier transform of the impulse response. This not only makes the experiment much quicker as only a single stimulus is required, but it also extends this identification approach to a broader class of real systems. As mentioned earlier, for many real systems it is not possible to produce and effective a sinusoidal stimulus, but it is still possible to stimulate the system with an impulse. In this example, we will compare the frequency curves produced by both the sinusoidal approach of Section 6.4.2 and impulse response approach to the system used in the last example.

Example 9.4:   Find and plot the frequency characteristics of the transfer function given in Example 9.3 using an empirical approach based on simulation. Probe the simulated system with both a series of sinusoids and an impulse function. Of course, since we know the Laplace transfer function we could easily convert it to the frequency domain ($s \rightarrow j\omega$) and use Bode plot techniques or MATLAB to construct the frequency curves. Here we pretend that the system is unknown and only its inputs and outputs are accessible to us. The approaches used here mimic a situation where the system itself is unknown but is available for empirical evaluation. In such a situation, you could evaluate the system's frequency response using a series of sinusoids as in Example 6.10, or a single impulse input and Fourier transform the resulting impulse response. The latter approach is much faster since it requires only a single stimulus signal, but may be more sensitive to noise from the system or in the measurement process.

Solution: The following MATLAB program uses convolution and the system impulse response (taken directly from the last example) to simulate the response of the system to a series of sinusoids and an impulse input. With the sinusoids, some trial and error may be required to find the most interesting frequency range. This would also need to be done in a real-world situation. You simply try a range of frequencies, plotting the magnitude (and possibly the phase) as you go until you have a respectable looking Bode plot.

```
% Example 9.4 Example to simulate a second-order Transfer
% Function
```

```
% and evaluate the responses to a sinusoidal series and an
% impulse.
%
close all; clear all;
fs = 2000;                      % Sample frequency
t = 0:1/fs:1.5;                 % Generate a time vector
ln = length(t);
% Now construct the impulse response
h = .05*(exp(-3.21*t) - exp(-21.79*t));
%
% Stimulate with sinusoids
for f = 1:100
  f1(f) = f;                    % Set frequency (1-100 Hz)
  input = 1.414 *...            % Generate sinusoid
  cos(2*pi*f*t);
  x = filter(input,1,h);        % Simulate response (use filter)
  Mag(f) = sqrt(mean(x.^2));% Calculate RMS value
end
subplot(1,2,1);
semilogx(f,20*log10(Mag),...% Plot frequency response in
'k');                         % dB
....... labels and axis .......
%
% Now determine the Transfer Function from the impulse
  response
Mag = abs(fft(h));              % Take Fourier Transform
subplot(1,2,2);
  f = (1:length(Mag))*fs/...% Frequency vector for plotting
  length(Mag);
  semilogx(f,20*log10(Mag), 'k');
  ......labels and axis.......
```

Analysis: The two magnitude frequency curves are shown in the figure on the next page and are seen to be similar and have the same slopes but differ slightly in vertical values and the sinusoidal method appears to have a lower limit of around −40 dB. These differences are likely due to small computational errors.

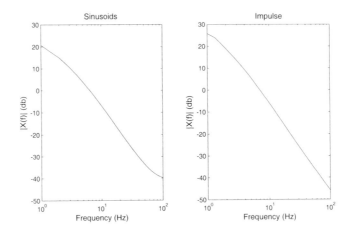

The impulse response was constructed as in the last example using the same sampling frequency. In this program, the MATLAB `filter` routine was used to implement convolution just for variety. The root-mean-squared (RMS) value of the output sinusoid was taken as the magnitude ratio, V_{out}/V_{in}, because the input sinusoid has an RMS value of 1.0. In the section that involved the impulse response, it was not necessary to use convolution to get the impulse response since we already have the impulse response. In a practical situation with a real system, you would need to stimulate the system with an impulse to find this response. One of the problems presents a more realistic situation where the system input and output are available, and the impulse response must be found in order to determine the system's frequency characteristics.

9.3   RESONANCE

Resonance is a behavior that may occur in almost any system. It is a frequency-dependent behavior characterized by a sharp increase (or decrease) in some system variables over a limited range of frequencies. It occurs frequently in mechanical and electrical systems and is a particularly useful behavior in chemical and molecular systems. Sometimes it is beneficial and exploited: Proton resonance is used in magnetic resonance imaging and optical resonance is used in spectroscopy to identifying molecular systems. However, it can also be undesirable, particularly in mechanical systems: Shock absorbers are placed on cars to increase damping and reduce the resonant properties of automotive suspension systems. Resonance will be discussed in terms of mechanical and electrical systems because these systems have already been well studied here, but most of the concepts are broadly applicable.

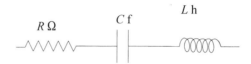

Figure 9.11 A series RLC circuit.

9.3.1 Resonant Frequency

In electrical and mechanical systems, resonance occurs when the impedance of an inertial-type element (mass or inductor) equals and cancels the impedance of a capacitive-type element. Consider the impedance of a series RLC circuit shown in Figure 9.11.

In the frequency (phasor) domain, the impedance of this series combination is:

$$Z(\omega) = R + j\omega L + \frac{1}{j\omega C} = R + j\left(\omega L - \frac{1}{\omega C}\right)\Omega \qquad \text{[Eq. 9.17]}$$

At some value of ω, the capacitor's impedance will be equal to the inductor's impedance and the two impedances will cancel. This will leave only the resistor to contribute to the total impedance. To determine the frequency at which this cancellation takes place, simply set the impedances equal and solve for frequency:

$$\omega_o L = \frac{1}{\omega_o C}; \quad \omega_o L(\omega_o C) = 1; \quad \omega_o^2 = \frac{1}{LC}$$

$$\omega_o = \frac{1}{\sqrt{LC}} \, \text{rad/sec} \qquad \text{[Eq. 9.18]}$$

where ω_o is the *resonant frequency*. Note that this is the same equation as for the undamped natural frequency, ω_n, in a second-order representation of the RLC circuit (see Eq. 8.47). If we were to plot the magnitude of the impedance in Eq 9.17, we would get a curve that reached a minimum value of R at $\omega = \omega_o$ with the curve increasing on either side (Figure 9.12). The sharpness of the curve would relate to the bandwidth of the resonant system as discussed in the next system.

9.3.2 Resonant Bandwidth, Q

When a system approaches the resonant frequency, the system variables (voltage–current or force–velocity) will increase (or decrease) to a maximum (or minimum). The sharpness of that curve will depend on the energy dissipation element (resistance or friction). Figure 9.13 shows an RLC circuit configured as an input–output system and Figure 9.14 shows the magnitude frequency characteristics for four values of resistance: 0.1, 1, 10, and 100 Ω. Figure 9.14 shows that the transfer function peak occurs at the same frequency for all values of R. This is expected since the resonant frequency is a function of only L and C (Eq. 9.18). Specifically, the peak occurs at $\omega_o = 1/\sqrt{LC} = 1/\sqrt{10^{-4}10^{-6}} = 100,000 \, \text{rad/sec}$.

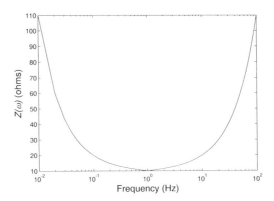

Figure 9.12   The impedance of series RLC combination as a function of frequency. $R = 10\,\Omega$, $L = 1$ h, and $C = 1$ f.

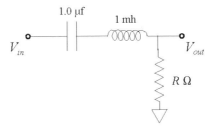

Figure 9.13   RCL circuit with realistic values for L and C. The transfer function frequency characteristics of this circuit are shown in Figure 9.14 for four different values of R.

Peak sharpness increases as R decreases. In the configuration shown in Figure 9.13, the circuit is a bandpass filter; however, if the resistor and capacitor were interchanged, the circuit would be a lowpass filter as used in previous examples. Nonetheless, it would still have a resonant peak at the same frequency and the sharpness of the peak would vary in the same manner. This is demonstrated in a problem at the end of this chapter.

The sharpness of the frequency curve around the resonant frequency is a very important property of a resonant system. This characteristic is often described by a number know as Q, which is defined as the resonant frequency divided by the bandwidth:

$$Q = \frac{\omega_o}{BW} \qquad\qquad \text{[Eq. 9.19]}$$

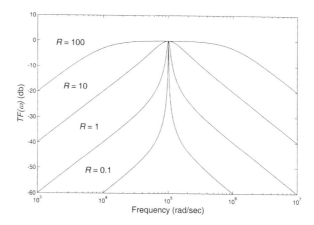

Figure 9.14 Magnitude frequency characteristics of the transfer function of the RLC circuit shown in Figure 9.13 with four different resistor values. The resonant frequency 100,000 rad/sec depends only on L and C, but sharpness of the curve depends strongly on R.

where ω_o is the resonant frequency and BW is the bandwidth, the difference between the high frequency cutoff and the low frequency cutoff (−3-dB points). This definition is also referred to as *selectivity* as Q has an alternate, classical definition described below.

Classically, Q was defined in terms of energy storage and energy loss to describe how close energy storage elements such as inductors and capacitors approach the ideal. Ideally, such elements should only store energy, but real elements also dissipate energy due to parasitic resistance. In this context, Q is defined as the energy stored over the energy lost in one cycle:

$$Q = 2\pi \frac{\text{Energy stored at resonance}}{\text{Energy dissipated at resonance}} \qquad \text{[Eq. 9.20]}$$

To calculate this for an inductor, assume an inductor having inductance L also has a parasitic resistance of $R\ \Omega$. The energy lost in this resistor over one sinusoidal cycle would be equal to the power integrated over the cycle, and power is just $V\ i$, or for a resistor, R_i^2. Assuming the current through the resistor is $i_R(t) = I \sin(\omega_o t)$, the energy lost becomes:

$$E_{lost} = \int_{Cyc} vi\,dt = \int_0^{2\pi} R(I\sin\omega_o t)^2\,dt = \frac{2\pi R I^2}{2\omega_o}\text{joules}$$

The energy stored in an inductor is also the integral of $v\ i$ over one cycle. The current through the inductor is the same as through the resistor, $i_L(t) = I \sin(\omega_{ot})$, and the voltage is L times the derivative of the current:

$$v_L(t) = L\frac{di}{dt} = \omega_{oL} I \cos(\omega_o t)$$

$$E_{stored} = \int_{Cyc} vidt = \int_0^{2\pi} \omega_{oL} I^2 \cos(\omega_o t)\sin(\omega_o t)dt = \frac{\omega_o L I^2}{2\omega_o}$$

Plugging in the two energies into Eq. 9.20, the Q of an inductor becomes:

$$Q_L = \frac{2\pi\left(\dfrac{\omega_o L I^2}{2\omega_o}\right)}{\left(\dfrac{2\pi R I^2}{2\omega_o}\right)} = \frac{\omega_o L}{R} \qquad\qquad \text{[Eq. 9.21]}$$

Similarly, it is possible to derive the Q of a capacitor having a capacitance value C and a parasitic resistance R as:

$$Q_C = \frac{1}{\omega_o RC} \qquad\qquad \text{[Eq. 9.22]}$$

However, in an LC circuit, most of the parasitic resistance will be from the inductor.

Based on these definitions, it is possible to derive Eq. 9.19 from the definition of bandwidth. Returning to the RLC circuit in Figure 9.13, the transfer function of this circuit can be written as:

$$TF(\omega) = \frac{R}{R + j\omega L + 1/j\omega C} = \frac{R}{Z(\omega)} = \frac{1}{Z(\omega)/R}$$

where $Z(\omega)$ is just the series R, L, C impedance. $Z(\omega)/R$ can also be written as:

$$\frac{Z}{R} = \frac{R + j\omega L - \dfrac{j}{\omega C}}{R} = 1 + j\left(\frac{\omega L}{R} - \frac{1}{\omega RC}\right); \quad \text{multiplying both imaginary terms by } \frac{\omega_o}{\omega_o}$$

$$\frac{Z}{R} = 1 + j\left(\frac{\omega}{\omega_o}\left(\frac{\omega_o L}{R}\right) - \frac{\omega_o}{\omega}\left(\frac{1}{\omega_o RC}\right)\right)$$

This allows us to substitute in the definitions of Q for an L and C into the equation for Z/R:

$$\frac{Z}{R} = 1 + j\left(\frac{\omega Q}{\omega_o} - \frac{\omega_o Q}{\omega}\right) = 1 + jQ\left(\frac{\omega}{\omega_o} - \frac{\omega_o}{\omega}\right) \qquad\qquad \text{[Eq. 9.2]}$$

The bandpass frequency curves shown in Figure 9.14 have two cutoff frequencies, ω_{low} and ω_{high}. From the definition of bandwidth, at these two cutoff frequencies the Transfer function is reduced by $0.707|TF(\omega_{high\ and\ low})| = 0.707\ |TF(\omega = \omega_o)|$. However, at the resonant frequency, $Z(\omega = \omega_o) = R$, and $TF(\omega = \omega_o) = 1.0$. Hence, at the cutoff frequencies $|TF(\omega_{high\ and\ low})| = 0.707$. So to find the bandwidth (which is just the difference between the cutoff frequencies), set $\left|\dfrac{Z}{R}\right| = \dfrac{1}{0.707} = \sqrt{2}$ and solve for ω_{high} and ω_{low}. Setting the magnitude of Eq. 9.23 to $\sqrt{2}$, we get the following:

$$\left|1 + jQ\left(\frac{\omega}{\omega_o} - \frac{\omega_o}{\omega}\right)\right| = \sqrt{2}; \quad \text{For } |1 + jB| = \sqrt{2}, B = \pm 1$$

$$Q\left(\frac{\omega}{\omega_o} - \frac{\omega_o}{\omega}\right) = \pm 1$$

There are two solutions to the equation: one for $+1$ and the other for -1:

$$\omega_{low} = \omega_0\left(1 - \frac{1}{2Q}\right); \quad \omega_{high} = \omega_0\left(1 + \frac{1}{2Q}\right)$$

$$BW = \omega_{high} - \omega_{low} = \frac{\omega_0}{Q}; \quad Q = \frac{\omega_0}{BW}$$

We can also relate Q to the standard coefficients of a second-order underdamped equation. Again referring to the RLC circuit in Figure 9.13, the transfer function in terms of the Laplace variable s is:

$$TF(s) = \frac{s\dfrac{R}{L}}{s^2 + \dfrac{R}{L}s + \dfrac{1}{LC}} = \frac{s\dfrac{R}{L}}{s^2 + 2\delta\omega_n s + \omega_n^2}$$

where $\omega_n \equiv \omega_o$. Equating coefficients:

$$2\delta\omega_o = \frac{R}{L}; \quad \delta = \frac{R}{2\omega_o L}; \quad \text{but } Q = \frac{\omega_o L}{R} \text{ so}$$

$$\delta = \frac{1}{2Q}; \quad Q = \frac{1}{2\delta} \qquad \text{[Eq. 9.24]}$$

The inverse relationship between Q and δ is straightforward.

Example 9.5: Find the Q of the mechanical system below. The system coefficients are $k_f = 6$ dyne-sec/cm; $m = 8$ gm; $k_e = 10$ dyne/cm.

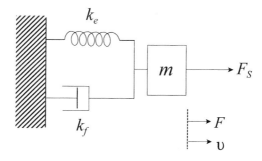

Solution: Q can be determined directly from δ in transfer function using Eq. 9.24. (Alternatively, we could determine δ directly from the Eq. 8.55.) Applying standard analysis:

$$F_S(s) - v(s)(k_f + ms + k_e/s) = 0; \quad v(s) = \frac{F_S(s)}{k_f + ms + k_e/s}$$

$$\frac{v(s)}{F_S(s)} = \frac{s/m}{s^2 + \dfrac{k_f}{m}s + \dfrac{k_e}{m}} = \frac{s/m}{s^2 + 2\delta\omega_n s + \omega_n^2}$$

Equating coefficients:

$$\omega_n = \sqrt{\frac{k_e}{m}}; \quad 2\delta\omega_n = \frac{k_f}{m}; \quad \delta = \frac{k_f}{2m\omega_n} = \frac{k_f}{2\sqrt{k_e m}} = \frac{6}{2\sqrt{10(8)}} = 0.335$$

$$Q = \frac{1}{2\delta} = 1.49$$

Example 9.6:   Plot the frequency characteristics and impulse response of a second-order system in which Q = 1, 10, and 100. Assume a resonant frequency of 1,000 rad/sec and use MATLAB.

Solution: Begin by using the standard second-order transfer function, but substitute Q for δ. Convert to phasor domain to plot the frequency response and to a time function for the time responses.

$$TF(s) = \frac{\omega_n^2}{s^2 + 2\delta\omega_n s + \omega_n^2};$$

$$\text{Substituting in } \delta = \frac{1}{2Q} \quad \text{and} \quad \omega_n = 1,000$$

$$TF(s) = \frac{\omega_n^2}{s^2 + \dfrac{\omega_n}{Q}s + \omega_n^2} = \frac{10^6}{s^2 + \dfrac{1,000}{Q}s + 10^6}$$

To find the frequency response, convert to phasor and plot for the requested values of Q:

$$TF(\omega) = \frac{10^6}{(j\omega)^2 + j\omega\dfrac{1,000}{Q} + 10^6} = \frac{1}{1 - \left(\dfrac{\omega}{10^3}\right)^2 + \dfrac{j\omega}{Q10^3}}$$

To find the impulse response, take the inverse Laplace transform of the transfer function. Since Q is equal to, or greater than 1.0, δ will be less than, or equal to

0.5, so the system is underdamped and entry no. 15 in the Laplace transform table can be used:

$$x(t) = \frac{\omega_n}{\sqrt{1-\delta^2}} \left[e^{-\delta\omega_n t} \sin(\omega_n \sqrt{1-\delta^2}\, t) \right] = \frac{10^3}{\sqrt{1-\frac{1}{4Q^2}}} \left[e^{-1,000 t/2Q} \sin\left(1,000\sqrt{1-\frac{1}{4Q^2}}\, t\right) \right]$$

The frequency and time domain equations are incorporated into the program below.

```
% Example 9.6
%
clear all; close all;
wn = 1000;                          % Define resonant frequency
w = (100:10:10000);                 % Define a frequency vector
t = (10^-5:10^-5:.2);               % Define a time vector
Q = [1 10 100];                     % Define Q's
for k = 1:3                         % Calc and plot the
                                    % frequency plots
  TF = 1./(1-(w/1000).^2 + j*w/(Q(k)*1000));
  TF = 20*log10(abs(TF));
  semilogx(w,TF,'k'); hold on;
end
xlabel('Frequency (rad/sec)'); ylabel('TF(w) (dB)');
axis([100 10000 -40 50]);
%
% Now construct the impulse responses
figure;
for k = 1:3                         % Cal. and plot the time
                                    % response
  d =sqrt(1-1/(4*Q(k)^2));          % Define √1-δ2
  x = (wn/d)*(exp(-wn*t/(2*Q(k)))).*sin(wn*d*t);
  subplot(3,1,k);                   % Plot separately for
                                    % clarity
      plot(t,x,'k');
      ylabel('x(t)');
end
xlabel('Time (sec)');
```

This program generates the plots shown in Figures 9.15 and 9.16. Figure 9.15 reiterates the message in Figure 9.14; that high-Q systems have sharp resonance peaks.

In the time domain, high Q corresponds to sustained oscillations, a sinusoid at the resonance frequency that diminishes very slowly (Figure 9.16). A common example of a high-Q mechanical system is a large church bell where the tone continues to sound long after it is stuck with something very like an impulse. In fact, this sustained oscillation is a characteristic of a high-Q system and is sometimes referred to as *ringing*.

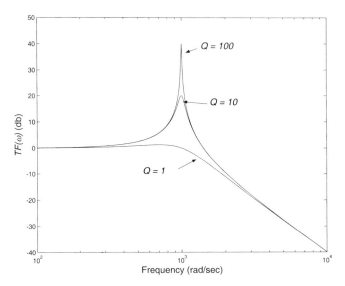

Figure 9.15   Frequency characteristic of a second-order system with a Q of 1, 10, and 100.

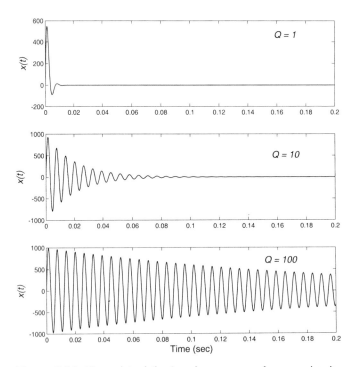

Figure 9.16   Time plot of the impulse response of a second-order system with a Q of 1, 10, and 100.

If a high-Q circuit is used as a filter, it can be quite selective with regard to the frequencies it allows to pass though. This is demonstrated in Example 9.7.

Example 9.7: Generate a sinusoidal waveform that continually increases in frequency with time. The frequency should increase from 1 to 500 rad/sec over a 5-second period. Assume a sampling frequency of 1 kHz. Pass this signal though the RLC circuit of Figure 9.13 with a Q of 10 and plot the output. Repeat for three values of resonant frequency: $\omega_o = 50$, 125, and 250 rad/sec. Use convolution in conjunction with the filter's impulse response to implement the filter. (To make life easier for us, we can use the impulse response from Example 9.6 setting ω_o and Q as desired.)

Solution: One of the challenges of this problem is generating the input signal. We need a sinusoid that linearly increases with frequency over a 5-second period between 1 and 1,000 rad/sec. Such a signal is termed a *chirp* signal because of the way it sounds if it is sent to an audio device. (If you have a sound system on your computer, try typing sound(xin,1000) in MATLAB after executing Example 9.7 from the disk.) The chirp signal can be very useful in certain signal processing operations since it contains a range of frequencies that vary in a well-defined manner with time.

To construct the chirp signal, we first define a time vector, t, that goes from 0 to 5 in steps of 0.001. This time vector is then used to construct a frequency vector of the same length that ranges from 0 to 250 (i.e., $\omega = 50\ t$) adding 1 to account for the beginning frequency of 1.0 rad/sec. The chirp signal is then constructed as the sine of the product of ω and t (i.e., sin(w.*t)) so that the frequency increases as t increases. We then define the impulse response used in Example 9.6 to have a Q of 10, but use three different resonant frequencies, ω_n. Convolve the input signal with the impulse response and remove the additional points produced by MATLAB's conv routine. (Alternatively, we could use filter.) This strategy is implemented in the code below.

```
% Example 9.7
%
clear all; close all;
wn = [50 125 250];              % Define resonant
                                % frequencies
Q = 10;                         % Define Q
t = (0:.001:5);                 % Define time vector (0-1
                                % sec; Ts = .001)
w = 50*t + 1;                   % Frequency goes from 1
                                % to .250 rad/sec
xin = sin(w.*t);                % Generate input signal
                                % (a "chirp" signal)
plot(t(1:1000),xin(1:1000),'k'); % Plot a segment of the
                                % chirp signal
```

```
xlabel('Time (sec)'); ylabel('xin(t)')
%
figure;
for k = 1:3                              % Calculate impulse
                                         % response

  d =sqrt(1-1/(4*Q^2));
  delta = (wn(k)/d)*(exp(-wn(k)*t/(2*Q))).*sin(wn(k)*d*t);
  xout = conv(xin, delta);               % Filter xin
  xout = xout(1:length(t));              % Remove extra points
  subplot(3,1,k);                        % Plot separately
    plot(t,xout,'k');
    ylabel('xout(t)');
end
xlabel('Time (sec)');
%
figure;
% Plot frequency characteristics of chirp
XOUT = abs(fft(xin));
f = (1:5000)/5;                          % Construct frequency
                                             vector for plotting

plot(2*pi*f(1:1000),XOUT(1:1000));
xlabel('Frequency (rad/sec)'); ylabel('Mag. Chirp');
```

Analysis: After defining the constants ω_n and Q, the time and frequency vectors are constructed. The chirp signal is generated by taking the $\sin$ of the (point-by-point) product of the frequency and time vectors. The first 1 second of the chirp signal is plotted in Figure 9.17 and shows the expected sinusoid that is increasing with frequency.

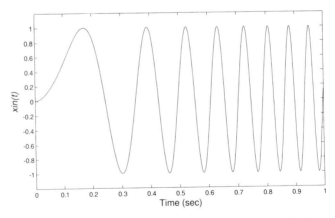

Figure 9.17 Time plot of a chirp signal, a sinusoidal signal whose frequency increases linearly with time.

The rest of the program is the similar to that of Example 9.6 except that Q is constant and the resonant frequency is modified. When the chirp passes thought this moderately high-Q filter, only that portion of the signal that is around the resonant signal is allowed to pass. Since the chirp signal's frequency depends on time, the signal will pass though the RLC filter at a time corresponding to the filter's resonant frequency. For example, the chirp signal's frequency approaches 50 rad/sec at 0.5 seconds so when the resonant frequency of the filter is 50 rad/sec, only the signal around 0.5 seconds passes through the filter to the output (Figure 9.18, upper curve). The chirp signal reaches a frequency of 250 rad/sec at 2.5 seconds so this portion of the signal is selected when the filter resonant frequency is 250 rad/sec (Figure 9.18, lower curve). The frequency selectivity of a high-Q filter is well demonstrated in Figure 9.18.

The last section of the programs plots the frequency spectrum of the chirp signal using the standard `fft` command. The spectrum is seen in Figure 9.19 to be relatively flat, containing a mixture of all frequencies between 0 and 100 rad/sec.

9.4 SUMMARY

Complex systems composed of many modules represented by transfer functions can easily be analyzed using algebra. The overall transfer function of two or more modules in sequence is just the product of the individual transfer functions. In feedback systems, the output of a module is connected to the input of a preceding module. Again, using algebra, it is easy to determine the transfer function for the feedback system (i.e., the closed-loop transfer function) as was done in Chapter 1 (Eq. 1.7). However, we should remember that the underling assumption regarding input and output impedances of connected systems must be met ($Z_{in} >> Z_{out}$).

Convolution is a technique for determining the output of a system to any general input without leaving the time domain. The approach is based on the impulse response: the impulse response is used as a representation of the system's response to an infinitesimal segment of the input. If the system is linear and superposition holds, the impulse response from each input segment can be summed to produce the system's output. The convolution integral (Eq. 9.12 and Eq. 9.13) is a running correlation between the input signal and the impulse response. This integration can be cumbersome for complicated input signals or impulse responses, but is easy to program on a computer. Basic MATLAB provides two routines, `conv` and `filter`, to perform convolution. (The `conv` routine actually converts to the frequency domain using the Fourier transformation, multiplies the two signals, then converts back to the time domain using the inverse Fourier transform, and this makes it faster than the `filter` routine.) Convolution is very commonly used in signal processing to implement digital filtering, and a simple example of such an application was found in Example 9.2.

Resonance is a phenomenon commonly found in nature. In electrical and mechanical systems, it occurs when two different energy storage devices have equal (but oppositely signed) impedances. During resonance, energy is passed back and

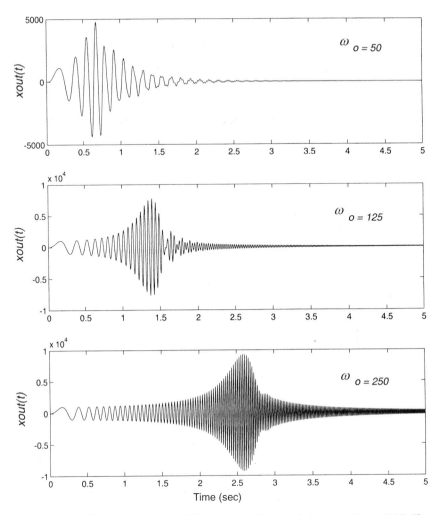

Figure 9.18   The output of the RLC filter to the chirp signal shown in Figure 9.17. The filter's ability to select a specific frequency range is shown for three different resonant frequencies: $\omega_o = 50$, 125, and 200 rad/sec.

forth between the two energy storage devices. For example, in an oscillating mechanical system, the moving mass stretches the spring transferring the kinetic energy of the mass to potential energy in the spring. Once the spring is appropriately compressed (or extended), the energy is passed back to the mass as the spring uncoils. Without friction, this process would continue forever, but friction removes energy from the system so the oscillation gradually decays. The quantity Q, is a measure of the ratio of energy storage capabilities of the mass and spring to the energy dissipation caused by the friction. The higher the Q, the longer the resonance will continue, before all the energy is removed. Electrical RLC circuits behave in exactly the same way passing energy between the inductor and capacitor while the resistor

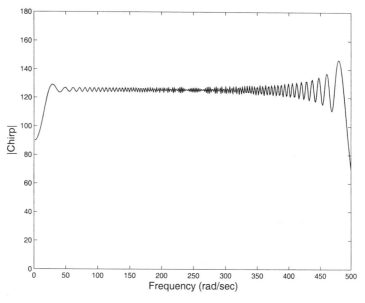

Figure 9.19 Frequency spectrum of the chirp signal shown in Figure 9.17. The spectrum is essentially flat between 0 and 500 rad/sec.

removes it from the system. (Recall both friction and resistance remove energy in the form of heat.)

PROBLEMS

1. A second-order system has an ω_n of 10 rad/sec and a δ of 0.5. This system becomes the feedforward element [i.e., $G(s)$] of a feedback system where the feedback element, $H(s)$, is a constant of k [i.e., $H(s) = k$]. What should be the value of k to double the resonant frequency of the overall feedback system? What is the value of δ in the feedback system? (*Hint:* For $G(s)$ use the standard (unity gain) second-order transfer function: $\dfrac{\omega_n^2}{s^2 + 2\delta\omega_n s + \omega_n^2}$.)

2. A second-order system has an ω_n of 1 rad/sec and a δ of 0.1. This system becomes the feedforward element [i.e., $G(s)$] of a feedback system where the feedback element, $H(s)$, has a transfer function of ks [i.e., $H(s) = ks$]. What should be the value of k to make the overall feedback system critically damped (i.e., $\delta = 0.707$)? Use the same transfer function for $G(s)$ as in Problem 9.1. (*Note:* Feedback of the form ks $\left(\text{or more practicably: } \dfrac{s}{s+a}\right)$ is referred to as *derivative feedback* for obvious reasons.)

3. In the RLC circuit shown below, what should be the values of C and R so the circuit has a resonant frequency of 1,000 rad/sec and a bandwidth of 50 rad/sec.

(*Hint:* To meet the bandwidth requirements, find Q, then the required value for R.)

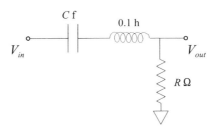

MATLAB Problems

4. Use convolution to solve for the step response, $v(t)$, of the mechanical system in Figure 9.13. First find the transfer function, $F(s)/v(s)$, then take the inverse Laplace transform to get the impulse response. Program this inverse impulse response in MATLAB as in Example 9.2. (*Hint:* You can use a modified version of the code in Example 9.3 to generate the step input and response.)

5. The impulse response of a filter is given by the four sequential data points below.

$$h[n] = \frac{1+\sqrt{3}}{8}, \quad \frac{1+\sqrt{3}}{8}, \quad \frac{1-\sqrt{3}}{8}, \quad \frac{1-\sqrt{3}}{8}$$

(A) Apply this filter to the same chirp signal used in Example 9.5 except make the maximum chirp frequency 1,500 rad/sec and plot the response.

(B) Determine and plot the frequency characteristics of this filter by taking the Fourier transform of the impulse response. To improve the frequency plot, pad the response to 256 points or more.

This filter is known as a four-coefficient *Daubechies filter*, and is one of a family of filters used in Wavelet analysis.

6. Repeat Problem 9.5 above using a simple three-coefficient *moving average*–type filter.

$$h[n] = \frac{1}{3}, \frac{1}{3}, \frac{1}{3}$$

Note that when this impulse response is applied to input data through convolution, it constructs an output that is an average of every three points in the input. Intuitively, this will reduce large point-to-point changes in the data. Such sharp changes are associated with high frequencies so this should act as a lowpass filter. Plotting the Fourier transfer in part B will confirm this. Finally, make note of the fact that even a very simple impulse response function can have a significant impact on the data when applied as a filter. (*Hint:* this problem and the next can be solved with only minor modifications of the code developed to solve Problem 9.5.)

7. Repeat Problem 9.4 for one of the simplest of all filters consisting of only two coefficients:

$$b[n] = \frac{1}{2}, \frac{1}{2}$$

This filter which acts as a two-point moving average is known as the *Haar filter*, and although it is not a very strong filter, it is useful in demonstrating some of the principles of Wavelet analysis.

8. Load the MATLAB data file "filter1.mat" and apply the impulse response variable h contained in that file using convolution to the EEG signal in eeg_data.mat. (Recall this file contains variable eeg.) Plot the EEG data before and after filtering as in Example 9.2. What does this mystery filter do?

9. Load the MATLAB file bp_filter that contains the impulse response of a band-pass filter in variable h. Apply this filter to a chirp signal used in Example 9.7. Apply the filter in two ways:
   A. In the time domain using convolution.
   B. In the frequency domain using multiplication. Get the transfer function of the filter, and the frequency domain representation of the chirp, using the fft. Multiply the two frequency domain functions (point-by-point), then take the inverse Fourier transform using MATLAB's ifft function to get the time domain output. Compare the two time-domain responses. (*Hint:* To multiply the two frequency domain functions together using point-by-point multiplication, they must be at the same length, so when you take the fft of the filter impulse response, pad it to be the same length as the chirp signal. In addition, ifft produces a very small imaginary component in the output, which will generate an error message in plotting, although the plot will still be correct. This error message can be eliminated by plotting only the real part of the output variable [e.g., real(xout)].)

10. Load the MATLAB data file "x_impulse," that contains the impulse response of an unknown system in variable h. Find the frequency characteristics of that unknown system in two ways: (a) use convolution to probe the system's response to a series of sinusoids; and (b) by taking the Fourier transform of the impulse response as in Example 9.4. When generating the sinusoids you will have to guess at how long to make them and what sampling rate to use. (*Suggestion:* A 1.0-second period and a sampling frequency of 2 kHz would appear to be a good place to start.)

11. Find the transfer function in the phasor domain of an RLC circuit similar to that of Figure 9.2 except reverse the positions of the resistor and capacitor. Plot the frequency characteristics for the same four values of R (0.1, 1.0, 10, 100 Ω).

12. Repeat Problem 9.11 above, but reverse the positions of the resistor and inductor.

10 BASIC ANALOG ELECTRONICS: OPERATIONAL AMPLIFIERS

Electronics circuits come in two basic varieties: analog and digital. Digital circuits feature electronic components that produce only two voltage levels, one high, and the other low. This limits the signals they can process to only two values such as zero and one. To transmit information, these high/low values, termed *bits*, are combined in groups to make up a binary number. Groupings of eight are common and an 8-bit binary number is called a byte. Bytes can be combined to make larger binary numbers or can be used to encode alphanumeric characters, most often in a coding scheme know as American Standard Code for Information Exchange (ASCII) code. Digital circuits form the basis of all modern computers and microprocessors. Although most medical instruments contain one or more small computers (i.e., microprocessors) along with related digital circuitry, bioengineers are not usually concerned with their design. They may be called on to develop some, or all, of the software, but not the actual electronics, except perhaps for some basic interface circuits.

Analog circuit elements support a continuous range of voltages and the information they carry is usually encoded as a time-varying continuous signal, similar to those used throughout this text. All of the circuits described thus far have been analog circuits. Analog circuitry is a necessary part of most medical instrumentation because the devices that perform physiological measurements, so-called transducers or biotransducers, usually produce analog electric signals. This includes devices that measure movement, pressure, bioelectric activity, sound and ultrasound, light, and other forms of electromagnetic energy. Bioelectric signals such as the electroencephalogram (EEG), electrocardiogram (ECG), and electromyelogram (EMG) are also considered analog signals. Before any of these analog signals can be processed by a digital computer, some type of manipulation is usually required while the signals are still in the analog domain. This analog signal processing may consist only of increasing the amplitude of the signal, but may also include filtering and other basic signal-processing operations. Unlike digital circuitry, the design of analog circuits is often the responsibility of the bioengineer. After analog signal processing, the signal is usually converted to a digital signal using an analog-to-digital converter. The components of a typical biomedical instrument are summarized in Figure 10.1.

Typical Bioengineering Measurement System

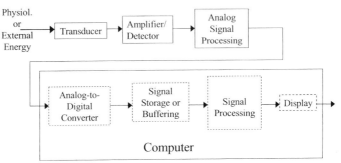

Figure 10.1 Basic elements of a typical biomedical instrument.

This chapter discusses the design and construction of some basic analog circuits such as amplifiers and filters. The design of biotransducers, often the most important biomedical instrument, can be found elsewhere (Northrop, 2004) along with the development of signal-processing software (Semmlow, 2004).

10.1   THE AMPLIFIER

Increasing the amplitude or gain of an analog signal is termed *amplification* and is achieved using an electronic device known as an *amplifier*. The properties of an amplifier are commonly introduced using a simplification called the *ideal amplifier*. Under this pedagogical scenario, the properties of a real amplifier are described as deviations from the idealization. In many practical situations, real amplifiers closely approximate the idealization in most of their properties: The limitations of real amplifiers, the deviations from the ideal, become important only in more challenging applications. Nevertheless, the bioengineer involved in circuit design must know these limitations to understand when a typical amplifier circuit is being challenged.

An ideal amplifier is one that has a well-defined gain at all frequencies (or at least over a specific range of frequencies), has an ideal source for an output (i.e., Z_{out} is zero), and presents an ideal load to the input (i.e., an infinite input impedance, Z_{in}). As a systems element, an ideal amplifier is simply a pure gain term, with ideal input and output properties. The electrical schematic and system representation of an ideal amplifier is shown in Figure 10.2.

The transfer function of this amplifier would be:

$$\frac{V_{out}(s)}{V_{in}(s)} = G$$

[Eq. 10.1]

where G would usually be a constant, or a function of frequency.

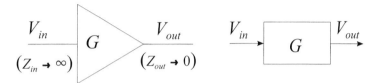

Figure 10.2 Schematic (*left*) and block diagram (*right*) of an ideal amplifier with gain of G.

Many amplifiers have a differential input configuration; that is, there are two separate inputs and the output is a constant times the difference between the two inputs. Stated mathematically:

$$V_{out} = G(V_{in2} - V_{in1}) \qquad \text{[Eq. 10.2]}$$

The schematic for such a *differential amplifier* is shown in Figure 10.3. Note that one of the outputs is labeled +, the other −, to indicate how the difference is taken. (It is common to draw the negative input above the positive input.) The + the terminal is known as the *noninverting input*, whereas the − terminal is referred to as the *inverting input*.

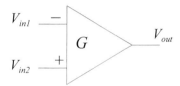

Figure 10.3 An amplifier with a differential input configuration. The output of this amplifier would be G times $V_{in2} - V_{in1}$; i.e., $V_{out} = G(V_{in2} - V_{in1})$.

Some transducers, and a few amplifiers, produce a differential signal, actually two signals that move in opposite directions with respect to the information they represent. For such differential signals, a differential amplifier is ideal, because it takes advantage of both input signals. Moreover, the subtraction tends to cancel any signal that is common to both inputs, and noise is often common (that is, similar) to both inputs. However, usually only a single signal is available. In these cases, a differential amplifier can still be used, but one of the inputs is set to zero by grounding. If the positive input is grounded and the signal is sent into the negative input (Figure 10.4, left side), the output will be the inverse of the input:

$$V_{out} = -GV_{in} \qquad \text{[Eq. 10.3]}$$

In this case the amplifier can be called an *inverting amplifier* for obvious reasons. If the opposite strategy is used and the signal is sent to the positive input while the

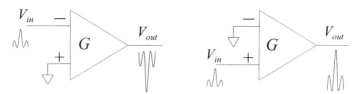

Figure 10.4   A differential amplifier configured as an inverting (*left*) and noninverting (*right*) amplifier.

negative input is grounded as in Figure 10.4 (right side), the output will have the same direction as the input. This amplifier is termed a *noninverting amplifier*. (Somewhat of a double negative.)

10.2   THE OPERATIONAL AMPLIFIER

The *operational amplifier*, or *op amp* for short, is a basic building block for a wide variety of analog circuits. One of its first uses was to perform mathematical operations, such as addition and integration in analog computers, hence the name *operational* amplifier. Although the functions provided by analog computers are now performed by digital computers, the op amp remains a valuable, perhaps the most valuable, tool in analog circuit design.

In its idealized form, the op amp has the same properties as the ideal amplifier described above except for one curious departure: it has *infinite* gain. Thus, an ideal op amp has infinite input impedance (i.e., an ideal load), zero output impedance (i.e., an ideal source), and a gain, A_v, of infinity. (The symbols A_v or A_{VOL} are commonly used to represent the gain of an operational amplifier.) Obviously, an amplifier with a gain of infinity is of limited value, so an op amp is rarely used alone but usually used in conjunction with other elements that reduce its gain to a finite level. Negative feedback can be used to limit the gain.

Consider the feedback system in Figure 10.5. The gain of the system can be found from the basic feedback equation first introduced in Chapter 1 (Eq. 1.7) and again in the last chapter (Eq. 9.3). Inserting A_V and β into the feedback equation, the overall system gain, G, becomes:

$$G = \frac{A_V}{1 + A_V \beta} \qquad \text{[Eq. 10.4]}$$

Letting the feedforward gain, A_V (the gain of the operational amplifier) go to infinity:

$$G = \lim_{A_V \to \infty} \frac{A_V}{1 + A_V \beta} = \lim_{A_V \to \infty} \frac{A_V}{A_V \beta} = \frac{1}{\beta} \qquad \text{[Eq. 10.5]}$$

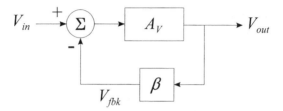

Figure 10.5 A basic feedback control system used to illustrate the use of feedback to set a finite gain in a system that has infinite feedforward gain, A_V.

If the overall gain is expressed in decibels (dB), then:

$$G_{dB} = 20 \log G = 20 \log\left(\frac{1}{\beta}\right) = -20 \log \beta \qquad \text{[Eq. 10.6]}$$

If $\beta < 1$, then the gain G will be >1 and if $\beta = 1$, $G = 1$. A $\beta > 1$ would lead to a gain <1 and a reduction in signal amplitude. If a reduction is gain is desired, it is easier to use a passive voltage divider, so in operational amplifier circuits, the feedback gain, β, will always be ≤1 to achieve a gain ≥1. The need for a feedback gain that is ≤1 turns out to be advantageous since gains <1 can be easily achieved using a simple voltage-divider network. A voltage divider network is just a pair of resistor in series where one of the resistors is attached to ground and the reduced voltage is taken from the point between the two resistors (details will be given in the next section). A feedback gain of $\beta = 1$ can be obtained even easier, just feedback all of the output back to the input.

The approach of beginning with an amplifier that has infinite gain, then reducing that gain to a finite level with the addition of feedback, would appear to be needlessly convoluted. Why not design the amplifier to have a fixed, finite gain to begin with? The answer to this question can be summarized in two words: flexibility and stability. If feedback is used to set the gain of an op amp circuit, only one basic amplifier needs to be produced and the desired gain for a specific application can be achieved by the feedback network. More important, the feedback network is usually implemented using passive resistors (sometimes a capacitor may also be involved) and such passive components are more stable than transistor-based amplifiers—that is, they are more immune to fluctuations due to temperature, age, and other environmental factors than active elements. Passive elements can also be more easily manufactured to tighter tolerances than active elements. For example, it is easy to buy resistors that have a 1% error in their values while most common transistors have variations in their gain by a factor of two or more (i.e., 100% error). Finally, a wide variety of different feedback configurations can be used allowing one operational amplifier chip to perform many different signal-processing operations. Some of these different functions are explored in the section on op amp circuits at the end of this chapter.

10.3   THE NONINVERTING AMPLIFIER

Negative feedback can be achieved by feeding the output back to the inverting (i.e., negative) input of an op amp using a simple voltage divider. Consider the voltage divider network in Figure 10.6.

The feedback voltage can be found by the simple application of KVL. Assuming that V_{out} is an ideal source (which it will be because it will be the output of an ideal amplifier):

$$V_{out} - i(R_f + R_1) = 0; \quad i = \frac{V_{out}}{R_f + R_1}:$$

$$V_{fbk} = iR_1 = V_{out}\left(\frac{R_1}{R_1 + R_f}\right) \qquad \text{[Eq. 10.7]}$$

For the system diagram in Figure 10.5, we see that $\beta = V_{fbk}/V_{out}$. Using Eq. 10.7 we can solve for β in terms of the voltage divider network:

$$\beta = \frac{V_{fbk}}{V_{out}} = \frac{R_1}{R_1 + R_f} \qquad \text{[Eq. 10.8]}$$

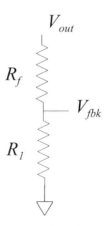

Figure 10.6   A voltage divider network that can be used to feed back a portion of the output signal to the inverting or negative input of an operational amplifier.

The transfer function, or gain, of an operational amplifier circuit that uses feedback to set the gain is just $1/\beta$ (Eq. 10.5):

$$G \equiv \frac{V_{out}}{V_{in}} = \frac{1}{\beta} = \frac{1}{\dfrac{R_1}{R_1 + R_f}} = \frac{R_1 + R_f}{R_1} \qquad \text{[Eq. 10.9]}$$

An op amp circuit using this feedback network is shown in Figure 10.7. The gain of this amplifier is given in Eq. 10.10. Because the input signal is fed to the positive side of the op amp, this circuit is a *noninverting amplifier*.

The transfer function for this circuit for the circuit in Figure 10.7 can also be found by circuit analysis, but a couple of helpful rules are needed.

1. Since the input resistance of the op amp approaches infinity, there will be no current flowing into, or out of, either of the op amp's input terminals.
2. Since the gain of the op amp approaches infinity, the only way the output can be finite is if the input is zero—that is, the difference between the plus input and the minus input must be zero. Stated yet another way, the voltage on the plus input terminal must be the same as the voltage on the minus input terminal of the op amp and vice versa.

In practical op amps, the gain is large (up to 10^6) but not infinite, so the voltage difference in a practical op amp circuit might be a few millivolts, but this small difference can generally be ignored. Similarly, the input resistance, while not infinite, is quite large: values of r_{in} (resistances internal to the op amp are denoted in lower case) are usually greater than $10^{12}\ \Omega$ so that any input current will be very small and can be disregarded (especially since the input voltage must be zero or at least very small). Note that the input characteristics of an op amp with feedback are a little peculiar: no current flows into its inputs and its differential input voltage is zero. We use these two observations to solve for the transfer function of a noninverting amplifier in the following example.

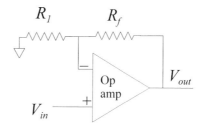

Figure 10.7 Noninverting op amp circuit. The gain of this op amp circuit in terms of the feedback resistors is given in Eq. 10.9.

Example 10.1:   Find the transfer function of the noninverting op amp circuit in Figure 10.7 using network analysis.

Solution: First note that by Rule 2, the voltage between the two resistors must be V_{in} since V_{in} is applied to the lower terminal and the voltage difference between the two terminals is zero. Next, define the three currents in and out of the node between the two resistors and apply KCL to that node. Substitute in the voltages for the currents and solve for V_{out}/V_{in}.

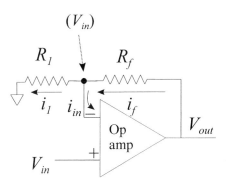

by KCL: $-i_1 - i_{in} + i_f = 0$, but $i_{in} = 0$ according to Rule 1.
   Hence, if $-i_1 - i_{in} + i_f = 0$; then $i_1 = i_f$.
   Applying Ohm's law:

$$i_1 = \frac{V_{in}}{R_1} \quad \text{and} \quad i_f = \frac{V_{out} - V_{in}}{R_f}$$

$$\text{Then since } i_1 = i_f: \quad \frac{V_{in}}{R_1} = \frac{V_{out} - V_{in}}{R_f}$$

Solving for V_{out}:

$$V_{out} - V_{in} = V_{in}\left(\frac{R_f}{R_1}\right); \quad V_{out} = V_{in}\left(1 + \frac{R_f}{R_1}\right) = V_{in}\left(\frac{R_f + R_1}{R_1}\right)$$

and the transfer function becomes:

$$\frac{V_{out}}{V_{in}} = \frac{R_f + R_1}{R_1}$$

This is the same transfer function that was found using the feedback equation (Eq. 10.9).

10.4 THE INVERTING AMPLIFIER

To construct an amplifier circuit that inverts the input signal, the ground and signal inputs of the noninverting amplifier are reversed as shown in Figure 10.8.

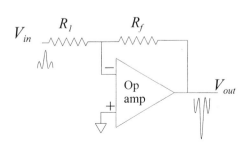

Figure 10.8 The operational amplifier circuit used to construct an inverting amplifier.

The transfer function of the inverting amplifier circuit is somewhat different from that of the noninverting amplifier, but can be easily found using the same approach (and tricks) used in Example 10.1.

Example 10.2: Find the transfer function, or gain, of the inverting amplifier circuit shown in Figure 10.8.

Solution: Define the currents and apply KCL to the point between the two resistors. Note that in this circuit the point between the two resistors must be at 0 V according to Rule 2. Since the plus side is grounded and the difference between the plus and minus side must be zero, the minus side is effectively grounded. The inverting input terminal is sometimes referred to as a *virtual ground* in the inverting amplifier configuration.

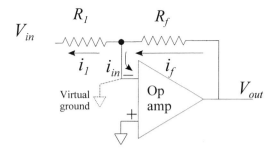

As in Example 10.1, we apply KCL to the inverting terminal and find that: $i_1 = i_f$ and by Ohm's law:

$$i_1 = \frac{0 - V_{in}}{R_1}; \quad i_f = \frac{V_{out} - 0}{R_f}; \quad \frac{-V_{in}}{R_1} = \frac{V_{out}}{R_f}$$

Again solving for V_{out}/V_{in}:

$$\frac{V_{out}}{V_{in}} = -\frac{R_f}{R_1} \qquad \text{[Eq. 10.10]}$$

Note the negative sign demonstrating that this is an inverting amplifier: The output is the negative, or inverse, of the input. The output is also larger than the input by a factor of R_f/R_1. If $R_1 > R_f$, the output will actually be reduced. The important factor is that the circuit designer has control over the gain of this circuit simply by varying the values of R_1 and/or R_f. If one of the resistors were variable (i.e., a potentiometer), the amplifier would have a variable gain. This holds for both inverting and noninverting amplifier circuits although the gain equations are different (Eq. 10.9 versus Eq. 10.10).

Example 10.3:   Design an inverting amplifier circuit with a variable gain between 10 and 100. Assume you have a variable 1-MΩ potentiometer; that is, a resistor that can be varied between 0 and 1 MΩ. Also assume you have available a wide range of fixed resistors.

Solution: The amplifier circuit will have the general configuration of Figure 10.8. It is possible to put the variable resistance as part of either R_{in} or R_f, but let us assume that the potentiometer is part of the latter along with a fixed series resistance. (Later we will see that there is some advantage to putting the variable resistor in the feedback path, R_f. The circuit then becomes:

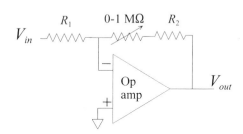

Because the variable resistor is in the feedback path, we will vary feedback resistance, R_f, to get the desired gain variation. Assume that the variable resistor will be 0 Ω when the gain is 10 and 1 MΩ when the gain is 100. We can write two equations based on Eq. 10.10 for the gain limits and solve for our two unknowns, R_1 and R_2.

For G = 10 the variable resistor is 0 Ω: $\dfrac{R_f}{R_1} = 10; \quad \dfrac{R_2 + 0}{R_1} = 10; \quad R_2 = 10R_1$

For $G = 100$ the variable resistor is $10^6 \, \Omega$:

$$\frac{R_2 + 10^6}{R_1} = 100; \quad \text{Substituting for } R_2: \; 10R_1 + 10^6 = 100R_1$$

$$90R_1 = 10^6; \quad R_1 = 11.1 \text{ k}\Omega \quad R_2 = 111 \text{ k}\Omega \text{ and}$$

The final circuit becomes:

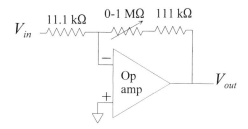

The equation for gain of noninverting and inverting op amp circuits can be extended to include feedback networks that contain capacitors and inductors. To modify these equations to include components other than resistors simply substitute impedances for resistors. So the gain equation for a noninverting op amp circuit becomes:

$$\frac{V_{out}}{V_{in}} = \frac{Z_f + Z_1}{Z_1} \qquad \text{[Eq. 10.11]}$$

and the equation for an inverting op amp circuit becomes:

$$\frac{V_{out}}{V_{in}} = -\frac{Z_f}{Z_1} \qquad \text{[Eq. 10.12]}$$

10.5 PRACTICAL OPERATIONAL AMPLIFIERS

Practical op amps, the kind that you buy from electronics supply houses, differ in a number of ways from the idealizations used above. In many applications, perhaps most applications, the limitations inherent in real devices can be ignored. The problem is that the bioengineer designing analog circuitry must know when the limitations are important and when they are not, and to do this, it is necessary to understand the characteristics of real devices. Only the topics that involve the type of circuits the bioengineer is likely to encounter are covered here. Several excellent references can be found to extend these concepts (particularly Horowitz and Hill, 1989).

Deviations of real op amps from the ideal op amp can be classified into three general areas: deviations in input characteristics, deviations in output characteristics, and deviations in transfer characteristics. Each of these areas is discussed in turn, beginning with the area likely to be of utmost concern to biomedical engineers, transfer characteristics.

10.5.1 Limitations in Transfer Characteristics of Real Operational Amplifiers

The most important limitations in the transfer characteristics of real op amps are bandwidth limitations and stability. In addition, real op amps have large, but not infinite, gain. Bandwidth limitations occur because an op amp's magnitude gain is not only finite, but decreases with increasing frequency. Stability, or rather the lack of stability that results in oscillations, is due to the op amp's increased phase shift with increasing frequency.

10.5.1.1   Bandwidth

The magnitude frequency characteristics of a popular op amp, the LF 356, are shown in Figure 10.9. Not surprisingly, even at low frequencies, the gain of this op amp is less than infinity. This in itself would not be a cause for much concern because the gain is still quite high: approximately 106 dB or 199,530. The problem is that

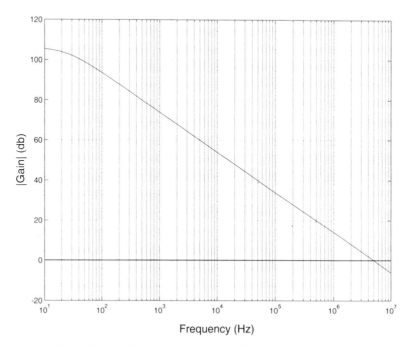

Figure 10.9 The open-loop magnitude gain characteristics of a popular operational amplifier, the LF 356.

this gain is also a function of frequency, so that at higher frequencies, the gain becomes quite small. In fact, there is a frequency above which the gain is actually less than 1. Thus, at higher frequencies the transfer function equations no longer hold because they were based on the assumption that op amp gain was infinite. Because the bandwidth of a real op amp is limited, the bandwidth of an amplifier using such an op amp must also be limited. Essentially, the gain of an op amp circuit is limited by the bandwidth limitations of the op amp or the feedback, *whichever is lower*. An easy technique for determining the bandwidth of an op amp circuit is to plot the gain that should be produced by the feedback circuit superimposed on the bandwidth curve of the real op amp. The former is referred to as the *closed-loop gain* because it includes the feedback, whereas the latter is termed the *open-loop gain* because it is the gain of the op amp without feedback.

To determine the transfer function of a real op amp circuit, we start with the fact that the theoretical closed-loop transfer function is $1/\beta$ (Eq. 10.5). Since the real transfer function gain is either this value or the op amp's open-loop gain, we plot $1/\beta$ superimposed on the open-loop curve. The real gain is simply the lower of the two curves. If the feedback network consists only of resistors, β will be constant for all frequencies, so $1/\beta$ will plot as a straight line on the frequency curve. (Although real resistors have some small inductance and capacitance, the effect of these *parasitic elements* can be ignored except at very high frequencies.)

For example, assume that 1/10 of the signal is fed back to the inverting terminal of a real op amp. Then, feedback gain is:

$$\beta = 0.1 = -20 \text{ dB and } 1/\beta = 10 = 20 \text{ dB}$$

Figure 10.10 shows the open-loop gain characteristics of a typical op amp (LF 356) with the plot of $1/\beta$ superimposed (dashed line). The overall gain will follow the dashed line until it intersects the op amp's open-loop curve (solid line) where it will follow that curve (solid line) since this is less. Hence, the amplifier's magnitude frequency characteristic will follow the heavy dash-dot lines seen Figure 10.10. Given this particular op amp and this value of β, the bandwidth of the amplifier circuit is approximately 50 kHz. The feedback gain, β, is the same for both inverting and noninverting op amp circuits, so this approach for determining amplifier bandwidth is the same in both configurations. The value $1/\beta$ is sometimes referred to as the *noise gain* because it is also the gain factor for input noise and errors, again irrespective of the specific configuration.

Example 10.4:   Find the bandwidth of the inverting amplifier circuit below.

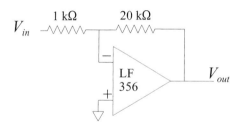

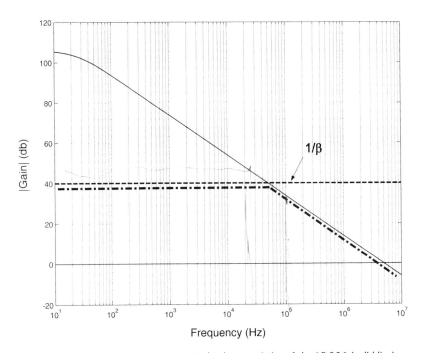

Figure 10.10 The open-loop magnitude characteristics of the LF 356 (*solid line*) as shown in Figure 10.9, with a plot of $1/\beta = 20$ dB superimposed (*dashed line*). The overall gain follows the dashed line until it intersects the solid open-loop curve then follows the open-loop curve.

Solution: First determine the feedback gain, β, then plot the inverse on top of the open-loop gain curve obtained from the op amp's specification sheets.

$$1/\beta = \frac{R_f + R_{in}}{R_{in}} = \frac{20 + 1}{1} \approx 20 = 26 \text{ dB}$$

From the superimposed plot, we see that the bandwidth is approximately 200 kHz.

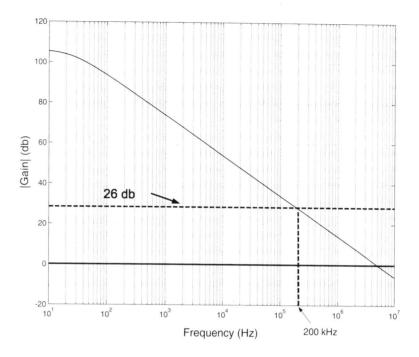

As shown in Figure 10.9, the magnitude curve of a typical op amp has very high values at low frequencies, but a fairly low cutoff frequency, followed by a rolloff of 20 dB/decade as in a first-order (or one-pole) lowpass filter. For example, the LF 356 has a low frequency gain of around 106 dB, but a cutoff frequency of only around 25 Hz. The addition of feedback dramatically lowers the overall gain, but just as dramatically increases the cutoff frequency. Indeed, the idea of using negative feedback in amplifier circuits was first introduced to improve bandwidth by trading reduced gain for increased cutoff frequency.

If the feedback gain is a multiple of 10, as is often the case, the bandwidth can be determined without actually plotting the two curves. The bandwidth can be determined directly from the amplifier gain and the frequency at which the open-loop curve intersects 0 dB. This is the frequency at which the op amp has a gain of 1.0 and is termed the *gain bandwidth product* (GBP). The GBP is given in the op amp specifications. For example, the GBP of the LF 356 is 5.0 MHz. Because the high frequency slope of the op amp magnitude curve is 20 dB/decade, for every 20 dB in gain, the closed-loop bandwidth decreases 1 decade. At a closed-loop gain of 0 dB, the bandwidth equals the GBP. For every 20 dB (or factor of 10) above 0, the bandwidth is reduced 1 decade. Hence, if an LF 356 with a GBP of 5 MHz is used to build an amplifier with a gain of 10 (20 dB), the bandwidth will be 5/10 MHz or 500 kHz. If the gain were 100 (40 dB), the bandwidth would be 50.0 kHz, whereas if the gain were 1,000 (60 dB), the bandwidth would be 5 kHz. Thus, it is often possible to determine the bandwidth directly, from the GBP, without resorting to the open-loop and $1/\beta$ plots. If the $1/\beta$ gain is not a power of 10, a logarithmic interpolation is required and it may be easier to revert to the plotting technique.

Example 10.5:   Using LF 356, what is the maximum amplifier gain (i.e., close-loop gain) that can be obtained with a bandwidth of 100 kHz?

Solution: From the open-loop curved given in the Figure 10.9, the open-loop gain at 100 kHz is approximately 30 dB. Hence, this is the maximum close-looped gain that will reach the desired cutoff frequency. Designing the appropriate feedback network to attain this gain (and bandwidth) is straightforward using Eq. 10.9 or Eq. 10.10.

10.5.1.2   Stability

Most op amp circuits use negative feedback, which is why the feedback voltage, V_{fbk}, is fed to the inverting (i.e., minus) input of the op amp. Except in very special situations, positive feedback is to be avoided. Positive feedback creates a *vicious circle*: The feedback signal enhances the feedforward signal which enhances the feedback signal which enhances . . . , and so forth. A number of things can happen to a positive feedback network, most of them bad. The two most likely outcomes are that the output is driven into saturation, locked into producing the maximum (or minimum) output possible, or the output can oscillate often between the output extremes. When the word *stability* is used in context with an op amp circuit, it means the absence of oscillation or other deleterious effects associated with positive feedback. The oscillation that may be produced by positive feedback is a sustained repetitive waveform, which could be a sinusoid, but it may also be more complicated.

Positive feedback oscillation occurs in a feedback circuit where the overall gain or *loop gain* (i.e., the gain of the feedback and feedforward circuits) is greater than or equal to one and has a phase shift of 360 degrees:

Loop gain (for oscillation) $\equiv$ Feedforward gain $\times$ Feedback gain $\geq 1.0 \angle 360$ degrees
[Eq. 10.13]

When this condition occurs, any small signal will feedback positively and grow to produce a sustained oscillation. Sometimes this oscillation will ride on top of the signal, sometimes it will overwhelm the signal, but in either case, it is unacceptable.

Since the feedback signal is sent to the inverting input of the op amp, positive feedback should not occur. The noninverting input induces a phase shift of 180 degrees, so the feedback signal is negative ($v \angle 180 \equiv -v$). However, if the op amp induces an additional phase shift of 180 degrees, the negative feedback becomes positive feedback because the *total* phase shift is 360 degrees ($v \angle 360 = +v$). If the loop gain happens to be ≥ 1 when this occurs, the circuit will oscillate: A repetitive waveform will be generated or saturation will occur. The base frequency of that oscillation will be equal to the frequency where the *overall* phase shift becomes 360 degrees; that is, the frequency where the op amp contributes a phase shift of an additional 180 degrees. Hence, oscillation is a result of phase shifts that occur at higher frequencies in real op amps.

A rigorous analysis of stability is beyond the scope of this text. However, because stability is so often a problem in op amp circuits, some discussion is warranted. Since the inverting input contributes 180-phase shift (to make the feedback negative), to ensure stability we must make sure that everything else in the feedback loop contributes a phase shift that is less than 180 degrees. If β is a constant, any additional phase shift must come from the op amp, so all we need are op amps that never approach a phase shift of 180 degrees. Unfortunately, all op amps will reach an internal phase shift of 180 degrees if we go high enough in frequency. The alternative strategy then is to ensure that when the op amp reaches a phase shift of 180 degrees, the overall loop gain is less than 1.0.

The overall loop gain is just $A_V(\omega)\beta(\omega)$. Putting the condition for stability in terms of the gain symbols we have used thus far, the condition for stability is:

$$\text{Loop-gain stability} = A_V(\omega)\beta(\omega) < 1.0 \angle 360 \text{ degrees} \qquad \text{[Eq. 10.14]}$$

where A_V is the gain of the op amp at a specific frequency and β is the feedback gain. Alternatively, if we want to build an oscillator, the condition for oscillation would be:

$$\text{Loop-gain oscillation} = A_V(\omega)\beta(\omega) \geq 1.0 \angle 360 \text{ degrees} \qquad \text{[Eq. 10.15]}$$

With respect to β, the worst case for stability occurs when β is the largest. In most op amp circuits β is less than 1, but can sometimes be as large as 1. A feedback gain of $\beta = 1$ corresponds to the lowest op amp gain: a gain of 1 ($V_{out} = V_{in}$). While it is somewhat counterintuitive, this means that stability is more likely to be a problem in *low gain* amplifier circuits where β is large, and most likely to be a problem when the gain is 1.0 since $\beta = 1$ in this case.

If the op amp gain, A_V, is less than 1.0 when its phase shift hits 180 degrees, and β is at most 1.0, $A_V \beta$ will be less than 1, and the conditions for stability are met (Eq. 10.14). Stated in terms of phase, the op amp's phase shift should be less than 180 degrees for all frequencies where its gain is 1 or more. In fact, most op amps have a maximum phase shift that is less than 120 degrees to be on the safe side for gains greater than, or equal to, 1.0. Such op amps are said to be *unity gain stable* because they will not oscillate even when $\beta = 1$, and the noninverting gain is 1. However, to achieve this criterion requires some compromise on the part of the op amp manufacture, usually some form of phase compensation that reduces the GBP. In many op amp applications where gain is high so that β is low, unity gain stability is overkill and results in a needless reduction of bandwidth. Op amp manufactures have come up with two strategies to overcome the problem: Produce different versions of the same basic op amp, one that has higher bandwidth but requires a minimum gain while another is unity gain stable with a lower bandwidth; or produce a single version but have the user supply the compensation (usually as an external capacitor) to suit the needs of the application. The former has become more popular because it does not require additional circuit components. The LF

356 is an example of this strategy. The LF 356 with a GBP of 5 MHz is unity gain stable while its "sister" chip, the LF 357, requires a $1/\beta$ gain of 5 or more, but has a GBP of 20 MHz.

Even if using an op amp that is unity gain stable (such as the LF 356), stability problems can still occur if the feedback network introduces a phase shift. A feedback network containing only resistors might be considered safe, but parasitic elements, small inductances, and capacitances can create an undesirable phase shift. Consider the feedback circuit in Figure 10.11 in which a capacitor is placed in parallel with one of the resistors. Will this additional capacitance present a problem with regard to stability?

To answer this question, we need to find the phase shift of the network at the frequency when the loop gain is one. The loop gain will be 1 when:

$$A_V \beta = 1; \quad A_V = \frac{1}{\beta} \quad \text{[Eq. 10.16]}$$

Hence the loop gain is 1 when $1/\beta$ equals the op amp gain A_V. On the plot of A_V and $1/\beta$, this occurs when the two curves intersect. In the circuit presented in Figure 10.11, the $1/\beta$ curve will not be a straight line, but can be easily found using the phasor techniques presented in Chapter 4.

For the voltage divider:

$$V_{fbk} = \frac{Z_1}{Z_1 + Z_f} V_{out}; \quad \text{Solving for } \beta: \frac{V_{fbk}}{V_{out}} = \beta = \frac{Z_1}{Z_1 + Z_f}$$

$$\text{where } Z_1 = \frac{R_1(1/j\omega C)}{R_1 + (1/j\omega C)} = \frac{10^4(10^9/j\omega)}{10^4 + (10^9/j\omega)} = \frac{10^4}{1 + j\omega/10^5}$$

$$Z_f = 10^5 \, \Omega$$

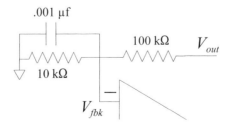

Figure 10.11 A feedback network that includes some capacitance either due to the intentional addition of a capacitor or due to parasitic elements.

Substituting in Z_1 and Z_f into the inverse of Eq. 10.9 and solving for β:

$$\beta = \frac{\dfrac{10^4}{1+j\omega/10^5}}{10^5 + \dfrac{10^4}{1+j\omega/10^5}} = \frac{10^4}{10^5 + j\omega + 10^4} = \frac{.09}{1+j\omega/1.1\times10^5}$$

This is just a low-pass filter with a cutoff frequency of 1.1×10^5 rad/sec or 17.5 kHz. Figure 10.12A shows the magnitude frequency plot of β while Figure 10.12B shows the $1/\beta$ curve plotted superimposed on the open-loop curve of the LF 356 op amp. From Figure 10.12B, the two curves intersect at around 85 kHz.

The phase shift contributed by the feedback network at 85 kHz can be readily determined from the equation for β. At 85 kHz, $\omega_1 = 2\pi(85 \times 10^3) = 5.34 \times 10^5$.

$$\beta = \frac{0.09}{1+j\omega/1.1\times10^5} = \frac{0.09}{1+j2\pi85\times10^3/1.1\times10^5} = \frac{0.09}{4.96 \angle 78} = 0.018\angle-78$$

Thus, the feedback network contributes 78 degrees phase shift to the overall loop gain. While the phase shift of the op amp at 85 kHz is not known (detailed phase information is not often provided in op amp specifications), we can only be confident that the phase shift is no more than 120 degrees. Adding the worst-case op amp phase shift to the feedback network phase shift (120 + 78) results in a total phase shift that is more than 180 degrees, and this circuit is likely to oscillate. A good rule of thumb is that the circuit will be unstable if the $1/\beta$ curve breaks upward before intersecting the A_V line of the op amp. The reverse is also true: the circuit will be stable if the $1/\beta$ line intersects the A_V line at a point where it is flat or going downward.

If the feedback network can make the circuit unstable, it stands to reason that it can also make the amplifier less unstable. This occurs when capacitance is added to the feedback circuit. In fact, the "quick fix" approach to oscillations is to add a capacitor to the feedback network in parallel with the feedback resistor. This usually works although the capacitor might have to be large. As shown in the section on filters, adding feedback capacitance reduces bandwidth and the larger the capacitance the greater the reduction in bandwidth. Sometimes a reduction in bandwidth is desired to reduce noise, but often it is disadvantageous. The influence of feedback capacitance on bandwidth is explored in the section on filters, whereas its influence on stability is demonstrated in Problem 10.6.

10.5.2 Input Characteristics

The input characteristics of a real op amp can best be described as involved, but not complicated. In addition to a large, but finite, input resistance, r_{in}, several voltage and current sources are found (Figure 10.13). The values of these elements are given for the LF 356 in parenthesis. (The curious units for the voltage and current noise sources, nV/$\sqrt{\text{Hz}}$ and pA/$\sqrt{\text{Hz}}$, are explained later.) These sources have very small

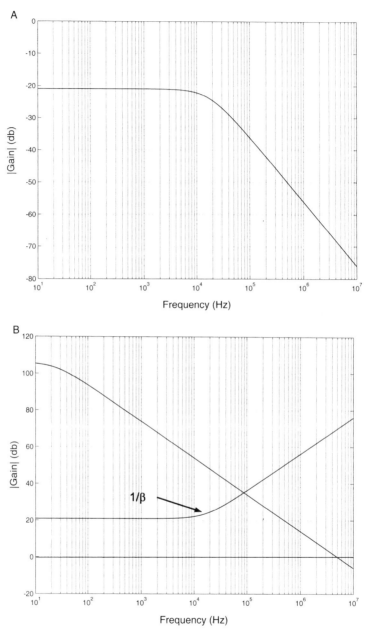

Figure 10.12 **A:** The magnitude frequency plot of the feedback gain, β, of the network shown in Figure 10.11. **B:** The inverse of the feedback gain (i.e., $1/\beta$) plotted superimposed in the A_v curve of the LF 356 op amp.

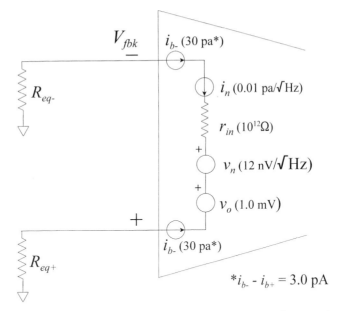

Figure 10.13 A schematic representation of the input elements of a practical operational amplifier. The input can be thought of as containing several voltage and current sources as well as a large input impedance.

values and can often be ignored, but again, it is important for the bioengineer to understand the importance of these elements to make intelligent decisions.

10.5.2.1 Input Voltages Sources

The voltage source, v_o, is a constant voltage source termed the *input offset voltage*. It indicates that the op amp will have a nonzero output even if the input is zero $(V + -V = 0)$. The output voltage produced by this small voltage source depends on the gain of the circuit. To find the output voltage under zero-input conditions (the *output* offset voltage), simply multiply the input offset voltage by the $1/\beta$ gain, also known as the *noise gain* for reasons given later. This is demonstrated in the following example.

Example 10.6: Find the offset voltage at the output of the amplifier circuit shown below. This is the same as asking for V_{out} when $V_{in} = 0$.

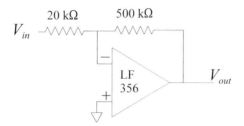

Solution: First find the noise gain, $1/\beta$, then multiply this by the input offset voltage, v_o (1.0 mV). The value of $1/\beta$ is:

$$\frac{1}{\beta} = \frac{R_f + R_1}{R_1} = \frac{500 + 20}{20} = 26$$

Given the typical value of v_o as 1.0 mV, V_{out} for zero input becomes:

$$V_{out} = v_o(26) = 26 \text{ mV}$$

The input offset voltage given in the LF 356 specifications (see Appendix F) is a *typical value*, not necessarily the value of any individual operational amplifier chip. This means that the input offset voltage will be around the value shown, but for any given chip, it could be larger or smaller. It could also be either positive or negative leading to a positive or negative output offset voltage. Often a maximum or *worst case* value is also specified.

   The noise voltage source, v_n, specifies the noise normalized for bandwidth—more precisely, normalized to the square root of the bandwidth—which accounts for the strange units nV/$\sqrt{\text{Hz}}$. Like Johnson noise and shot noise described in Chapter 1 (Section 1.4), noise in an op amp is distributed over the entire bandwidth of the op amp. To determine the actual noise in an amplifier it is necessary to multiply v_n by the square root of the circuit bandwidth as determined using the methods described above. This value should then be multiplied by the noise gain (i.e., $1/\beta$) to find the noise at the output. (We finally see why $1/\beta$ is also referred to as the noise gain.) This procedure is demonstrated in Example 10.7.

Example 10.7:   Find the noise at the output of the amplifier used in Example 10.6 that is due only to the op amp's noise voltage. (The resistors in the feedback network will also contribute Johnson noise as will the input current noise source, i_n.)

Solution: Find the noise gain, then determine the bandwidth from $1/\beta$ using the open-loop gain curve in Figure 10.9. Multiply the input noise voltage v_n, by the square root of the bandwidth to find the noise at the input. Then multiply the result by the noise gain to find the value of noise at the output.

From Example 10.6, the noise gain is 26, or 28 dB. Referring to Figure 10.9, a $1/\beta$ line at 28 dB will intersect A_V at approximately 200 kHz. Hence, the bandwidth can be taken as 200 kHz and the input noise voltage becomes:

$$v_{n\,input} = 12 \times 10^{-9} \sqrt{200 \times 10^3} = 5.37\ \mu V$$

The noise at the output is:

$$V_{n\,output} = v_{n\,input}(1/\beta) = 5.37(26) = 139.6\ \mu V$$

The value of input voltage noise used in Example 10.7 was again a typical value for frequencies above 100 Hz. Op amp noise generally increases at the lower frequencies and many op amp specifications include this information, including those of the LF 356. For the LF 356 the input voltage noise increases from 12 nV/√Hz at 200 Hz and up, to 60 nV/√Hz at 10 Hz (see Appendix F). Presumably, the voltage noise becomes even higher at lower frequencies, but values below 10 Hz are not given for this chip.

10.5.2.2 Input Current Sources

To evaluate the influence of input current sources, it is easiest to convert them to input voltages by multiplying by the equivalent resistance at the terminals. Figure 10.13 represents the equivalent input resistances at the plus (+) and minus (−) terminals as R_{eq+} and R_{eq-}. These would have been determined by using the network reduction methods described in Chapter 7 and will be illustrated in a subsequent example. The two current sources shown at the two inputs, i_{b+} and i_{b-}, are known as the bias currents and will contribute to the overall offset voltage. It may seem curious to show two current sources rather than one for both inputs, but this has to do with the fact that the offset currents at the two terminals are not exactly equal, although they do tend to be similar. In addition, as with the bias voltage, the bias currents could be in either direction, in or out of their respective terminals.

To determine this contribution, convert the bias current to voltages by multiplying by the appropriate R_{eq}'s, then multiplying by the noise gain. In most operational amplifiers, the two bias currents are approximately the same, so the influence on output offset voltage tends to cancel *if* the equivalent resistances at the two terminals are the same. Sometimes the op amp circuit designer will try to make the equivalent resistances at the two terminals the same just to achieve this cancellation. The amount that the two bias currents are different, the imbalance between the two currents, is called the *offset current* and is usually much less than the bias current. For example, in the LF 356 typical bias currents are 30 pA while the offset current is only 3.0 pA, an order of magnitude less.

Figure 10.14 shows an inverting op amp circuit where a resistor has been added between the positive terminal and ground. The current flowing through this resistor is essentially zero (if you ignore the small bias currents) because the op amp's input impedance is quite large. So there is negligible voltage drop across the resistor, and the positive terminal is still at ground potential. This resistor performs no

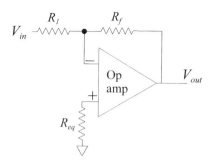

Figure 10.14 An inverting operational amplifier with a resistor added to the noninverting terminal to balance the bias currents. R_{eq} should be set to equal the equivalent resistance on the inverting terminal.

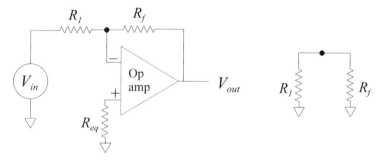

Figure 10.15 The left circuit is a typical inverting operational amplifier. If the op amp output, V_{out} and the input, V_{in}, can be considered ideal sources, then the equivalent resistance at the inverting terminal is the parallel combination of R_f and R_1 as shown on the right side.

function in the circuit except to balance the bias currents. To achieve this balance, the resistor should be equal to the equivalent resistance at the op amp's negative terminal.

To determine the equivalent resistance at the negative terminal, we use the approaches described in Chapter 7 and make the usual assumption that the input to the op amp is an ideal source (Figure 10.15A). We also assume that the output of the op amp is essentially an ideal source. (Op amp output characteristics are covered in the next section.)

Since the equivalent resistance of an ideal source is 0.0 Ω, the two resistors go from the negative terminal to ground and are in parallel. Hence, the equivalent resistance at the negative terminal of an op amp is:

$$R_{eq-} = \frac{R_f R_1}{R_f + R_1}; \quad \text{or more generally: } Z_{eq-} = \frac{Z_f Z_1}{Z_f + Z_1} \qquad \text{[Eq. 10.17]}$$

So to balance the resistances (or impedances) at the input terminals, the resistance at the positive terminal should be set to the parallel combination of the feedback and input resistors (Eq. 10.17). Sometimes, as an approximation, the resistance at the positive terminal is set to equal the lower of the two feedback network resistors (usually R_1). Another strategy is to make this resistor variable, then adjust the resistance to cancel the output offset voltage from a given op amp. This has the advantage of removing any output offset voltage due to the input offset voltage, v_o. The primary downside to this approach is that the resistor must be carefully adjusted after the circuit is built.

The current noise source is treated the same way as the bias currents: It is multiplied by the two equivalent resistances to find the input current noise, then multiplied by the noise gain to find the output noise. To find the total noise at the output, it is necessary to add in the voltage noise. Because the noise sources are independent, they add as the square root of the sum of the squares (see Eq. 1.11). In addition to the voltage and current noise of the op amp, the resistors will produce voltage noise as well. To repeat the equation for Johnson noise for a resistor from Chapter 1 (Eq. 1.8):

$$V_J = \sqrt{4kT\,R\,BW}\ \text{V} \qquad \text{[Eq. 10.18]}$$

The three different noise sources associated with an operational amplifier circuit are all dependent on bandwidth. The easiest way to deal with these three different sources is to combine them in one equation that includes the bandwidth:

$$V_{n\,in} = \left[\left(v_n^2 + (i_n(R_{eq+} + R_{eq-}))\right)^2 + 4kT(R_{eq+} + R_{eq-})\right)BW\right]^{1/2} \qquad \text{[Eq. 10.19]}$$

This equation gives the summed input noise. To find the output noise, multiply by the noise gain.

$$V_{n\,out} = V_{n\,in}(Noise\ Gain) = V_{n\,in}(1/\beta) \qquad \text{[Eq. 10.20]}$$

Use of this approach to calculate the noise out of a typical op amp amplifier circuit is shown in Example 10.8.

Example 10.8: Find the noise at the output of the operational amplifier circuit shown in Figure 10.14 where: $R_f = 500$ kΩ, $R_1 = 10$ kΩ, and $R_{eq} = 9.8$ kΩ.

Solution: First find the noise gain, $1/\beta$. From $1/\beta$ determine the bandwidth of the amplifier using the open-loop gain curves in Figure 10.9. Apply Eq. 10.19 to find the total input voltage noise including both current and voltage noise. Then multiply this voltage by the noise gain to find output voltage noise.

The noise gain is:

$$1/\beta = \frac{R_f + R_1}{R_1} = \frac{500 + 10}{10} = 51$$

From the specifications of the LF 356, $v_n = 12$ nV/$\sqrt{\text{Hz}}$ and $i_n = 0.01$ pA/$\sqrt{\text{Hz}}$. The equivalent resistance at the negative terminal is found from Eq. 10.17 to be 9.8 kΩ (the parallel combination of 500 kΩ and 10 kΩ). For a $1/\beta$ of 51, the bandwidth is approximately 100 kHz (Figure 10.9). Using $T = 310°$K, $4\,kT$ is 1.7×10^{20} J, Eq. 10.17 becomes:

$$V_{n\,in} = \left[\left((12 \times 10^{-9})^2 + (0.01 \times 10^{-12}(19.6 \times 10^3))^2 + 1.7 \times 10^{-20}(19.6 \times 10^3)\right)10^5\right]^{1/2}$$

$$V_{n\,in} = \left[(1.44 \times 10^{-16} + 3.8 \times 10^{-20} + 3.33 \times 10^{-16})10^5\right]^{1/2}$$

$$V_{n\,in} = \left[(4.77 \times 10^{-16})10^5\right]^{1/2} = 6.9\ \mu V$$

The noise at the output is found by multiplying by the noise gain:

$$V_{n\,out} = V_{n\,in}(Noise\ Gain) = 6.9 \times 10^{-6}(51) = 0.35\ \text{mV}$$

One advantage to including all the sources in a single equation is that the relative contributions of each source can be compared. After converting to a voltage, the current noise source is approximately four orders of magnitude less than the other two voltage noise sources, so its contribution is negligible. The op amp's voltage noise does contribute to the overall noise, but most of the noise is coming from the resistors. Finding an op amp with a lower noise voltage would lower the noise, but of the 0.35 mV noise at the output, 0.29 mV is from the resistors. Of course, this is using the value of noise voltage for frequencies above 200 Hz. The noise voltage of the op amp at 10 Hz is 60 nV/$\sqrt{\text{Hz}}$, four times the value used in this example. If noise at the lower frequencies is a concern, another op amp may be of value. (For example, the Op-27 op amp features a noise voltage of only 5.5 nV/$\sqrt{\text{Hz}}$ at 10 Hz.)

10.5.2.3   Input Impedance

Although the input impedance of most op amps is quite large, the actual input impedance of the circuit depends on the configuration. The noninverting op amp has the highest input impedance, that of the op amp itself. In practice, it may be difficult to attain the high impedance of many op amps due to leakage currents in the circuit board or wiring. In addition, the bias currents of an op amp will decrease its effective input impedance.

For an inverting amplifier, the input impedance is approximately equal to the input resistance, R_1 (Figure 10.8). This is because the input resistor is connected to *virtual ground* in the inverting configuration. While this will be much lower than the input impedance of the noninverting configuration, it is usually large enough for most applications. Where very high input impedance is required, the noninverting configuration should be used. If even higher input impedances are required, op amps with particularly high input impedances are available, but the limitations on impedance are usually set by other components of the circuit such as the lead-in wires and circuit board.

10.5.3   Output Characteristics

Compared to the input characteristics, the output characteristics of an op amp are quite simple: A Thévenin source where the ideal source is A_V ($V_{in+} - V_{in-}$) and the

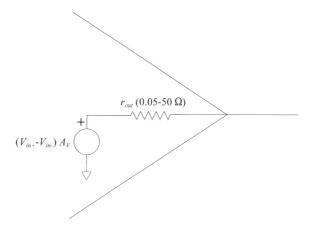

Figure 10.16 The output characteristics of an op amp are those of a Thévenin source including an ideal voltage source, $A_V(V_{in+} - V_{in-})$, and an output resistor, r_{out}.

resistance is r_{out} (Figure 10.16). For the LF 356, A_V is given as a function of frequency in Figure 10.9. The Thévenin output resistance r_{out} is quite low, approximately 0.05 Ω at low frequencies and for low values of the loop gain when $\beta = 1$. The output resistance increases to as large as 50 Ω at higher frequencies when β is small and amplifier gain is large.

In some circumstances, the output characteristics can become more complicated. Maximum voltage swing at the output must always be a few volts less than the voltage that powers the op amp (see next section), but the output signal range is also limited at higher frequencies. In addition, many op amps have stability problems when driving a capacitive load. Figure 10.17 shows a circuit taken from the LF 356 specification sheet that can be used to drive a large capacitive load. (A

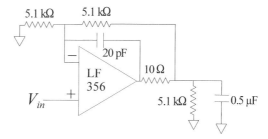

Figure 10.17 An operational amplifier circuit that can be used to drive a large capacitive load. Note that this is a noninverting amplifier with a gain of 2. The main purpose of this circuit is to provide a drive to the capacitive load, not to amplify the signal.

0.5-μf capacitor is considered fairly large in electronic circuits.) One strategy illustrated in this figure is to place a resistor in parallel with the capacitor load. Another strategy that is shown is to place a small resistor at the output of the op amp before the feedback resistor. A final strategy is to add a small feedback capacitor, which, as mentioned above, improves stability. These strategies are often implemented on an ad hoc basis, but the design engineer should anticipate possible problems when capacitive (or inductive) loads are involved.

10.6 POWER SUPPLY

Op amps are active devices and require external power to operate. This external power is delivered as a constant voltage or voltages from a device known, logically, as a *power supply*. Power supplies are commercially available in a wide range of voltages and current capabilities. Many operational amplifiers are *bipolar*, that is, they can handle both positive and negative voltages. (Unlike its use in psychology, in electronics the term bipolar has nothing to do with stability.) Bipolar applications require both positive and negative power supply voltages, and values of ±12 V or ±15 V are common. The higher the power supply voltage, the larger the output voltages the op amp can produce, but all op amps have a maximum voltage limit. The maximum voltage for the LF 356 is ±18 V with a special version, the LF 356B, that can handle ±22 V. Higher voltage op amps are available as are low voltage op amps for battery use. The latter also feature lower current consumption. (The LF 356 uses a nominal 5 to 10 mA and a number of them in a circuit will *eat* through 9-V batteries fairly quickly.)

The power supply connections are indicated on the op amp schematic by vertical lines coming from the side of the amplifier icon as shown in Figure 10.18. Sometimes the actual chip pins are indicated on the schematic as in this figure. [Pin numbers are for the 8 pin DIP (dual inline package) configuration of the LF 356.] Figure 10.18 also shows a curious collection of capacitors attached to the two supply voltages. Power supply lines often go to a number of different op amps or other analog circuitry and make great pathways for spreading signal artifacts, noise, and other undesirable fluctuations among the op amps in a circuit. One op amp circuit might induce fluctuations on the power line, and these fluctuations could then pass to all the other circuits. Practical op amps do have some immunity to power supply fluctuations, but this immunity falls off significantly with the frequency of the fluctuations. For example, the LF 356 will attenuate power supply variations at 100 Hz by 90 dB (a factor of 31,623), but this attenuation falls to 10 dB (a factor of 3) at 1 MHz. A capacitor placed right at the power supply pin will reduce these fluctuations. In a sense, such a capacitor isolates or disconnects the op amp from power line noise, so this capacitor is often called a *decoupling capacitor*. Figure 10.18 shows two capacitors on each supply line: a large 10-μf capacitor and a small 0.01-μf capacitor. Since the two capacitors are in parallel, they are in theory equivalent to a single 10.01 μf capacitor. The small capacitor would appear to be contributing very little. In fact, the small capacitor is there

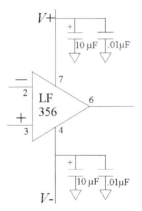

Figure 10.18 An operational amplifier (op amp) with the power supply connections shown. The capacitors attached to the positive and negative leads smooth the supply voltage to the op amp. They are an attempt to *decouple* this circuit from the other circuits attached to the same power lines. This schematic also shows the pin connections for the LF 356 op amp chip (8-pin DIP package).

because large capacitors have very poor high-frequency performance: They look more like inductors than capacitors at higher frequencies. The small capacitor serves to reduce high-frequency fluctuations while the large capacitor does the same at low frequencies. While a given op amp circuit may not need both these decoupling capacitors, the 0.01 μf capacitor is routinely included in the op amp circuits by most design engineers. The larger capacitor is added if strong low-frequency signals are present in the network.

10.7 OPERATIONAL AMPLIFIER CIRCUITS, OR 101 THINGS TO DO WITH AN OPERATIONAL AMPLIFIER

Although there are more than 101 different signal-processing operations that can be performed by op amp circuits, this is an introductory course, so only a handful will be presented. For a look at the other 90+, see the *Art of Electronics* by Horowitz and Hill (1989).

10.7.1   The Differential Amplifier

We have already shown how to construct inverting and noninverting amplifiers. Why not throw the two together to produce an amplifier that does both—a differential amplifier? As shown in Figure 10.19, a differential amplifier is a combination of the inverting and noninverting amplifier. To derive the transfer function of the circuit in Figure 10.19, we will once again use the principle of superposition. Setting V_{in2} to zero effectively grounds the lower input resistor, R_1, and the circuit becomes a standard inverting op amp with a resistance between the positive terminal and ground (Figure 10.20, left side). As stated previously, this resistance does not alter the voltage at V_t. For this partial circuit, the transfer function is:

$$V_{out} = -\frac{R_f}{R_1} V_{in1}$$

Setting V_{in1} to zero grounds the upper R_1 resistor, and the circuit becomes a noninverting amplifier with a voltage divider on the input. With respect to the voltage V' (Figure 10.20, right side), the circuit is a standard noninverting op amp:

$$V_{out} = \frac{R_f + R_1}{R_1} V'$$

The relationship between V_{in2} and V' is given by the voltage divider equation:

$$V' = \frac{R_f}{R_f + R_1} V_{in2}$$

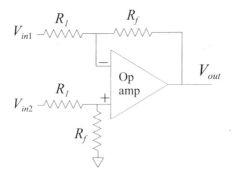

Figure   10.19 A   differential   amplifier circuit. This amplifier combines both inverting and noninverting amplifier into a single circuit. As described in the text, this circuit amplifies the difference between the two input voltages: $V_{out} = R_f/R_{in}(V_{in2} - V_{in1})$.

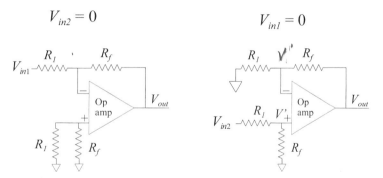

Figure 10.20 Superposition applied to the differential amplifier. **Left circuit:** The V_{in2} input is set to zero (i.e., grounded) leaving a standard inverting amplifier circuit. **Right circuit:** The V_{in1} input is grounded leaving a noninverting operational amplifier with a voltage divider circuit on the input.

Substituting and solving for V_{in2}:

$$V_{out} = \frac{R_f + R_1}{R_1}V' = \frac{R_f + R_1}{R_1}\frac{R_f}{R_f + R_1}V_{in2} = \frac{R_f}{R_1}V_{in2}$$

By superposition, the two partial solutions can be combined to find the transfer function when both voltages are present.

$$V_{out} = \frac{R_f}{R_1}V_{in1} + \frac{R_f}{R_1}V_{in2} = \frac{R_f}{R_1}(V_{in2} - V_{in1}) \qquad \text{[Eq. 10.21]}$$

Thus, the circuit shown in Figure 10.19 amplifies the *difference* between the two input voltages.

10.7.2 The Adder

If the sum of two or more voltages is desired, the circuit shown in Figure 10.21 can be used.

It is easy to show, using an extension of the approach used in Example 10.2, that the transfer function of this circuit is:

$$V_{out} = \left(\frac{R_f}{R_1}\right)V_{in1} + \left(\frac{R_f}{R_2}\right)V_{in2} + \left(\frac{R_f}{R_3}\right)V_{in3} \qquad \text{[Eq. 10.20]}$$

The derivation of this equation is left as an exercise at the end of the chapter. This circuit can be extended to any number of inputs by adding more input resistors. If $R_1 = R_2 = R_3$, then the output is the straight sum of the three input signals amplified by R_f/R_1. Otherwise the summation is weighted by the various resistor combinations as given in Eq. 10.20.

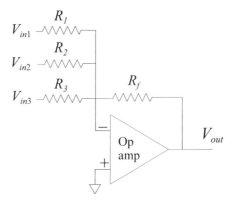

Figure 10.21 An operational amplifier circuit that takes a weighted sum of three input voltages.

10.7.3 The Buffer Amplifier

At first glance, the circuit in Figure 10.22 appears to be of little value. In this circuit all of the output is feedback to the inverting input terminal, so the feedback gain, β, equals 1. Because the gain of a noninverting amplifier is $1/\beta$, the gain of this amplifier is 1 and $V_{out} = V_{in}$. (This can also be shown using circuit analysis, and this is an exercise in the problem section.) Although this amplifier does nothing to enhance the amplitude of the signal, it does a great deal when it comes to impedance. Specifically, the incoming signal sees a large impedance, the input impedance of the operational amplifier ($>10^{12}$ Ω for the LF 356) making it a near ideal load, whereas the output impedance is very low (0.02 Ω at 10 kHz for the LF 356), approaching that of an ideal source. This circuit can take a signal from a high-

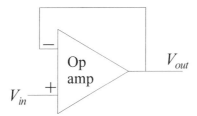

Figure 10.22 A buffer amplifier circuit. This amplifier provides no gain ($G = 1/\beta = 1$), but presents a very high impedance to the signal source (nearly an ideal load) and generates a low-impedance signal that looks like it comes from a near ideal source.

impedance Thévenin source and provide a low-impedance, nearly ideal, source that can be used to drive several other devices. While all noninverting op amp circuits will have this impedance transformation function, the unity gain circuits is particularly effective and will have the highest bandwidth since $1/\beta = 1$. Because this circuit provides a buffer between the high-impedance source and the other devices it drives, it is sometimes referred to as a *buffer amplifier*. The low-output impedance also reduces noise pick up, and this circuit is invaluable whenever a signal is sent over long wires or off the circuit board.

10.7.4 The Transconductance Amplifier

Figure 10.23 shows another simple circuit that looks like an inverting op amp circuit except the input resistor is missing.

The input to this circuit is a current, not a voltage, and the circuit is used to convert this current into a voltage. Applying KCL to the negative input terminal, the transfer function for this circuit is easily determined.

$$i_n = i_f = \frac{V_{out}}{R_f};$$

Solving for V_{out}:

$$V_{out} = R_f i_n \qquad \text{[Eq. 10.23]}$$

Some transducers produce a current and this circuit can be used as the first stage to convert that current signal to a voltage. An example transducer common to many medical instruments is the photodiode, which produces a current proportional to the light falling on it. Usually these currents are very small (in the nanoamps or picoamps) and R_f is chosen to be very large (≈ 100 MΩ) so that a reasonable output voltage is produced.

Since the input to the op amp is current, noise depends only on current noise. This would include the noise generated by the op amp and the feedback resistor.

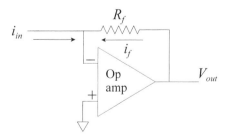

Figure 10.23 An operational amplifier circuit used to convert a current to a voltage. Because of this current-to-voltage transformation, this circuit is also referred to as a *transconductance amplifier*.

The net input current noise is then multiplied by R_f to find the output voltage noise. This approach is illustrated in the very practical problem posed in Example 10.9

Example 10.9:   A transconductance amplifier shown in Figure 10.24 is used to convert the output of a photodetector into a voltage. The photodetector has a sensitivity of: $\mathcal{R} = 0.01\ \mu A/\mu W$. (The $\mathcal{R}$ stands for *responsivity*, which is essentially the same as sensitivity, the output for a given input.) The photodiode also has a noise current, which is called *dark current*, of $i_d = 0.05\ pA/\sqrt{Hz}$. What is the minimum light flux, φ in μW, that can be detected with an SNR (signal-to-noise ratio) of 20 dB with a bandwidth of 1 kHz?

Solution: This problem requires a number of steps, but the heart of the problem is how much current noise is generated at the input of the op amp. Once this is determined, the minimum signal current can be determined as 20 dB, or a factor of 10, times this current noise. Once the minimum signal current is found, the minimum light flux, φ_{min} can be calculated as $i_{min}/\mathcal{R}$. To find the noise current, use a version of Eq. 10.19 for current rather than voltage:

$$i_{n\,Total} = \left[\left(i_d^2 + i_n^2 + \frac{4kT}{R_f} \right) BW \right]^{1/2} \qquad \text{[Eq. 10.24]}$$

Note that the value of current noise *decreases* for increased values of R_f so a good design would use as large a value of R_f as possibly needed.

Solving Eq. 10.24 for a bandwidth of 1 kHz, a value of i_n from the LF 356, the value of i_d for the photodetector, and a value of R_f of 10 MΩ:

$$i_{n\,Total} = \left[\left((0.05 \times 10^{-12})^2 + (0.01 \times 10^{-12})^2 + \frac{1.7 \times 10^{-20}}{10^7} \right) 10^3 \right]^{1/2}$$

$$= \left[(2.5 \times 10^{-27} + 1 \times 10^{-28} + 1.7 \times 10^{-27}) 10^3 \right]^{1/2} = 2.07\ pA$$

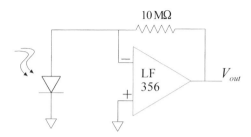

10 MΩ

LF 356

V_{out}

Figure 10.24   The output of a photodetector is fed to a transconductance amplifier. This circuit is used in Example 10.9 to determine the minimum light flux that can be detected with a signal-to-noise ratio of 20 dB.

Thus the minimum signal required for a SNR of 20 dB is: 10×2.07 pA $= 20.7$ pA. From the sensitivity of the photodetector, the minimum light flux that can be detected is:

$$\varphi_{min} = \frac{i_{min}}{\mathcal{R}} = \frac{20.7 \times 10^{-12}}{0.01} = 2.07 \times 10^{-9}\,\text{W} = 0.00207\ \mu\text{W}$$

Note that since $\mathcal{R}$ is in μA/μW this is the same dimensions as A/W so it is not necessary to scale this number. To determine the output voltage produced by this minimum signal, or any other signal for that matter, simply multiply this input current by R_f:

$$V_{out} = i_{in}R_f = 20.7 \times 10^{-12}(10^7) = 0.207\ \text{mV}$$

This is a small signal, so it would be a good idea to increase the value of R_f. Because R_f is the second largest contributor of noise, increasing its value by a factor of 10 would both increase the output signal by that amount and decrease the noise (since noise current is inversely related to the resistance). Further improvement could be obtained by using an op amp with a lower current noise voltage.

10.7.5 Analog Filters

As first shown in Chapter 6, a simple single-pole low pass filter can be constructed using an RC circuit. An op amp can also be used to construct a low pass filter with improved input and output characteristics and provide increased signal amplitude. The easiest way to construct an op amp lowpass filter is to add a capacitor in parallel to the feedback resistor as shown in Figure 10.25.

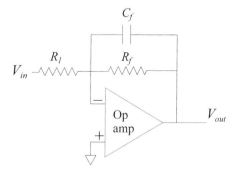

Figure 10.25 An inverting amplifier that also functions as a lowpass filter. As described in the text, the low-frequency gain of this amplifier is R_f/R_1 and the cutoff frequency is $\omega = 1/R_fC_f$.

The equation for the transfer function of an inverting op amp with impedances in the feedback circuit was give in Eq.10.12 and repeated here:

$$\frac{V_{out}}{V_{in}} = -\frac{Z_f}{Z_1} \qquad\qquad \text{[Eq. 10.25]}$$

Applying Eq. 10.25 to the circuit in Figure 10.25:

$$\frac{V_{out}}{V_{in}} = -\frac{Z_f}{Z_1} = -\frac{\dfrac{R_f\, 1/j\omega C_f}{R_f + 1/j\omega C_f}}{R_1} = \left(\frac{R_f}{R_1}\right)\frac{1}{(1 + j\omega R_f C_f)} \qquad \text{[Eq. 10.26]}$$

At frequencies well below the cutoff frequency, the second term goes to 1/1, and the gain is R_f/R_1. The second term is a lowpass filter with a cutoff frequency of $\omega = 1/R_f C_f$ rad/sec or $f = 1/2\pi R_f C_f$ Hz. Design of an active lowpass filter is given in Problem 10.12.

It is also possible to construct a second-order filter using a single op amp. A popular second-order op amp circuit is shown in Figure 10.26.

Derivation of the transfer function requires applying KCL to two nodes and is provided in Appendix A.3. The transfer function will simply be given here as:

$$\frac{V_{out}}{V_{in}} = \frac{G/(RC)^2}{s^2 + \dfrac{3-G}{RC}s + \dfrac{1}{(RC)^2}} \quad \text{where } G = \frac{R_f + R_1}{R_1} \qquad \text{[Eq. 10.27]}$$

where G is the gain of the noninverting amplifier and equals $1/\beta$. Equating coefficients of Eq. 10.27 with the standard Laplace transfer function of a second-order system (Eq. 8.46):

$$\omega_o = \frac{1}{RC}; \quad \delta = \frac{3-G}{2} \qquad\qquad \text{[Eq. 10.28]}$$

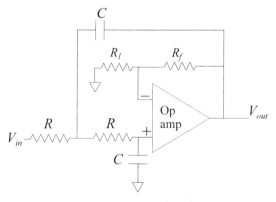

Figure 10.26 A two-pole active lowpass filter constructed using a single operational amplifier.

Example 10.10: Design a second-order filter with a cutoff frequency of 5 kHz and a damping of 1.0.

Solution: Because there are more unknowns than equations, several component values may be set arbitrarily with the rest determined by Eq. 10.28. To find R and C, it is common to pick a value for C that is easy to obtain, then calculate the value for R (capacitor values are more limited than resistor values). Assume $C = 0.001 \, \mu f$, a common value. Then the value of R is:

$$\omega_o = 2\pi f = \frac{1}{RC} = 2\pi(5,000) = 31,416 \, \text{rad/sec}$$

$$R = \frac{1}{31,416(0.001 \times 10^{-6})} = 31.8 \, \text{k}\Omega$$

The value for G, the gain of the noninverting amplifier would be:

$$\delta = \frac{3-G}{2} = 1.0; \quad G = 3 - 2\delta = 3 - 2 = 1$$

Hence, in this particular damping, $G = 1/\beta = 1$ and $\beta = 1.0$. So all of the output is feedback to the noninverting input and a resistor divider network is not required as in a buffer amplifier. Other values of damping would require the standard resistor divider network to achieve the desired gain. A second-order active filter having the desired cutoff frequency and damping is shown in Figure 10.27.

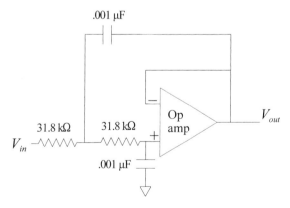

Figure 10.27 A two-pole lowpass filter with a cutoff frequency of 5 kHz and a damping factor of 1.0.

10.7.6 Instrumentation Amplifier

The differential amplifier shown in Figure 10.19 can be very useful in certain biomedical engineering applications, particularly as amplifiers for biotransducers that produce a differential output. Such transducers actually produce two voltages that move in opposite directions to a given input. An example of such a transducer is the strain gage bridge shown in Figure 10.28.

Here the strain gages are arranged in such a way that when a force is applied to the gages, two of them (A-B and C-D) undergo tension while the other two (B-C and D-A) undergo compression. The two gages under tension decrease their resistance while the two under compression increase their resistance. The net effect is that the voltage at B increases whereas the voltage at D decreases in response to the applied force. If the difference between these voltages is amplified using a differential amplifier such as shown in Figure 10.19, the output voltage will be the difference between the two voltages and reflect the force applied. If the force reverses, the output voltage will change sign. One of the significant advantages of this arrangement is that much of the noise, particularly that picked up by the wires leading to the differential amplifier, will be common to both of the inputs and will tend to cancel. To optimize this kind of noise cancellation, the gain of each of the two inputs should be exactly equal in magnitude (but opposite in sign, of course). Not only must the two inputs be balanced, but the input impedance should also be balanced and often it is desirable that the input impedance be quite high. An *instrumentation amplifier* is a differential amplifier circuit that meets these criteria: balanced gain along with balanced and high input impedance. In addition, low noise characteristics are generally a desirable feature.

A circuit that fulfills this role is shown in Figure 10.29. The output op amp performs the differential operation, and the two leading op amps configured as unity gain buffer amplifier provide similar high-impedance inputs. If the requirements for balanced gain were high, one of the resistors would be adjusted until the two channels had equal but opposite gains. It is common to adjust the lower R_2 resistor. Since

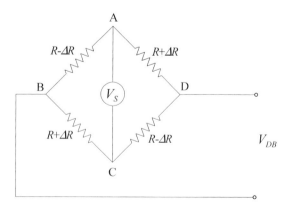

Figure 10.28 A bridge circuit that produces a differential output. The voltage at node D moves in opposition to the voltage at node B. The output voltage is best amplified by a differential amplifier.

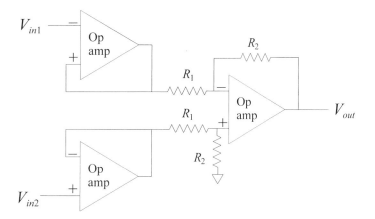

Figure 10.29 A deferential amplifier circuit with high input impedance. One R_2 resistor could be adjusted to balance the differential gain so that the two channels have equal but opposite gains. In the interest of symmetry, it is common to reverse the position of the positive and negative operational amplifier inputs in the upper input operational amplifier.

the two input op amps provide no gain, the transfer function of this circuit is just the transfer function of the second stage, which is shown in Eq. 10.21 to be:

$$V_{out} = \frac{R_1 + R_2}{R_1}(V_{in2} - V_{in1}) \qquad \text{[Eq. 10.27]}$$

There is one serious drawback to the circuit in Figure 10.29. To increase or decrease the gain it is necessary to change two resistors simultaneously: either both R_1's or both R_2's. Moreover, to maintain balance, they would both have to be changed exactly the same amount. This can present practical difficulties. A differential amplifier circuit that requires only one resistor change for gain adjustment is shown in Figure 10.30. The derivation for the input–output relationship of this circuit is more complicated than for the previous circuit, and is given in Appendix A.4.

$$V_{out} = \frac{R_4}{R_3}\left(\frac{R_1 + 2R_2}{R_1}\right)(V_{in2} - V_{in1}) \qquad \text{[Eq. 10.28]}$$

Since R_1 is a now single resistor, the gain can be adjusted by modifying this resistor. As this resistor is common to both channels, changing its value affects the gain of each channel equally and does not alter the balance between the gains of the two channels. The instrumentation amplifier is a popular first stage whenever a differential input signal is available. It is possible to obtain integrated circuit instrumentation amplifiers that place all the components of Figure 10.30 on a single chip. Such packages generally have excellent balance between the two channels, very high input impedance, and low noise. For example, an instrumentation amplifier made by Analog Devices, Inc. (Noorwood, MA), the ADC624 has an input impedance of $10^9\ \Omega$ and a noise voltage of 4.0 nV/√Hz at 1.0 kHz.

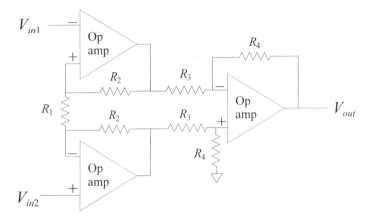

Figure 10.30 An instrumentation amplifier circuit. This circuit has all the advantages of the one shown in Figure 10.29 (i.e., balanced channel gains and high input impedance), but with the added advantage that the gain can be adjusted by modifying a single resistor, R_1.

The balance between the channels is measured in terms of V_{out} when the two inputs are at the same voltage. Voltage that is common (i.e., the same) to both input terminals is termed the *common mode voltage*. In theory, the output should be zero no matter what the input voltage so long as it is the same at both inputs. However, any imbalance between the gains of the two channels will produce some output voltage, and this voltage will be proportional to the common mode voltage. Because the idea is to have the most cancellation and the smallest output voltage to a common mode signal, the common mode voltage is specified as an inverse gain. This inverse gain is called the common mode rejection ratio (CMRR), and usually given in decibels.

$$V_{out} = \frac{V_{CM}}{CMRR} \qquad \text{[Eq. 10.30]}$$

Hence the higher the CMRR, the less voltage at the output from a common mode signal. The ADC624 has a CMRR of 120 dB. This means that the common mode gain is −120 dB. For example, if 10 V were applied to each of the input terminals (i.e., $V_{in1} = V_{in2} = 10$ v), V_{out} would be:

$$V_{out} = \frac{V_{CM}}{CMRR} = \frac{10}{120 dB} = \frac{10}{10^{120/20}} = \frac{10}{10^6} = 10 \; \mu V$$

While not zero, this value is quite small and would be close to the noise level for most applications.

10.8 SUMMARY

There is more to analog electronics than just op amp circuits, but they do encompass most analog applications. The ideal op amp is an extension of the concept of an ideal amplifier. An ideal amplifier has infinite input impedance, zero output impedance, and a fixed gain at all frequencies. An ideal op amp also has infinite input and output impedance, but has infinite gain. The actual gain of an op amp circuit to is determined by the feedback network, which is usually constructed from passive devices. This provides great flexibility with a wide variety of design options and the inherent robustness and long-term stability of passive elements.

Real op amps come reasonably close to the idealization. They have very high input impedances and quite low output impedances. Deviations from the ideal can be fall three categories: deviations in transfer characteristics, deviations in input characteristics, and deviations in output characteristics. The two most important transfer characteristics are bandwidth and stability. Stability in this context refers to the ability to avoid oscillation. The bandwidth of an op amp circuit can be determined by combining the frequency characteristics of the feedback network with the frequency characteristics of the op amp itself. The immunity of an op amp circuit from oscillation can also be estimated from the frequency characteristics of the operational amplifier and feedback network. Input errors include bias voltages and currents, and noise voltages and currents. The bias and noise currents are usually converted to voltages by multiplying them by the equivalent resistance at each of the input terminals. The effect of these input errors on the output can be determined by multiplying all the input voltage errors by the noise gain, $1/\beta$.

A wide variety of very useful analog circuits are based on the op amp. These include inverting and noninverting amplifiers, filters, buffers, adders, subtractors (including differential amplifiers), transconductance amplifiers, and many more circuits not covered. The design and construction of real circuits that use op amps is straightforward, although some care may be necessary to prevent noise and artifact from spreading through the power supply lines. Decoupling capacitors, capacitors running from the power supply lines to ground, are often placed at the op amp's power supply feed to reduce the spread to of noise through the power lines.

PROBLEMS

1. Design an noninverting amplifier circuit with a gain of 500.
2. Design an inverting amplifier with a variable gain from 50 to 250.
3. What is the bandwidth of the noninverting amplifier below? If the same feedback network were used to design an inverting amplifier, what would be the bandwidth of this circuit?

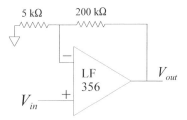

4. An amplifier has a GBP of 10 MHz. It is used in a noninverting amplifier where $\beta = 0.01$. What is the gain of the amplifier? What is the bandwidth?
5. An LF 356 is used to implement the variable gain amplifier in Example 10.3. What are the bandwidths of this circuit at the two extremes of gain?
6. A 0.001-μf capacitor is added to the feedback circuit of the inverting op amp circuit shown below. You can assume that before the capacitor was added, the phase shift due to the amplifier when $A_v\beta = 1$ was 120 degrees. (The criterion for stability is that the phase shift induced by the op amp and the feedback network must be less than 180 degrees when $A_v\beta = 1$.) After the capacitor is added, what is the phase shift of the op amp plus feedback network at the frequency where $A_v\beta = 1$. Follow the example given in the section on stability.

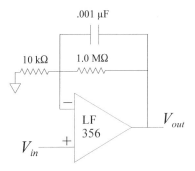

7. For the circuit below, what is the total offset voltage at the output (due to both offset voltage and current)? ($R_f = 500\,\text{Kr}$; $R_1 = 20\,\text{Kr}$; $R_{eq} = 19\,\text{Kr}$.) How much is this offset voltage increased if the 19-kΩ ground resistor on the positive terminal is replaced with a short circuit?

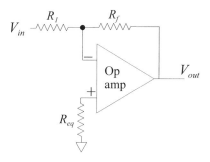

8. Derive the transfer function of the adder circuit below. Use KVL applied to node A.

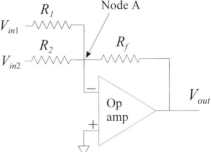

9. Find the total noise at the output of the circuit below. Identify the major source(s) of noise. What would the output noise be if the 50-kΩ ground resistor at the positive terminal was replaced with a short circuit? (Note: this resistor adds to both the Johnson noise from the resistors and to the voltage noise generated by the op amp's noise current.)

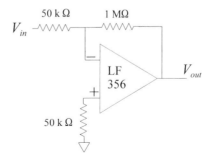

10. For the circuit in Problem 10.9, what is the minimum signal that can be detected with an SNR ratio of 10 dB? What will be the voltage produced by such a signal at V_{out}?

11. An op amp has a noise current of 0.1 pA$\sqrt{Hz}$. This op amp is used as a transconductance amplifier (Figure 10.23). What should the minimum value of feedback resistance be so that the noise contribution from the feedback resistor is less than the noise contribution from the op amp?

12. Design a one-pole lowpass filter with a bandwidth of 1 kHz. Assume you have capacitor values of 0.001 μf, 0.01 μf, 0.05 μf, and 0.1 μf, and a wide range of resistors.

13. Design a two-pole lowpass filter with a cutoff frequency of 500 Hz and a damping factor of 0.8. Assume the same component availability as in Problem 10.12.

14. Design a two-pole highpass filter with a cutoff frequency of 10 kHz and a damping factor of 0.707. (The circuit for a highpass filter is the same as for a lowpass filter except that the capacitors and resistors are reversed as shown in the figure.)

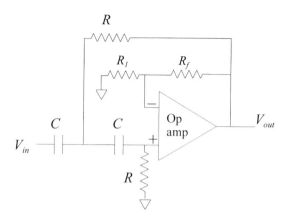

15. Design an instrumentation amplifier with a switchable gain of 10, 100, and 1,000. (*Hint:* Switch the necessary resistors in our out of the circuit as needed.)

APPENDIX A

A.1 DERIVATION OF EULER'S FORMULA

Assume a sinusoidal function:

$$x(\theta) = \cos\theta + j\sin\theta \qquad\qquad \text{[Eq. A.1]}$$

(where $j = \sqrt{-1}$ as usual)
Differentiating with respect to θ produces:

$$\frac{dx}{d\theta} = -\sin\theta + j\cos\theta = j(\cos\theta + j\sin\theta) = jx \qquad\qquad \text{[Eq. A.2]}$$

Separating the variables gives:

$$\frac{dx}{x} = jd\theta \quad \text{and integrating both sides gives:} \quad \ln x = j\theta + K \qquad \text{[Eq. A.3]}$$

where K is the constant of integration. To solve for this constant, note that in Eq. A.1: $x = 1$ when $\theta = 0$. Applying this condition to Eq. A.3: $\ln 1 = 0 = 0 + K$; hence, $K = 0$; and Eq. A.3 becomes $\ln x = j\theta$

or

$$x = e^{j\theta} \qquad\qquad \text{[Eq. A.4]}$$

but since x is defined in Eq. A.1 as $\cos\theta + j\sin\theta$

$$e^{j\theta} = \cos\theta + j\sin\theta \qquad\qquad \text{[Eq. A.5]}$$

Alternatively,

$$e^{-j\theta} = \cos\theta - j\sin\theta \qquad\qquad \text{[Eq. A.6]}$$

A.2   CONFIRMATION OF THE FOURIER SERIES

Fourier showed that a periodic function of period T can be represented by a series, possibly infinite, of sinusoids, or sine and cosines:

$$x(t) = \frac{a_0}{2} + \sum_{n=1}^{\infty}(a_n \cos n\omega_o t + b_n \sin n\omega_o t) \qquad \text{[Eq. A.7]}$$

where $\omega_o = 2\pi/T$ and a_n and b_n are the Fourier coefficients.

To derive the Fourier coefficients, let us begin with the a_0 or the direct current (DC) term. Integrating both sides of Eq. A.7 over a full period:

$$\int_0^T x(t)dt = \int_0^T \frac{a_0}{2}dt + \sum_{n=1}^{\infty}\int_0^T (a_n \cos n\omega_o t + b_n \sin n\omega_o t)dt \qquad \text{[Eq A.8]}$$

For all $n > 0$, the second term on the right hand side is zero because we will be integrating the sine and cosine over a full period ($n = 1$) or multiple periods ($n > 1$). For ($n = 0$), the summation is still zero since it begins at one and Eq. A.8 becomes.

$$\int_0^T x(t)dt = \int_0^T \frac{a_0}{2}dt; \quad \int_0^T x(t)dt = \frac{a_0 T}{2};$$

$$a_0 = \frac{2}{T}\int_0^T x(t)dt \qquad \text{[Eq. A.9]}$$

To find the other coefficients, multiply both sides of Eq. A.7 by $\cos(m\omega_o t)$, where m is an integer and again integrate both sides.

$$\int_0^T x(t)\cos(m\omega_o t)dt = \int_0^T \frac{a_0}{2}\cos(m\omega_o t)dt$$

$$+ \sum_{n=1}^{\infty}\int_0^T (a_n \cos(m\omega_o t)\cos n\omega_o t + b_n \cos(m\omega_o t)\sin n\omega_o t)dt \qquad \text{[Eq. A.10]}$$

and rearranging:

$$\int_0^T x(t)\cos(m\omega_o t)dt = \int_0^T \frac{a_0}{2}\cos(m\omega_o t)dt + \sum_{n=1}^{\infty}a_n\int_0^T \cos(m\omega_o t)\cos n\omega_o t\, dt$$

$$+ \sum_{n=1}^{\infty}b_n\int_0^T \cos(m\omega_o t)\sin n\omega_o t\, dt \quad \text{[Eq. A.11]}$$

Because m is an integer, the first term and third terms on the right-hand side integrate to zero for all m. The middle term integrates to zero for all m and n except $m = n$. At $m = n$, this term becomes:

$$a_n\int_0^T \cos^2(n\omega_o t)dt = \frac{\pi}{\omega_o}a_n = \frac{T}{2}a_n; \quad \text{so that} \quad \frac{T}{2}a_n = \int_0^T x(t)\cos(m\omega_o t)dt$$

$$a_n = \frac{2}{T}\int_0^T \cos(n\omega_o t)dt; \quad m = 1, 2, 3, \ldots \qquad \text{[Eq. A.12]}$$

The b_n coefficients are found in a similar fashion except Eq. A.7 is multiplied by $\sin(m\omega_o t)$, then integrated. In this case, all but the third term integrates to zero and the third term is nonzero only for $m = n$.

$$b_n \int_0^T \sin^2(n\omega_o t)dt = \frac{\pi}{\omega_o}b_n = \frac{T}{2}b_n; \quad \text{so that} \quad \frac{T}{2}b_n = \int_0^T x(t)\sin(n\omega_o t)dt$$

$x(t)$

$$b_n = \frac{2}{T}\int_0^T \sin(n\omega_o t)dt; \quad m = 1, 2, 3, \ldots \qquad \text{[Eq. A.13]}$$

A.3 DERIVATION OF THE TRANSFER FUNCTION OF A SECOND-ORDER OP AMP FILTER

The operational amplifier (op amp) circuit for a second-order lowpass filter is shown in Figure A.1.

This derivation is developed for the lowpass version shown in Figure A.1, but also applies to the highpass version where the positions of R and C are reversed.

Note that at node V' by KCL: $i_1 - i_2 - i_3 = 0$. This allows us to write a nodal equation around that node:

$$\frac{V_{in} - V'}{R} - \frac{V' - Vout}{1/CS} - i_2 = 0$$

$$\text{where} \quad i_2 = \frac{V^+}{1/CS} = V^+ Cs \qquad \text{[Eq. A.13]}$$

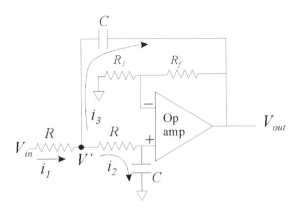

Figure A.1 Circuit diagram of a second-order lowpass filter circuit using a single operational amplifier. The equations developed here refer to this figure.

But, because the two terminals of an op amp must be at the same voltage, the voltage V^+ must be equal to V^-. Applying the voltage divider equation to the two feedback resistors, both V^- and V^+ can be found in terms of V_{out}:

$$V^+ = V^- = V_{out}/G \quad \text{where } G = \frac{R_f + R_1}{R_1}$$

So i_2 becomes $V_{out}(Cs)/G$. Substituting i_2 into the nodal equation Eq. A.13:

$$\frac{V_{in} - V'}{R} - (V' - V_{out})Cs - \frac{V_{out}(Cs)}{G} = 0$$

$$\frac{V_{in}}{R} - \frac{V'}{R} - V'Cs + V_{out}Cs - \frac{V_{out}Cs}{G} = 0$$

$$\frac{V_{in}}{R} - V'\left(\frac{1}{R} - V'Cs\right) + V_{out}Cs\left(1 - \frac{1}{G}\right) = 0 \qquad \text{[Eq. A.14]}$$

Note that V' can also be written in terms of just i_2:

$$V' = i_2\left(R + \frac{1}{CS}\right) = \frac{V_{out}Cs}{G}\left(R + \frac{1}{CS}\right)$$

Substituting this in for V' in Eq. A.14:

$$\frac{V_{in}}{R} - \frac{V_{out}Cs}{G}\left(\frac{1}{R} + Cs\right)\left(R + \frac{1}{Cs}\right) + V_{out}Cs\left(1 - \frac{1}{G}\right) = 0$$

$$\frac{V_{in}}{R} = \frac{V_{out}Cs}{G}\left(3 + \frac{1}{RCs} + RCs - G\right)$$

Solving for V_{out}/V_{in}:

$$\frac{V_{out}}{V_{in}} = \frac{G}{RCs(3 + 1/RCs + RCs - G)} = \frac{G}{(RCs)^2 + (3-G)RCs + 1}$$

$$\frac{V_{out}}{V_{in}} = \frac{G/(RC)^2}{S^2 + \dfrac{3-G}{RC}s + \dfrac{1}{(RC)^2}} \qquad \text{[Eq. A.15]}$$

A.4   DERIVATION OF THE TRANSFER FUNCTION OF AN INSTRUMENTATION AMPLIFIER

The classic circuit for a 3–op amp instrumentation amplifier is shown in Figure A.2.

To determine the transfer function, note that the voltage V_{in1} appears on both terminals of op amp 1, whereas V_{in2} appears on both terminals of op amp 2. (Recall that the voltage difference between op amp input terminals must be zero.) The

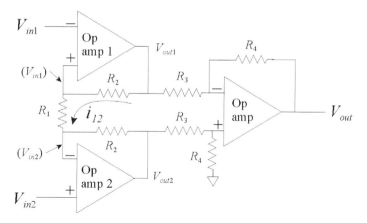

Figure A.2   A circuit diagram of an instrumentation amplifier.

voltage out of op amp 1 will be equal to V_{in2} after accounting for the voltage drop across the two resistors, R_2 and R_1:

$$V_{out1} = i_{12}(R_1 + R_2) + V_{in2}; \quad \text{but: } i_{12} = \frac{V_{in1} - V_{in2}}{R_1}$$

$$V_{out1} = \frac{V_{in1} - V_{in2}}{R_1}(R_1 + R_2) + V_{in2} \qquad \text{[Eq. A.16]}$$

Applying the same logic to op amp 2, its output, V_{out2} can be written as:

$$V_{out2} = \frac{V_{in2} - V_{in1}}{R_1}(R_1 + R_2) + V_{in1} \qquad \text{[Eq. A.17]}$$

The overall output, V_{out}, will be equal to the difference of $V_{out2} - V_{out1}$ times the gain of the differential amplifier: R_4/R_3. The difference voltage can be found by subtracting the two equations above and rearranging:

$$V_{out2} - V_{out1} = \frac{V_{in2} - V_{in1}}{R_1}(R_1 + R_2)2 - (V_{in2} - V_{in1})$$

$$= \left[\frac{2(R_1 + R_2)}{R_1} - 1\right](V_{in2} - V_{in1}) = \left[\frac{2R_1 + 2R_2 - R_1}{R_1}\right](V_{in2} - V_{in1})$$

$$V_{out2} - V_{out1} = \left(\frac{R_1 + 2R_2}{R_1}\right)(V_{in2} - V_{in1})$$

$$V_{out} = \left(\frac{R_4}{R_3}\right)(V_{out2} - V_{out1}) = \left(\frac{R_4}{R_3}\right)\left(\frac{R_1 + 2R_2}{R_1}\right)(V_{in2} - V_{in1}) \qquad \text{[Eq. A.18]}$$

APPENDIX B
LAPLACE TRANSFORMS

1) 1 $\dfrac{1}{s}$

2) t^n $\dfrac{n!}{s^{n+1}}$

3) $e^{-\alpha t}$ $\dfrac{1}{s+\alpha}$

4) $1 - e^{-\alpha t}$ $\dfrac{\alpha}{s(s+\alpha)}$

5) $\cos \beta t$ $\dfrac{s}{s^2 + \beta^2}$

6) $\sin \beta t$ $\dfrac{\beta}{s^2 + \beta^2}$

7) $\cosh \beta t$ $\dfrac{s}{s^2 - \beta^2}$

8) $\sinh \beta t$ $\dfrac{\beta}{s^2 - \beta^2}$

9) $e^{-\alpha t} - e^{-\gamma t}$ $\dfrac{\gamma - \alpha}{(s+\alpha)(s+\gamma)}$

10) $t - \dfrac{1}{\alpha}(1 - e^{-\alpha t})$ $\dfrac{\alpha}{s^2(s+\alpha)}$

11) $\left(\dfrac{b\beta - b\alpha + c}{2\beta}\right)e^{-(\alpha-\beta)t} + \left(\dfrac{b\beta + b\alpha - c}{2\beta}\right)e^{-(\alpha+\beta)t}$ $\dfrac{bs+c}{s^2 + 2\alpha s + \alpha^2 - \beta^2}$

12) $e^{-\alpha t}[b + (c - b\alpha)t]$ $\dfrac{bs+c}{(s+\alpha)^2}$

13)* $e^{-\alpha t}\left[\left(\dfrac{c-b\alpha}{\beta}\right)\sin\beta t + b\cos\beta t\right]$

$\dfrac{bs+c}{s^2+2\alpha s+\alpha^2+\beta^2}$

14)* $1-e^{-\alpha t}\left[\left(\dfrac{\alpha-b}{\beta}\right)\sin\beta t + \cos\beta t\right]$

$\dfrac{bs+\alpha^2+\beta^2}{s(s^2+2\alpha s+\alpha^2+\beta^2)}$

15)* $\dfrac{\omega_n}{\sqrt{1-\delta^2}}\left[e^{-\delta\omega_n t}\sin\!\left(\omega_n\sqrt{1-\delta^2}\,t\right)\right]$

$\dfrac{\omega_n^2}{s^2+2\delta\omega_n+\omega_n^2}$

16)* $1-\dfrac{e^{-\delta\omega_n t}}{\sqrt{1-\delta^2}}\sin\!\left(\omega_n\sqrt{1-\delta^2}\,t+\theta\right)$

$\dfrac{\omega_n^2}{s(s^2+2\delta\omega_n+\omega_n^2)}$

where $\theta=\tan^{-1}\!\left(\dfrac{\sqrt{1-\delta^2}}{\delta}\right)$

*Complex roots ($\delta<1.0$).

APPENDIX C
TRIGONOMETRIC AND
OTHER FORMULAS

$$\sin(-x) = -\sin x$$

$$\cos(-x) = \cos x$$

$$\tan x = \frac{\sin x}{\cos x}$$

$$\sin(\omega t) = \cos(\omega t - 90)$$

$$\cos(\omega t) = \sin(\omega t + 90)$$

$$\sin(x \pm y) = \sin x \cos x \pm \cos x \sin y$$

$$\cos(x \pm y) = \cos x \cos y \mp \sin x \sin y$$

$$\sin 2x = 2 \sin x \cos x$$

$$\cos 2x = \cos^2 x - \sin^2 x = 2 \cos^2 x - 1 = 1 - 2 \sin^2 x$$

$$\sin^2 x = \frac{1 - \cos 2x}{2}$$

$$\cos^2 x = \frac{1 + \cos 2x}{2}$$

$$\sin^2 x + \cos^2 x = 1$$

$$\sin x \pm \sin y = 2 \sin \frac{1}{2}(x \pm y) \cos \frac{1}{2}(x \mp y)$$

$$\cos x + \cos y = 2 \cos \frac{1}{2}(x + y) \cos \frac{1}{2}(x - y)$$

$$\cos x - \cos y = -2 \sin \frac{1}{2}(x + y) \sin \frac{1}{2}(x - y)$$

$$\sin x \cos y = \frac{1}{2}[\sin(x + y) + \sin(x - y)]$$

$$\cos x \cos y = \frac{1}{2}[\cos(x + y) + \cos(x - y)]$$

$$\sin x \sin y = \frac{1}{2}[\cos(x - y) - \cos(x + y)]$$

$$e^{\pm jx} = \cos x \pm j \sin x$$

$$\cos x = \frac{e^{jx} + e^{-jx}}{2}$$

$$\sin x = \frac{e^{jx} - e^{-jx}}{2j}$$

APPENDIX D
UNITS

METRIC CONVERSIONS

1 cc (1,000 mm^3)	1×10^{-6} m^3
1 kg wt	9.80665 N
1 kg wt	9.80665×10^5 dyne
1 gm wt	980.665 dyne
1 kg-m	9.80665 joule
1 kg-m	2.3427 gm-cal
1 gm	0.001 kg
1 J	1 watt-sec
1 J	10^7 ergs
1 J	2.778×10^{-7} kW hr
1 Pa	10 dyne/cm^2
1 PA	1 N/m^2
1 rad	57.296 deg
1 rad/sec	57.296 deg/sec
1 N	10^5 dyne
1 dyne	0.0010197 gm wt
1 erg (1 dyne-cm)	1×10^{-7} J
1 km	10^5 cm
1 m	10^{-9} nanometers
1 m	10^{-6} microns
1 L	1,000.027 cm^3 (cc)
1 W	1 J/sec
1 W	10^7 erg/sec
1 W	10^7 dyne-cm/sec
1 A	1 coulomb/sec
1 coulomb (A/sec)	$6.281 - 10^{18}$ electronic charges
1 coulomb	3×10^9 electrostatic units
1 ohm	1 V/A
1 V	1 J/coulomb

1 h	1 V-sec/A
1 f	1 col/V

METRIC TO ENGLISH CONVERSIONS

1 cm	0.3937 in
1 m	3.281 ft
1 km	0.6214 mile
1 km	3,280.8 ft
1 km	1.0567×10^{-13} light year
1 gm	0.035 oz (adps)
1 kg	2.2046 lb (adps)
1 deg/sec	0.1667 rpm
1 cm/sec	0.02237 mi/hr

ENGLISH TO METRIC CONVERSIONS

1 in (U.S.)	2.540 cm
1 ft	0.3048 m
1 yd	0.9144 m
1 fathom (6 ft)	1.829 m
1 mile	1.609 km
1 in^3	16.387 cm^3
1 in^2	6.451 cm^2
1 pt (0.5 qt)	0.47 L
1 gal (0.013368 ft^3)	3.7853 L
1 gal	3,785.4 cm^3
1 gal wt (8.337 lb)	3.111 kg
1 carat	6.2 gm
1 oz (troy)	31.1035 gm
1 oz (avdps)	28.35 gm
1 lb (troy)	373.24 gm
1 lb (troy)	0.3732 kg
1 lb (avdps)	453.59 gm
1 lb (avdps)	0.45359 kg
1 lb wt	4.448×10^5 dyne
1 lb wt	4.448 N
1 BTU	1,054.8 J
1 hp	0.7452 kW
1 rpm	6 deg/sec
1 rpm	0.10472 rad/sec

PRESSURE CONVERSIONS

1 cm Hg (0°C)	1.333×10^4 dyne/cm^2
1 cm Hg (0°C)	135.95 kg/m^2
1 cm H$_2$O (4°C)	980.638 dyne/cm^2
1 Pa	10 dyne/cm^2

CONSTANTS

Gravitational constant: g	32.174 ft/sec^2
Gravitational constant: g	980.665 cm/sec^2
Speed of light: v	2.9986×10^{10} cm/sec
Dielectric constant (vacuum): ε_0	8.85×10^{-12} coulomb^2/mm^2

AMERICAN STANDARD WIRE GAUGE (AWG) OR BROWN AND SHARPE (B&S) GAUGE

No.	Diameter (in)	No.	Diameter (in)
1	0.2893	17	0.0453
2	0.2576	18	0.0403
3	0.2294	19	0.0359
4	0.2043	20	0.0320
5	0.1819	21	0.0285
6	0.1620	22	0.0253
7	0.1443	23	0.0226
8	0.1285	24	0.0201
9	0.1144	25	0.0179
10	0.1019	26	0.0159
11	0.0907	27	0.0142
12	0.0808	28	0.0126
13	0.0720	29	0.0113
14	0.0641	30	0.100
15	0.0571	31	0.00893
16	0.0508	32	0.00795

SCALING PREFIXES

Scale	Prefix	Examples
10^{-15}	femto	femtoseconds
10^{-12}	pico	picoseconds, picoamps
10^{-9}	nano	nanoseconds, nanoamps, nanotechnology
10^{-6}	micro	microsecond, microvolts, microfarads
10^{-3}	milli	milliseconds, millivolts, milliamps, millimeters
10^{3}	kilo	kilohertz, kilohms, kilometers
10^{6}	mega	megahertz, megaohms, megabucks
10^{9}	giga	gigahertz, gigawatt
10^{12}	tera	terahertz

APPENDIX E
COMPLEX ARITHMETIC

Complex numbers and complex variables consist of two numbers rolled into one. However, they are also related in that most operations affect both numbers in some manner. The two components of a complex number are the real part and the imaginary part, the latter so-called because it is multiplied by $\sqrt{-1}$. It is given the symbol i in mathematic circles, but the symbol j is used in engineering because i is reserved for current. A typical complex number would be written as $a + jb$ where a is the real part and jb is the imaginary part.

Complex numbers are visualized as lying on a plane consisting of a real horizontal axis and an imaginary vertical axis (Figure E.1).

A complex number is one point on the real–imaginary plane and can be represented either in rectangular notation as a real and imaginary coordinate (Figure E.1, dashed lines), or in polar coordinates as a magnitude, C, and angle, θ (Figure E.1, solid line). In this text, the polar form is written using a shorthand notation: $C \angle \theta$.

To convert between the two representations, refer to the geometry of Figure E.1. To go from polar to rectangular, apply triangular trigonometry:

$$a = C\cos\theta \quad b = C\sin\theta \qquad \text{[Eq. E.1]}$$

To go in the reverse direction:

$$C^2 = a^2 + b^2$$
$$C = \sqrt{a^2 + b^2} \qquad \text{[Eq. E.2]}$$

$$\tan\theta = \frac{b}{a} \quad \theta = \tan^{-1}\left(\frac{b}{a}\right) \qquad \text{[Eq. E.3]}$$

These operations are very useful in complex arithmetic because addition and subtraction are done using the rectangular representation while multiplication and division are easier using the polar form. Care must be taken with regard to signs and the quadrant of the angle, θ.

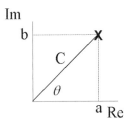

Figure E.1 A complex number can visualized as one point on a plane.

E.1.1 Addition and Subtraction

To add two or more complex numbers, add real numbers to real, and imaginary numbers to imaginary:

$$(a+jb)+(c+jd)=(a+c)+j(b+d)$$

Subtraction follows the same strategy:

$$(a+jb)-(c+jd)=(a-c)+j(b-d)$$

If the numbers are in polar form (or mixed), covert them to rectangular form first.

Example E.1:   Add the following complex numbers: $8\angle-30 + 6\angle60$

Solution: Convert both numbers to rectangular form following the rules above. Note that the first term is in the fourth quadrant so it will have the general form $a - jb$.

For the first term:

$$a = 8\cos(-30); \quad b = 8\sin(-30);$$

$$a = 6.9; \quad b = -4$$

For the second term:

$$c = 6\cos(60) = 3; \quad d = 6\sin(60) = 5.2$$

$$c = 3; \quad d = 5.2$$

$$\text{Sum} = 6.9 - j4 + 3 + j5.2 = 9.9 + j1.2$$

This could then be converted back to polar form if desired.

E.1.2 Multiplication and Division

These arithmetic operations are best done in polar form although they can be carried out in rectangular notation. For multiplication, multiply magnitudes and add angles:

$$C\angle\theta(D\angle\varnothing) = CD\angle(\theta+\varnothing)$$

For division, divide the magnitudes and subtract the angles:

$$\frac{C \angle \theta}{D \angle \varnothing} = \frac{C}{D} \angle (\theta - \varnothing)$$

Example E.2: Perform the indicated multiplications or divisions:

A. $8\angle - 30$ $(6 \angle 120)$;   C. $\dfrac{7 \angle 305}{6 \angle -80}$;

B. $(10 - j6)(1 + j10)$;   D. $\dfrac{6 \angle 50}{8 - j6}$

Solution: For the numbers already in polar coordinates, simply follow the rules given above. Otherwise, convert to polar form where necessary.

A. $8\angle - 30(6\angle 120) = 48\angle 90$

B. $(10 - j6)(5 + j10) = 11.66\angle -31(11.2\angle +63) = 130.6\angle 32$

C. $\dfrac{7 \angle 305}{6 \angle -80} = 1.166 \angle 385 = 1.166 \angle 25$

D. $\dfrac{6 \angle 50}{8 - j6} = \dfrac{6 \angle 50}{10 \angle -36.9} = 0.6 \angle 86.9$

More involved arithmetic operations can call for combinations of these conversions.

Example E.3: Add the two complex fractions:

$$\frac{5(j6)}{3 - j7} + \frac{-8 + j6}{-3 - j4}$$

Solution: Convert all the rectangular representations to polar form, carry out the division, then convert back to rectangular form for the addition. When converting from rectangular to polar form, note that each number is in a different quadrant, so care must be taken with the angles.

Evaluate each fraction in turn:

$$\frac{5 + j6}{3 - j7} = \frac{7.8 \angle 50}{7.6 \angle -66} = 1.03 \angle 116 = -0.45 + j0.93$$

$$\frac{-8 + j6}{-3 - j8} = \frac{10 \angle 143}{8.5 \angle 249} = 1.18 \angle -106 = -0.33 - j1.13$$

So the sum becomes:

$$-0.45 + j0.93 - 0.33 - j1.13 = -0.78 - j0.2 = 0.8 \angle 194$$

It is a good idea to visualize where each number falls in the real–imaginary plane, or at least what quadrant of the plane, to help keep angles and signs straight.

Multiplication or division by the number j has the effect of rotation of the complex point by ± 90 degrees. This is apparent if the number j, which is in rectangular form, is converted to polar form: $j = 1\angle 90$. So multiplying or dividing by j adds or subtracts 90 degrees from a number:

$$j(C \angle \theta) = 1 \angle 90(C \angle \theta) = C \angle (\theta + 90)$$

$$\frac{C \angle \theta}{j} = \frac{C \angle \theta}{1 \angle 90} = C \angle (\theta - 90)$$

APPENDIX F
LF 356 SPECIFICATIONS

Only those specifications useful for analyzing circuits and problems in Chapter 10 are given here. For a detailed set of specifications, see the various manufacturers' specification sheets. These specifications are for the LF 356 except as noted, and typical values are given.

Description	Specification
Bias current	30 pA (picoamps)
Offset current	3.0 pA
Offset voltage	3.0 mV
Gain bandwidth product	5.0 MHz (LF 356)
Gain bandwidth product	20 MHz (LF 357 – minimum $1/\beta = 5$)
Input impedance	$10^{12}\ \Omega$
Open loop gain (direct current)	106 dB
Common mode rejection ratio (CMRR)	100 dB
Maximum voltage	±18 V (LF 356); ±22 V (LF 356B)
Noise current	0.01 pA/$\sqrt{\text{Hz}}$
Noise voltage (1,000 Hz)	12 nV/$\sqrt{\text{Hz}}$
Noise voltage (100 Hz)	15 nV/$\sqrt{\text{Hz}}$
Noise voltage (10 Hz)	60 nV/$\sqrt{\text{Hz}}$

APPENDIX G   DETERMINANTS AND CRAMER'S RULE

The solution of simultaneous equations can be greatly facilitated by matrix algebra. When the solutions must be done by hand, the use of determinants is helpful, at least when only two or three equations are involved. A determinant is a specific, single value defined for a square array of numbers. Given a 2×2 array, the determinant would be found by the application of the so-called *diagonal rule* where the product of the main diagonal (solid arrow) is subtracted by the product of the off diagonal (dotted arrow):

$$\det = \begin{vmatrix} a_{11} & a_{12} \\ a_{21} & a_{22} \end{vmatrix}$$

This gives rise to the equation:

$$\det = \begin{vmatrix} a_{11} & a_{12} \\ a_{21} & a_{22} \end{vmatrix} = a_{11}a_{22} - a_{12}a_{21} \qquad \text{[Eq. G.1]}$$

For a 3×3 array, the determinant is found by an extension of the diagonal rule. One way to visualize this extension is to repeat the first two columns at the right side of the array. Then the diagonals can be drawn directly:

$$\det = \begin{vmatrix} a_{11} & a_{12} & a_{13} \\ a_{21} & a_{22} & a_{23} \\ a_{31} & a_{32} & a_{33} \end{vmatrix} \begin{matrix} a_{11} & a_{12} \\ a_{21} & a_{22} \\ a_{31} & a_{32} \end{matrix}$$

This procedure produces the equation:

$$\det = \begin{vmatrix} a_{11} & a_{12} & a_{13} \\ a_{21} & a_{22} & a_{23} \\ a_{31} & a_{32} & a_{33} \end{vmatrix}$$

$$= (a_{11}a_{22}a_{33} + a_{12}a_{23}a_{31} + a_{13}a_{21}a_{32}) - (a_{31}a_{22}a_{13} + a_{32}a_{23}a_{11} + a_{33}a_{21}a_{12}) \qquad \text{[Eq. G.2]}$$

(In MATLAB the determinant is found using the command: det(A), where A is the array variable.)

Cramer's rule is used to solve simultaneous equations using determinants. The equations are first put in matrix format (shown here using electrical variables):

$$\begin{vmatrix} v_1 \\ v_2 \end{vmatrix} = \begin{vmatrix} Z_{11} & Z_{12} \\ Z_{21} & Z_{22} \end{vmatrix} \begin{vmatrix} i_1 \\ i_2 \end{vmatrix}$$ [Eq. G.3]

The current i_1 is found using:

$$i_1 = \frac{\det Z_1}{\det Z} = \frac{d\begin{vmatrix} v_1 & Z_{12} \\ v_2 & Z_{22} \end{vmatrix}}{\begin{vmatrix} Z_{11} & Z_{12} \\ Z_{21} & Z_{22} \end{vmatrix}} = \frac{v_1 Z_{22} - v_2 Z_{12}}{Z_{11} Z_{22} - Z_{21} Z_{12}}$$ [Eq. G.4]

And, in a similar fashion, the current i_2 is found by:

$$i_2 = \frac{\det Z_1}{\det Z} = \frac{\begin{vmatrix} Z_{11} & v_1 \\ Z_{21} & v_2 \end{vmatrix}}{\begin{vmatrix} Z_{11} & Z_{12} \\ Z_{21} & Z_{22} \end{vmatrix}} = \frac{Z_{11} v_2 - Z_{21} v_1}{Z_{11} Z_{22} - Z_{21} Z_{12}}$$ [Eq. G.5]

Extending Cramer's rule to a 3×3 matrix equation:

$$\begin{vmatrix} v_1 \\ v_2 \\ v_3 \end{vmatrix} = \begin{vmatrix} Z_{11} & Z_{12} & Z_{13} \\ Z_{21} & Z_{22} & Z_{23} \\ Z_{31} & Z_{32} & Z_{33} \end{vmatrix} \begin{vmatrix} i_1 \\ i_2 \\ i_3 \end{vmatrix}$$ [Eq. G.6]

and, the three currents are obtained as:

$$i_1 = \frac{\det\begin{vmatrix} v_1 & Z_{12} & Z_{13} \\ v_2 & Z_{22} & Z_{23} \\ v_3 & Z_{32} & Z_{33} \end{vmatrix}}{\det\begin{vmatrix} Z_{11} & Z_{12} & Z_{13} \\ Z_{21} & Z_{22} & Z_{23} \\ Z_{31} & Z_{32} & Z_{33} \end{vmatrix}} \quad i_2 = \frac{\det\begin{vmatrix} Z_{11} & v_1 & Z_{13} \\ Z_{21} & v_2 & Z_{23} \\ Z_{31} & v_3 & Z_{33} \end{vmatrix}}{\det\begin{vmatrix} Z_{11} & Z_{12} & Z_{13} \\ Z_{21} & Z_{22} & Z_{23} \\ Z_{31} & Z_{32} & Z_{33} \end{vmatrix}} \quad i_3 = \frac{\det\begin{vmatrix} Z_{11} & Z_{12} & v_1 \\ Z_{21} & Z_{22} & v_2 \\ Z_{31} & Z_{32} & v_3 \end{vmatrix}}{\det\begin{vmatrix} Z_{11} & Z_{12} & Z_{13} \\ Z_{21} & Z_{22} & Z_{23} \\ Z_{31} & Z_{32} & Z_{33} \end{vmatrix}}$$ [Eq. G.7]

where each determinant would be evaluated using Eq. G.2.

BIBLIOGRAPHY

Bahil AT, Stark LS. Trajectories of saccadic eye movements. *Scientific American* 1979;240:84–93.

Bruce EN. *Biomedical signal processing and signal modeling.* New York: John Wiley & Sons, 2001.

Davasahayam SR. *Signals and systems in biomedical engineering: signal processing and physiological systems modeling.* New York: Kluwer Academic/Plenum Publishing, 2000.

Dodge NS, Babbage C. *IEEE Annals of the History of Computing* 2000;22(4):22–43.

Horowitz P, Hill W. *The art of electronics*, 2nd ed. New York: Cambridge University Press, 1989.

Johnson DE, Johnson JR, Hilbun JL. *Electric circuit analysis.* Englewood Cliffs, NJ: Prentice Hall, 1989.

Koo MCK. *Physiological control systems: analysis, simulation, and estimation.* Piscataway, NJ: IEEE, 2000.

Lide DR. *Handbook of physics and chemistry*, 85th ed. Boca Raton, FL: CRC Press, 2004.

Maple SL. *Digital spectral analysis with applications.* Englewood Cliffs, NJ: Prentice Hall, 1987.

Northrop RB. *Signals and systems analysis in biomedical engineering.* Boca Raton, FL: CRC Press, 2003.

Northrop RB. *Analysis and application of analog electronic circuits to biomedical instrumentation.* Boca Raton, FL: CRC Press, 2004.

Raju TNK. The Nobel chronicles—1924: William Einthoven (1860–1927). *Lancet* 1998;352(9139):1560.

Semmlow JL. *Biosignal and biomedical image processing: MATLAB-based applications.* New York: Marcel Dekker, Inc., 2004.

Wodbury W, Gordon A, Conrad JT. Muscle. In: *Physiology and biophysics.* Ruch TC, Patton HD, eds. Philadelphia: WB Saunders, 1965.

INDEX